PARTICLE-LADEN FLOW

ERCOFTAC SERIES

VOLUME 11

Series Editors

Aims and Scope of the Series

ERCOFTAC (European Research Community on Flow, Turbulence and Combustion) was founded as an international association with scientific objectives in 1988. ERCOFTAC strongly promotes joint efforts of European research institutes and industries that are active in the field of flow, turbulence and combustion, in order to enhance the exchange of technical and scientific information on fundamental and applied research and design. Each year, ERCOFTAC organizes several meetings in the form of workshops, conferences and summerschools, where ERCOFTAC members and other researchers meet and exchange information.

The ERCOFTAC Series will publish the proceedings of ERCOFTAC meetings, which cover all aspects of fluid mechanics. The series will comprise proceedings of conferences and workshops, and of textbooks presenting the material taught at summerschools.

The series covers the entire domain of fluid mechanics, which includes physical modelling, computational fluid dynamics including grid generation and turbulence modelling, measuring-techniques, flow visualization as applied to industrial flows, aerodynamics, combustion, geophysical and environmental flows, hydraulics, multi-phase flows, non-Newtonian flows, astrophysical flows, laminar, turbulent and transitional flows.

The titles published in this series are listed at the end of this volume.

Particle-Laden Flow

From Geophysical to Kolmogorov Scales

Edited by

BERNARD J. GEURTS

Multiscale Modeling and Simulation, J.M. Burgers Center, Faculty EEMCS, University of Twente, Enschede, The Netherlands

HERMAN CLERCX

Vortex Dynamics and Turbulence, Department of Applied Physics, Eindhoven University of Technology, Eindhoven, The Netherlands

and

WIM UIJTTEWAAL

Environmental Fluid Mechanics, Delft University of Technology, Delft, The Netherlands

Springer

A C.I.P. Catalogue record for this book is available from the Library of Congress.

ISBN 978-1-4020-6217-9 (HB)
ISBN 978-1-4020-6218-6 (eBook)

Published by Springer,
P.O. Box 17, 3300 AA Dordrecht, The Netherlands.
www.springer.com

Printed on acid-free paper

Contents

Preface

The dispersion of particles in a flow is of central importance in various geophysical and environmental problems. The spreading of aerosols and soot in the air, the growth and dispersion of plankton blooms in seas and oceans, or the transport of sediment in rivers, estuaries and coastal regions are striking examples.

These problems are characterized by strong nonlinear coupling between several dynamical mechanisms such as convective sweeping, rotation, buoyancy, bio-physical influences and interactions between particles and fluid. As a result, processes on widely different length and time scales are simultaneously of importance. These range from Kolmogorov scales at which the flow at particle-scales is central, to much larger-scale structures that can be appreciated best via satellite observations. The multiscale nature of this challenging field motivated this colloquium that was organized by the recently established Dutch Platform for Geophysical and Environmental Fluid-mechanics (PGEF). The meeting took place at the University of Twente (the Netherlands), June 21-23, 2006.

In total 55 participants from 13 different countries and 4 different continents contributed to the colloquium. The six keynote speakers provided reviews and recent research findings of areas that were central to the theme of the colloquium. These keynote lectures constituted the framework for the rest of the program, which contained 33 contributed papers, several of which are collected in this book.

Issues related to the large-scale environmental aspects of particle-laden flows were addressed by considering turbulence modulation arising in high density clay-laden flows, and by focussing on transport processes in the stratosphere and its relevance to climate and weather predictions. Fundamental aspects of transport of particles formed the topic of the second day of the colloquium. Insights from experimental and computational research were combined to understand the distortion of flow in the neighborhood of embedded particles. Aspects of Lagrangian statistics in turbulence were discussed at length, addressing the dispersion of embedded point particles. Bridging the environmental and the fundamental aspects of particle-laden flows was the topic of the final day of the colloquium. The Lagrangian dispersion of particles in the context of their environmental setting was presented. The closing lecture provided a synopsis of transport processes in particle-laden flow in which possibilities of multi-resolution, multi-physics modeling and monitoring were discussed.

The colloquium on particle-laden flow was organized under the auspices of EUROMECH, the European Mechanics Society, and the Universities of Technology of Delft, Eindhoven and Twente. It was supported financially by a number of institutions: ERCOFTAC (European Research Council On Flow, Turbulence and Combustion), COST Action P20 'LES-AID' (COoperation in the field of Science and Technology), the Netherlands foundation for fundamental research of matter (FOM), the Netherlands Royal Academy of Arts and Sciences (KNAW), the J.M. Burgers Center for fluid mechanics (JMBC), the Netherlands science foundation (NWO), the foundation for technical sciences (STW), Water Research Center Delft, Eindhoven University of Technology, the Faculty of Applied Physics of Eindhoven University of Technology, the University of Twente and the Twente institute for Mechanics, Processes and Control (IMPACT). This support was crucial for the organization of this colloquium and is gratefully acknowledged.

We hope that these proceedings will lead to new insights and fruitful developments.

Enschede, Eindhoven, Delft
February, 2007

Bernard J. Geurts
Herman Clercx
Wim Uijttewaal

Part I

Dispersion in environmental flows

Sand motion induced by oscillatory flows: sheet flow and vortex ripples

Jan S. Ribberink[1], Jebbe J. van der Werf[1] and Tom O'Donoghue[2]

[1] University of Twente, Faculty of Engineering, Water Engineering and Manage-ment, PO Box 217, 7500 AE Enschede, The Netherlands
j.s.ribberink@utwente.nl
[2] University of Aberdeen, Department of Engineering, King's College, Aberdeen AB24 3UE, Scotland

1 Introduction

Shoaling short gravity waves at sea approaching the shore become asymmetric and are able to generate a net resulting sand transport in cross-shore direction (on-shore-offshore transport). The wave-related sand transport is still very difficult to predict due to the complexity of its underlying processes, which mainly take place in a thin layer near the sea bed in the wave boundary layer (thickness of order centimeters). The development of models for cross-shore sand transport heavily relies on experimental lab research, especially as taking place in large oscillating water tunnels (see, e.g., Nielsen, 1992). In oscillating water tunnels the near-bed horizontal orbital velocity, as induced by short gravity waves, can be simulated above fixed or mobile sandy beds (for a detailed description, see, e.g., Ribberink and Al-Salem, 1994). It should be realized that the vertical orbital flow and relatively small wave-induced residual flows as streaming and drift are not reproduced in flow tunnels. Research aimed at their contribution to the net sediment motion under surface waves is still ongoing (see Ribberink et al., 2000).

The present study is focused at the sediment motion as occurring under the influence of horizontal oscillatory flows and measuring results will be presented of the Large Oscillating Water Tunnel (LOWT) of WL—Delft Hydraulics and the Aberdeen Oscillating Flow Tunnel (AOFT). Due to their large size (length of test sections: 10-15 m) they belong to the few available facilities in which the near-bed flows of full-scale waves can be generated and scale effects can be avoided.

Based on an energetics-approach Bagnold and Bailard (see Bailard, 1981) developed sand transport formulas for short gravity waves, relating the time-dependent sand transport rate during a wave-cycle $q_s(t)$ in a quasi-steady way to a power n of the horizontal velocity above the wave boundary layer $U(t)$:

Bernard J. Geurts et al. (eds), Particle Laden Flow: From Geophysical to Kolmogorov Scales, 3–14.

$$q_s(t) = m|U(t)|^{n-1}U(t) \tag{1}$$

Ribberink (1998) developed a similar quasi-steady formula based on the time-dependent bed-shear stress. For asymmetric waves this type of transport-formulas always leads to a time-averaged (net) transport rate which is 'on-shore' directed.

All net transport rate measurements which were collected during the preceding years in the LOWT and the AOFT for asymmetric waves are depicted in Figure 1 as a function of the sediment mobility number $\Psi = 2U_{rms}^2/\Delta gD$ (with U_{rms} the root mean square velocity of the wave, Δ the relative density of sand, D the grain diameter and g the gravitational acceleration).

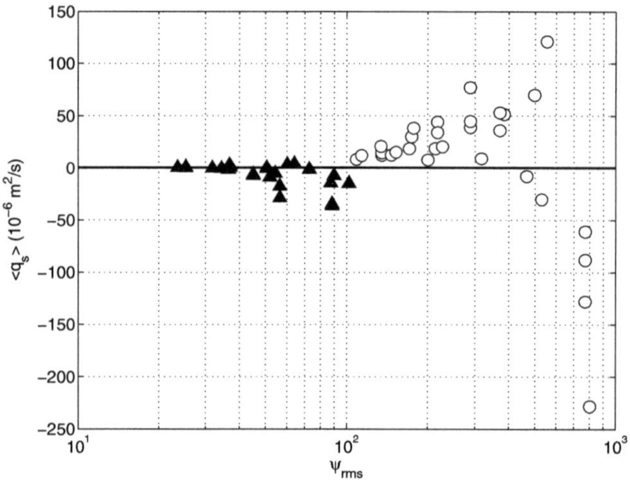

Fig. 1. Measured net sand transport rates $\langle q_s \rangle$ under asymmetric oscillatory flows as a function of the mobility number Ψ in the vortex ripple regime (triangles) and the sheet flow regime (circles). The data refer to flows with a constant degree of asymmetry $R = U_c/(U_c + U_t) = 0.62$, wave periods $T = 3 - 9$ s and sand with grain diameters $D = 0.13 - 0.45$ mm.

It is shown that - contrary to what transport model (1) suggests - the net transport rates can be 'on-shore' (> 0) as well as 'offshore' directed (< 0). Moreover, two data groups with different micro bed morphology can be observed : i) a vortex ripple regime ($\Psi < 100 - 200$) with mainly 'offshore' transport, and ii) a sheet flow regime with flat sea beds ($\Psi > 100 - 200$) with mainly 'on-shore' transport.

In order to obtain a better understanding of this variable behavior of the net sand transport, in the present paper the underlying boundary layer flow and sediment dynamics of these two bed regimes are discussed. Here-to insights as obtained during a series of Ph.D. studies in the preceding years (Dohmen-Janssen, 1999; Clubb, 2001; Wright, 2002; Hassan, 2003; Van der

Werf, 2006) are presented and different types of sand transport models are reviewed.

2 Oscillatory sheet flow

For large mobility numbers ($\Psi > 100 - 200$) ripples are washed out and the sea bed becomes plane. The oscillatory sand transport is now confined to a thin layer with a thickness of order $1cm$ near the bed, in which large sediment concentrations (10-50 volume percent) and large sand fluxes can occur (sheet flow layer).

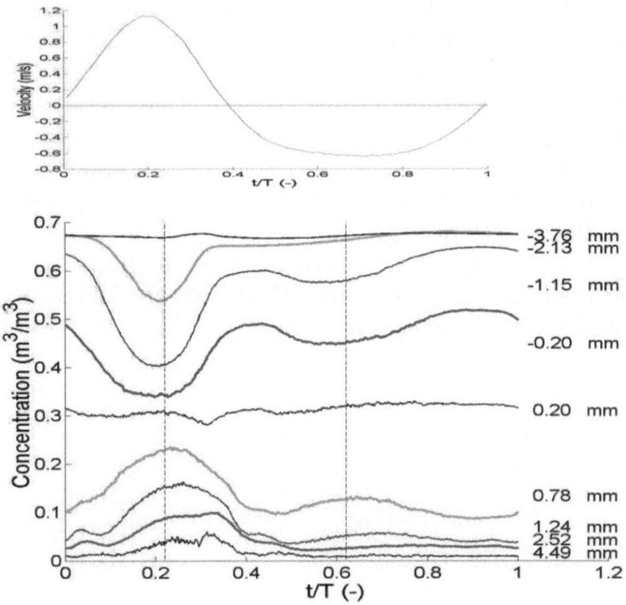

Fig. 2. Time-dependent flow velocity in the free stream (upper panel) and sand (volume) concentrations at different elevations in the sheet flow layer during 1 asymmetric wave cycle (experiment Mh, 0.2 mm sand).

Ribberink and Al-Salem (1995) and McLean et al. (2001) showed how small Conduction Concentration probes (CCM) can be used to visualize and measure the sand pick-up and redeposition processes in the sheet flow layer. The probes measure sand concentration and grain-velocity through electro-resistance of the sand water mixture (sensing volume of ca. 1 mm high).

Asymmetric gravity waves on the shore-face induce a horizontal oscillatory flow with a relatively large maximum velocity in on-shore direction velocity U_c (under the wave crest) and a relatively small maximum velocity in off-shore direction U_t (under the wave trough). Figure 2 shows time-dependent ensemble-averaged sand concentrations during an asymmetric wave cycle, as measured with CCM, at different elevations in the sheet flow layer ($z = 0$ *mm* refers to the original bed level without sand motion). The upper panel shows the horizontal asymmetric velocity in the free stream (above the wave boundary layer). The data were measured with 0.2 *mm* sand and reveal a two-layer structure of the near-bed sand transport layer, with a pick-up layer ($z < 0$) and an upper sheet flow layer ($z > 0$). During flow acceleration sand is picked up from the pick-up layer (decreasing concentrations) into the upper sheet flow layer (increasing concentrations). During flow deceleration the opposite occurs and sand settles back from the upper sheet flow layer into the pick-up layer. The upper elevations ($z = 0.78 - 4.49$ *mm*) show an increasing phase-lag of the maximum concentration with increasing elevation. These phase-lag effects play a crucial role in the magnitude and direction of the mean resulting horizontal transport of asymmetric waves in the sheet flow regime (Dohmen-Janssen et al, 2002; Hassan, 2003).

Phase-lags may occur in the pick-up process of sand grains, in the vertical upward transport of sand and in the resettling process. Figure 3 shows how the concentrations, measured at a fixed level in the upper sheet flow layer and scaled with the time-averaged concentration C_m, experience increasing phase-lags for decreasing wave periods T.

Further systematic experiments revealed that the phase-lags also increase for decreasing grain size D (slower resettling) and for increasing free stream velocities (entrainment to higher elevations).

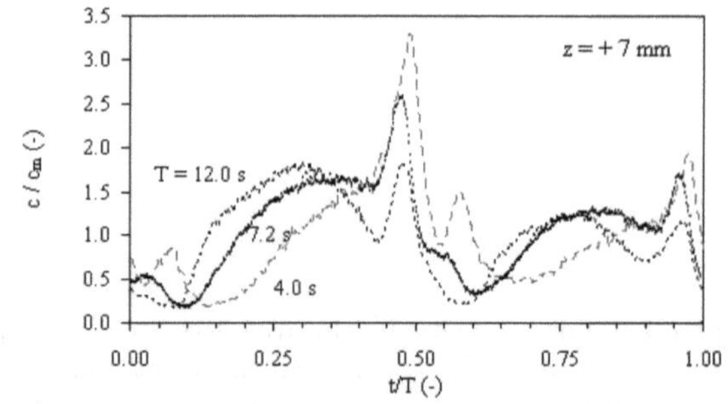

Fig. 3. Time-dependent volume concentrations at a fixed level $z = 7$ *mm* in the upper sheet flow layer during one wave cycle for three wave periods T. The concentrations are normalized with their mean values C_m.

Recently, O'Donoghue and Wright (2004) obtained further insight into the phase-lag phenomenon with high-resolution sand flux measurements using various sand sizes under asymmetric waves in the AOFT. They showed that for very fine sand (0.15 *mm*) the direction of the mean (horizontal) sand transport may even change sign (from 'on-shore' to 'offshore'), due to the fact that the (large) sand volumes stirred up during the (strong) 'on-shore' half wave cycle, are still suspended in the wave boundary layer during the 'offshore' half cycle. As an illustration Figure 4 shows the measured horizontal maximum 'on-shore', maximum 'offshore' and total mean flux profiles for two grain sizes, i.e., 0.28 (MA5010) and 0.15 *mm* (FA5010).

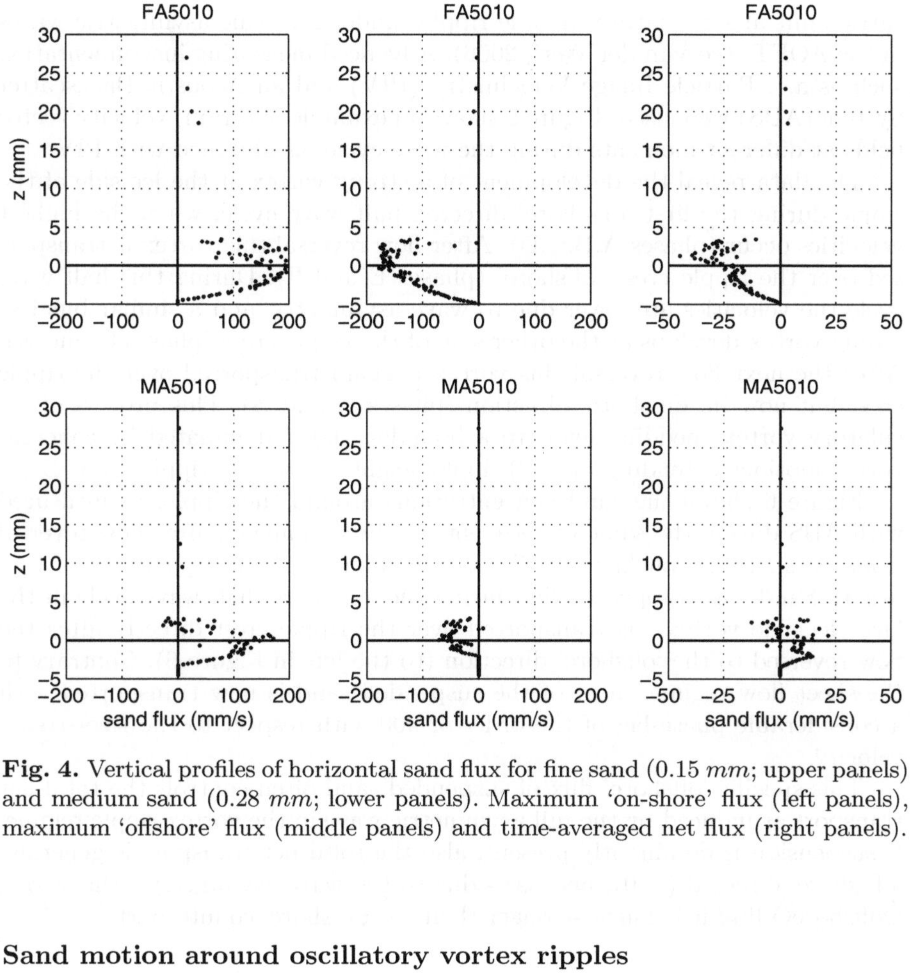

Fig. 4. Vertical profiles of horizontal sand flux for fine sand (0.15 *mm*; upper panels) and medium sand (0.28 *mm*; lower panels). Maximum 'on-shore' flux (left panels), maximum 'offshore' flux (middle panels) and time-averaged net flux (right panels).

Sand motion around oscillatory vortex ripples

Vortex ripples appear on the sea bed for mobility numbers $\Psi < 100 - 200$. Their dimensions, such as ripple height η (of the order $cm - dm$) and length λ (of the order $dm - m$), directly scale with the amplitude of the horizontal oscillatory motion near the sea bed, and show a variation with the mobility number (see Nielsen, 1992; O'Donoghue et al., 2006).

For an overview of present knowledge about this sand transport regime, reference is made to Van der Werf (2006). The flow dynamics in the vortex ripple regime differ strongly from the oscillatory sheet flow regime, mainly due to the fact that processes as flow separation and coherent vortex motions now dominate the entrainment, transport and resettling of sand grains. The turbulence associated with this vortex shedding process leads to much thicker wave boundary layers and to a more important role of suspended sediment in the sand transport process than in sheet flow conditions.

Recently, new flow velocity and sand concentration measurements were carried out around natural mobile ripples under full-scale asymmetric waves in the AOFT (see Van der Werf, 2006). Advanced measuring instrumentation such as a.o. Particle Image Velocimetry (PIV) and an Acoustic Backscatter System (ABS) were used. Figure 5 shows an example of (grain) velocity vector fields at different moments during the wave cycle, as obtained with PIV.

The data reveal the development of a strong vortex at the lee side of the ripple during the first 'on-shore' directed half wave cycle, when the highest velocities occur (phases A,B,C,D). After flow reversal this vortex is transported over the ripple crest 'offshore' (phases E and F). During this half wave cycle the velocities are lower due to wave asymmetry, and a similar but less strong vortex develops at the other side of the ripple crest (phases G and H). After the next flow reversal this vortex is again transported over the ripple crest but now in 'on-shore' direction (phases A and B). This process of oscillatory vortex shedding leads to a boundary layer dominated by coherent vortex motions extending up to 2 ripple heights above the ripple crest.

Figure 6 shows the sand concentrations around the ripple, as measured with ABS during the same experiment, at three moments after flow reversal from 'on-shore' to 'offshore' ($t/T = 0.5, 0.56$ and 0.61, see upper plot of Figure 5). Sand - as trapped earlier during the 'on-shore' half wave cycle in the large lee-side vortex - is transported over the ripple crest directly after the flow reversed to the 'offshore' direction (to the left in Figure 6). Contrary to the sheet flow regime, most of the suspended sand is now transported with a considerable phase-lag of the order of 900 with respect to the free-stream velocity.

This specific 'offshore' flux of suspended sand often controls the total net transport as induced by the full asymmetric wave in the vortex ripple regime. If suspension is dominantly present, also the total net transport is generally 'off-shore' directed (< 0), because - due to the wave asymmetry - the strong ÔoffshoreÕ flux it is much stronger than its 'on-shore' counterpart.

3 1DV RANS modeling

Reynolds-averaged 1DV Navier-Stokes equations with different turbulence closures are combined with an advection-diffusion equation for suspended sediment concentrations and solved numerically to simulate time-dependent suspended sediment fluxes and sand transport by waves in flat sea bed conditions with sheet flow (see Ribberink and Al-Salem, 1995; Uittenbogaard and Klopman, 2001; Malarkey et al., 2003). In general the flow is driven by the

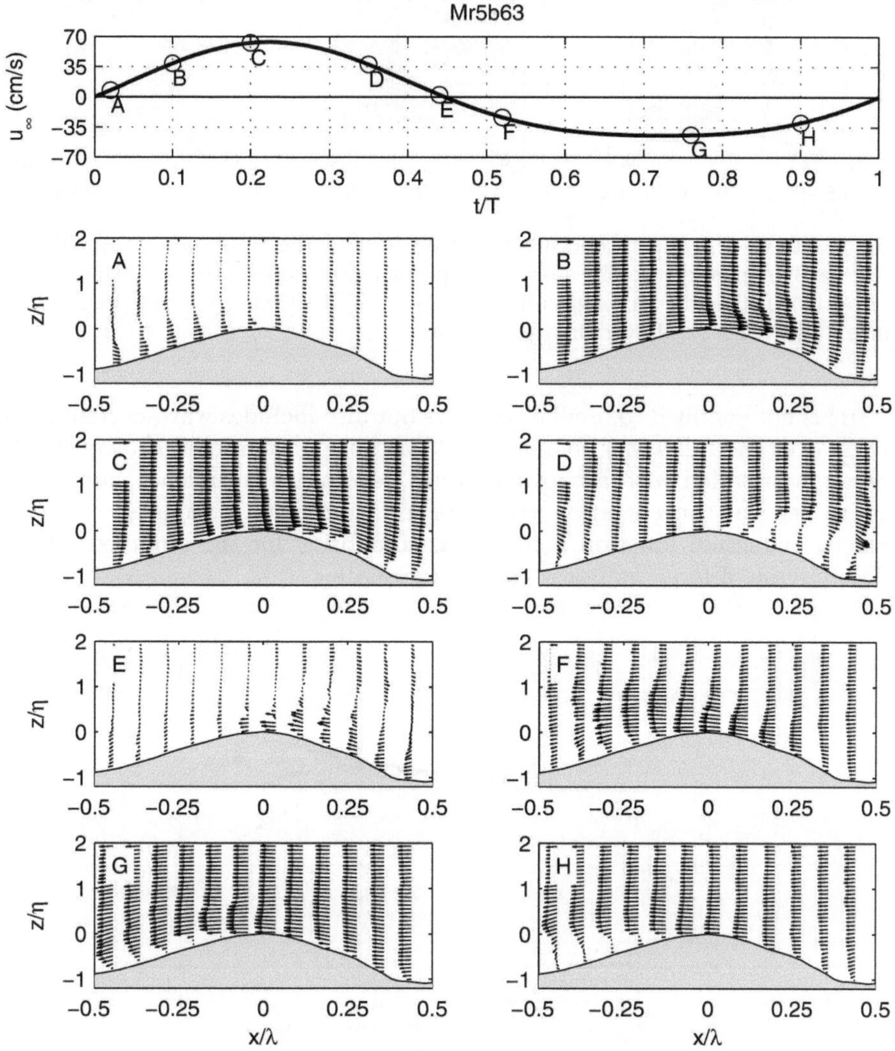

Fig. 5. Grain-velocity vector fields around a ripple at 8 phases during the wave cycle as measured with PIV (exp. Mr5b63) Positive, 'on-shore' flow is to the right. The top panel shows the free-stream orbital velocity u_∞ during the wave cycle. Vertical coordinate z and horizontal coordinate x are normalized with the ripple height η and respectively the ripple length λ.

time-dependent pressure-gradient near the sea bed and a rough wall boundary condition is assumed at the (fixed) bed level. A reference concentration at a level close to the bed is prescribed as a function of the time-dependent bed shear stress. The Point Sand Model (PSM) of Uittenbogaard and Klopman

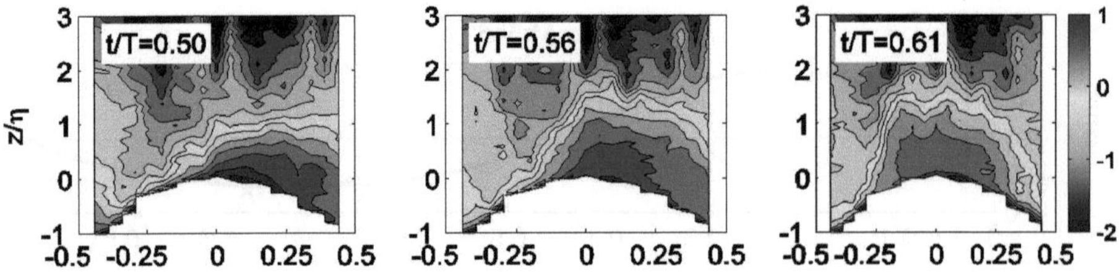

Fig. 6. Time-dependent concentration field around the ripple, as measured with ABS at 3 moments directly after flow reversal from 'on-shore' to 'offshore', i.e., $t/T = 0.5, 0.56$ and 0.61 (experiment Mr5b63), and showing the transport of the vortex filled with sand over the ripple crest.

(2001) is not confined to oscillatory flows but also includes wave-current interaction over the full water column. In PSM the equation for turbulent kinetic energy is provided with a buoyancy term, leading to turbulence suppression due to vertical concentration gradients (stratification). Moreover, a settling velocity reduction function is included, accounting for the hindered settling effect in case of large sediment concentrations.

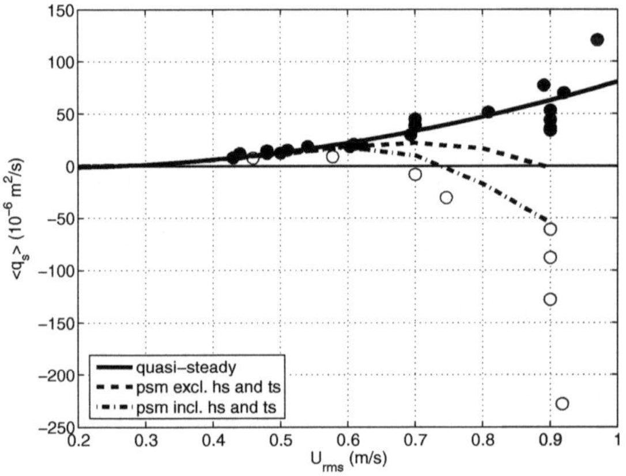

Fig. 7. Measured net transport rates (symbols) and predicted net transport rates with PSM (lines) for asymmetric waves as a function of U_{rms} (sheet flow regime). The solid line and black symbols refer to 3 medium sands (0.21, 0.32, 0.46 mm), the dashed lines and open symbols refer to 2 fine sands (0.13, 0.15 mm).

PSM is used to explain the phase-lag effects, occurring in oscillatory sheet flows, as discussed above. Figure 7 summarizes the results of this investigation by showing predicted and measured net transport rates as a function of U_{rms} (= root-mean-square velocity of the oscillatory flow velocity) for two fine sands with $D_{50} = 0.13$ and 0.15 mm and three medium sands with $D_{50} = 0.21$, 0.32 and 0.46 mm.

The measured transport rates are all from oscillatory flow experiments (LOWT and AOFT) with a small range in wave asymmetry ($R \approx 0.60 - 0.65$) and wave periods ($T = 5 - 9\ s$). In accordance with the data, PSM shows a distinct separation between positive ('on-shore') and negative ('offshore') transport rates for medium respectively fine sands. The medium sands behave in a quasi-steady way, showing increasing positive transport rates with increasing U_{rms} ($\langle q_s \rangle \sim U_{rms}^3$). The fine sands experience a substantial phase-lag effect, which is simulated by PSM ($D_{50} = 0.13\ mm$, $R = 0.62$, $T = 6.5\ s$) in a reasonable way. Inclusion of turbulence suppression and hindered settling effects in PSM leads to increased phase-lags and a better agreement between the data and the simulations. Nevertheless, it appears that the phase-lag effects as well as the negative transport rates are still underestimated by the model.

Similar conclusions follow from simulations with a two-layer model, provided with an empirical sheet flow layer description (Malarkey et al., 2003), and a two-phase flow model (see O'Donoghue et al., 2004). It is concluded that a better description of the time-dependent sand exchange between sea bed and sheet flow layer (pick-up and re-settling) is probably needed for further quantitative improvement of the models.

Using the new experimental tunnel data in the vortex ripple regime Van der Werf (2006) shows that - although this regime would require in principle a 2DV or 3D modeling approach - also a 1DV model with adjusted descriptions for eddy-viscosity and sediment entrainment is able to give a good description of the over-all time-dependent sand fluxes and net transport rates.

4 Unsteady sand transport formulas

The process-research as described above has stipulated the dominant role of phase-lag effects in the sand transport process by waves and their different character in the vortex ripple and sheet flow regime.

For the sheet flow regime the 'quasi-steady' transport model of Ribberink (1998) has therefore been adjusted to an 'unsteady' transport model. Hereto, the sheet flow concentrations are modeled in a schematic way as an advection-diffusion process with a constant eddy-viscosity ε_s and a simplified reference boundary condition. The equations are solved analytically for asymmetric oscillatory flows and a dimensionless phase-lag parameter is obtained, representing the ratio of sediment entrainment height above the bed (= sheet flow layer thickness) and the vertical settling distance of sediment during a wave period ($W_s T$). The sheet flow layer thickness δ_s is modeled as a function of the Shields parameter. Using this phase-lag model an unsteady transport formula is developed, which provides a strongly improved description of measured net transport rates in the sheet flow regime. For more details see Dohmen-Janssen, et al. (2002) and Hassan (2003).

In the vortex ripple-regime the phase-lag effects are generated in a different way than in sheet flow conditions, due to the vortex shedding process. Apart from the strength of the lee-side vortices and the moment of their upward injection during the wave cycle (near flow reversal), also the potential of the sediments to be entrained into suspension in the lee-side vortex plays an important role.

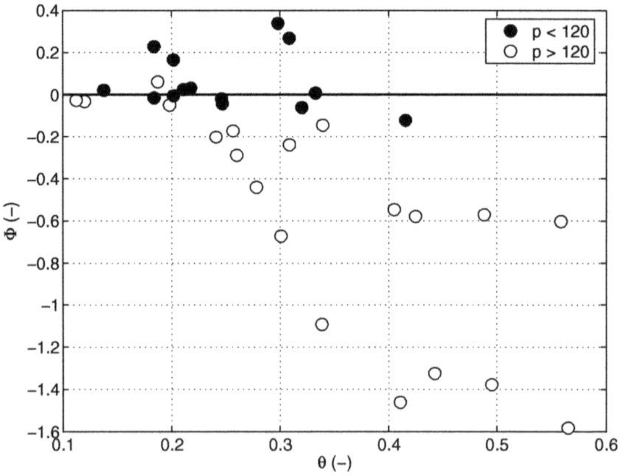

Fig. 8. Measured non-dimensional net transport rates Φ for asymmetric oscillatory flows in the vortex ripple regime as a function of the Shields parameter θ for 2 ranges of the vortex suspension parameter p. Positive Φ-values refer to 'on-shore' transport. The data are from oscillatory flow experiments with the same degree of asymmetry ($R = 0.6 - 0.63$), wave periods $T = 3 - 12\ s$ and grain-size $D_{50} = 0.21 - 0.44\ mm$.

Figure 8 shows how - during asymmetric wave experiments (in AOFT and LOWT) - the direction and magnitude of the dimensionless mean transport rate Φ $(= \langle q_s \rangle / \sqrt{\Delta g D^3})$ is influenced by two parameters, i.e., the Shields parameter θ $(= f_w U_{rms}^2 / (\Delta g D))$ and a new vortex-suspension parameter $p = \eta / D$. The latter parameter, representing the ratio of the ripple height η (dimension of the vortex) and the grain-size D, indicates to what extent suspended sediment and associated phase-lag processes are important. For small p-values (small vortex/ coarse sediment) bed-load transport dominates and the net transport rates are mainly 'on-shore'. For large p-values (large vortex / fine sediment) suspended sediment is dominant and phase-lag effects lead to 'offshore' directed transport rates. The transport rates also tend to become more negative ('offshore' directed) for increasing Shields parameter. Using the available dataset of tunnel measurements in the vortex ripple regime, Van der Werf (2006) developed a new empirical unsteady sand transport formula, in which the vortex suspension parameter p is one of the key parameters.

5 Conclusions

An overview is given of experimental research in oscillating water tunnels (LOWT and AOFT) aimed at a better understanding of sand motion in oscillatory flows. Two different transport regimes were investigated using advanced measuring techniques (CCM, UVP, PIV, ABS), i.e., the vortex ripple regime and the sheet flow regime. New insights and detailed data were obtained of the unsteady flow processes controlling the total net sand transport under asymmetric waves. Phase-lag effects or time-history effects during the wave cycle strongly control whether net sand transport under asymmetric oscillatory flows is 'on-shore' or 'offshore' directed. The time-history effects in the two transport regimes show large differences, mainly due to the large differences in sand entrainment and transport mechanisms. In the vortex ripple regime vortex shedding and suspended sediment generally lead to large phase-lags and 'offshore' directed transport. Oscillatory sheet flow is often a more quasi-steady process with 'on-shore' directed sand transport, except for fine sand and short wave periods when phase-lag effects may lead to 'offshore' transport. It was shown that a RANS model, which includes hindered settling and turbulence suppression is able to explain the direction change from 'on-shore' to 'offshore' transport in the sheet flow regime in a qualitative sense. Based on the new insights new unsteady sand transport formulas were developed for the two bed-form regimes.

References

[1] Bailard, J.A. (1981): An energetics total load sediment transport model for a plane sloping beach. J. Geoph. Research, Vol. 86, No. C11, pp. 10938-10954.

[2] Clubb, G.S. (2001): Experimental study of vortex ripples in full-scale sinusoidal and asymmetric flows, PhD thesis, University of Aberdeen, Scotland.

[3] Dohmen-Janssen (1999): Grain size influence on sediment transport in oscillatory sheet flow, phase lags and mobile-bed effects. Ph.D. Thesis, Delft Univ. of Technology, The Netherlands.

[4] Dohmen-Janssen, C.M., D.F. Kroekenstoel, W.N.M. Hassan, and J.S. Ribberink (2002): Phase lags in oscillatory sheet flow: experiments and bed load modelling. J. Coast. Eng., Vol. 46, pp. 61-87.

[5] Hassan, W.N.M.H. (2003): Transport of size-graded and uniform sediments under oscilla-tory sheet-flow conditions, Ph.D. Thesis, University of Twente, The Netherlands

[6] Malarkey, J., A.G. Davies, Z.Li (2003): A simple model of unsteady sheet flow sediment transport, Coastal Eng. Vol. 48, pp 171-188.

[7] McLean, S.R., J.S. Ribberink, C.M. Dohmen-Janssen and W.N.M. Hassan (2001): Sediment transport measurements within the sheet flow layer

under waves and currents. J. Waterway, Port, Coast. and Ocean Eng., ISSN 0733-950X.

[8] Nielsen, P. (1992): Coastal bottom boundary layers and sediment transport, World Scien-tific, Singapore, 324 p.

[9] O'Donoghue, T. and Wright, S. (2004) : Flow tunnel measurements of velocities and fluxes in oscillatory sheet flow for well-sorted and graded sands, Coastal Eng. Vol. 51, pp 1163-1184

[10] O'Donoghue, T., Ming Li, J. Malarkey, S. Pan, A.G. Davies, B.A. O'Connor (2004): Numerical and experimental study of wave generated sheet flow, Proc. 29th Int. Conf. On Coast. Eng., Lisbon, Portugal , pp. 1690-1702.

[11] O'Donoghue, T., Doucette, J.S., Van der Wef J.J. and J.S.Ribberink (2006): The dimensions of sand ripples in full-scale oscillatory flows, accepted for publication in Coastal Engineering.

[12] Ribberink, J.S. and A.A. Al-Salem (1994): Sediment transport in oscillatory boundary layers in cases of rippled beds and sheet flow. J. Geophysical Research, Vol. 99, No. C6, pp. 12,707-12,727.

[13] Ribberink, J.S. and A.A. Al-Salem (1995): Sheet flow and suspension in oscillatory boundary layers. Coast. Eng., Vol. 25, pp. 205-225.

[14] Ribberink, J.S. (1998): Bed-load transport for steady flows and unsteady oscillatory flows. Coast. Eng., Vol. 34, pp. 59-82.

[15] Ribberink, J.S., Dohmen-Janssen, C.M., Hanes, D.M., McLean, S.R. and C. Vincent (2000): Near-bed sand transport mechanisms under waves - a large-scale flume experiment (Sistex99), In Proc. 27th Int. Conf. On Coast. Eng., Sydney, Australia, pp. 3263-3276.

[16] Uittenbogaard, R.E. (2000): 1DV Simulation of Wave Current Interaction. In Proc. 27th Int. Conf. On Coast. Eng., Sydney, Australia, pp. 255-268.

[17] Uittenbogaard, R.E. and G. Klopman (2001): Numerical simulation of wave-current driven sediment transport, Proc. Coastal Dynamics'01, ASCE, Lund, pp 568-577.

[18] Van der Werf, J.J. (2006): Sand transport over rippled beds in oscillatory flow, Ph.D. Thesis, University of Twente, The Netherlands

[19] Wright, S. (2002): Well-sorted and graded sands in oscillatory sheet flow, PhD thesis, Uni-versity of Aberdeen, Scotland.

Sediment transport, ripple dynamics and object burial under shoaling waves

S.I.Voropayev[1,2], F.Y.Testik[1,3], H.J.S.Fernando[1], and S.Balasubramanian[1]

[1] Arizona State University, Environmental Fluid Dynamics Program, Department of Mechanical and Aerospace Engineering, Tempe, AZ, 85287-9809
s.voropayev@asu.edu; j.fernando@asu.edu; sridhar29@asu.edu
[2] Institute of Oceanology, Russian Academy of Sciences, Moscow, 117851, Russia
[3] Present address : Clemson University, Department of Civil Engineering, Clemson, SC, 29634-0911 ftestik@clemson.edu

Summary. A comprehensive study is being conducted with emphasis on establishing laboratory-based physical model for predicting ripple dynamics and scour, burial behavior of bottom objects in oceanic coastal environments. A brief description of this study as well as salient observations and quantitative parameterizations on ripples, scour and burial that resulted from it are presented. Comparisons with field data are also given. The results have important coastal engineering and naval applications related to sediment transport and coastal morphodynamics under shoaling waves that are typical of coastal zone.

1 Introduction and general analysis

This research is motivated by recent naval interests to predict the behavior and fate of cylindrical anti-ship mines that are deployed in the coastal shoaling zone and are heavy to repose on the ocean floor with little migration. The behavior of objects of different shape, which can move/drift under the wave action, was studied in (Voropayev et al., 1998, 2001; Luccio et al., 1998) and not considered here.

A solid object resting at the sand-water interface in the oceanic coastal region is subject to highly nonlinear processes associated with the sand-water-object interaction. A myriad of physical processes determine the fate of the object (scour, burial, migration, etc.) and morphodynamics at the immediate vicinity of the object. To develop predictive models, a fundamental understanding of the sand-water-object interaction under shoaling and breaking waves is required. However, nonlinear nature of these processes and particular complications associated with the interaction of three (sand-water-object) different phases, such as sediment transport modeling, changing boundaries, have been the bane for theoretical and numerical analysis (see e.g., Sleath, 1984; Blondeaux, 1990; Nielsen, 1992; Vittory and Blondeaux, 1990; Andersen, 2001; Zhao and Fernando, 2006 and references herein). Therefore, an

Bernard J. Geurts et al. (eds), Particle Laden Flow: From Geophysical to Kolmogorov Scales, 15–27.

experimental approach is used in our studies to establish a laboratory-based physical modeling for predicting the ripple dynamics and scour/burial behavior of bottom objects in the oceanic coastal environment.

Taking into account that large number of governing parameters significantly complicates the laboratory modeling, we first introduce some justified simplifications and identify the most important governing parameters. Following Voropayev et al., (2003a), consider a heavy cylinder (diameter D, length L_c) placed at time $t = 0$ on a layer of sand (grain size d, density ρ_s) placed on a slope (angle β) in oscillating water flow (oscillation frequency ω, amplitude of the maximum near-bottom horizontal wave orbital velocity U, water density ρ and viscosity ν) and the cylinder axis has an angle α relative to the flow direction. The full set of (twelve) primary external parameters determining a particular scour/burial characteristic, say B, of the cylinder is thus given by

$$B = F(D, L_c, d, \rho_s, t, U, \omega, \rho, \beta, \nu, g, \alpha) \quad (1)$$

(g - gravitational acceleration, F - function). Using dimensional analysis and general physical arguments, (1) can be simplified and written in dimensionless form as

$$B^* = F^*(Re, KC, \theta, \beta, \alpha, \tau, a). \quad (2)$$

Here we use: the Reynolds number, $Re = DU/\nu$, Keulegan-Carpenter number, $KC = UT/D = 2\pi\varepsilon/D$ (T - wave period, $\varepsilon = U/\omega$ - amplitude of near-bottom horizontal water particle excursion), Shields parameter, $\theta = (f/2)\Psi$ [$\Psi = U^2/(g^* d)$ - mobility parameter, $g^* = g(\rho_s/\rho - 1)$ and f - friction coefficient that is well parameterized (Nielsen 1992, Sleath 1984)], non-dimensional time, $\tau = \omega t$, and aspect ratio, $a = L_c/D$. For the simple case of $\alpha = 0$, $a = 5$, corresponding to field experiments, and assuming the Reynolds number similarity, (2) becomes

$$B^* = F^*(KC, \theta, \beta, \tau), \quad (3)$$

thus simplifying the problem considerably. For fixed bed slope and steady state, $\tau \gg 1$, only two parameters remain in (3).

To evaluate the functional form in (3) (see Voropayev et al., 2003a,b; Testik 2003; Testik et al., 2005a,b), laboratory experiments were conducted and the main dimensional and dimensionless governing parameters for laboratory and oceanic conditions are given in Table 1. As can be seen, it is possible to match KC and θ between typical oceanic and laboratory situations, but there is a large mismatch of Re. However, observations (Testik et al., 2005b) demonstrate that the Reynolds number flow similarity holds for $Re \geq 2000$, and recent field data show that the results obtained under similarity assumption are valid for very large Re (Quinn, 2006).

Table 1. Typical values of parameters

	Laboratory	Ocean
U (cm s^{-1})	10-46	20-100
T (s)	2.5-5	4-10
D (cm)	3-8	50
d (cm)	0.04-0.06	0.02-0.1
KC	6-60	1.6-16
θ	0.015-0.3	0.01-0.8
Re	(4-35)10^3	(1-5)10^5

2 Experimental set-up and flow conditions

A detailed description of the experimental facility is given in (Testik, 2003, Voropayev et al., 2003a,b) and only a brief description is given below. The oceanic coastal zone was modeled in a wave tank ($3200 \times 90 \times 180$ cm^3) with a sloping (slope, $\beta = 1/24$) sandy bottom and before each experiment the sand surface on the slope was made planar. A vertical paddle, driven by a piston, is used to generate sinusoidal periodic waves, with the water depth near the paddle being 100 cm (Fig. 1). The tank is equipped with measurement instruments, which include wave gages, a three-component acoustic Doppler velocimeter (ADV) and a structured light device (SLD) (Faraci et al., 2000).

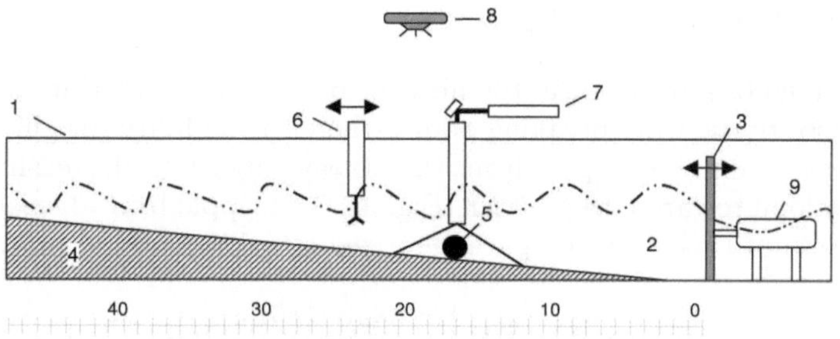

Fig. 1. Schematic of the experimental setup: 1 - tank, 2 - water, 3 - vertical wave-paddle, 4 - sloping bottom, 5 - bottom object, 6 - acoustic Doppler velocimeter (ADV), 7 - structured light device (SLD), 8 - photo/video camera, 9 - hydraulic system. Section numbers are also shown and the length of each section is 61 cm.

Although the wave paddle forcing is sinusoidal, with frequency ω and peak-to-peak horizontal displacements $2\varepsilon_0$, the waves steepen and change their height as they propagate from deep to shallow water. The underlying velocity field becomes increasingly nonlinear and more energetic (Fig. 2) with noticeable wave asymmetry and undertow velocity (Testik et al., 2006a). While this complicates somewhat the flow description (the wave characteristics should be measured/parameterized along entire slope), such geometry has obvious benefits. The advantage of the above experimental set-up with a slope, compared to the standard set-up with horizontal bed, is that in our geometry the flow intensity (e.g., U and ε) change along the slope, thus enabling one to collect data on bottom morphology under different local flow conditions (different dimensionless parameters) using a given experiment with fixed wave forcing. In the latter approach, however, a series of experiments is necessary for this purpose.

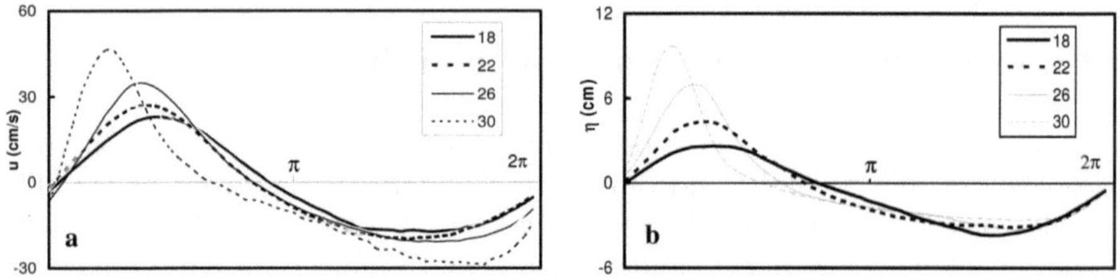

Fig. 2. Typical horizontal along-slope near-bed water velocity profiles (a) (at 10 cm above the bed) and wave elevation profiles (b) at four different sections (see legend for section numbers).

3 Ripple dynamics

3.1 Ripple formation and growth

Our experiments show that, in the absence of an object, the ripples first start to form close to the wave breaking point, where the mobility (or Shields) parameter is maximum. The ripple front then propagates with decreasing velocity down the slope towards deep water (Fig. 3). The ripple front displacement X as a function of time t can be parameterized as (Voropayev et al., 2003b)

$$X/X_0 = 1 - \exp(-t/t_0), \tag{4}$$

where X_0 - maximum displacement and $t_0 = C/(\omega\Psi_0^{1/2})$ - characteristic time ($\Psi_0 \approx 5.4$ is the critical value of the mobility parameter for ripple formation, $C \approx 2500$ - empirical constant).

Observations show that the height h and spacing Λ of ripples evolve with time t as (Voropayev et al., 1999, 2003b)

$$\Lambda(t)/\Lambda_0 = h(t)/h_0 = 1 - \exp(-t/t_1), \quad t_1 = C/\omega\Psi^{1/2}. \tag{5}$$

After a typical transition time interval (t $\approx 3t_1$, t_1 - characteristic time for ripple growth), the ripples reach an equilibrium state with a characteristic height, h_0, and spacing, Λ_0, given by (see Sleath, 1984; Nielsen, 1992; Voropayev et al., 1999, 2003b)

Fig. 3. Ripple front (shown by arrow) propagating (a, b) along initially flat slope in the offshore (from left to right) direction.

$$\Lambda_0 = \varepsilon(2.2 - 0.35\Psi^{1/3}), \quad h_0 = \Lambda_0(0.18 - 0.24\theta^{3/2}), \tag{6}$$

where ε, Ψ and θ are based on local values. Note, that (4) and (5) also describe sand ripple growth in an annular sand-water cell (Scherer et al., 1999).

3.2 Similarity of asymmetric ripples

Under asymmetric waves (Fig. 2) ripples are also asymmetric (Fig. 4). Results of measurements (Testik et al., 2006a) show that under nonlinear shoaling waves sand ripple profiles demonstrate self-similar behavior. When dimensional data on ripple sizes (Fig. 5a) are plotted in dimensionless form (Fig. 5b) this similarity becomes obvious. Measurements show that in a broad range of parameters studied, the resulting dimensionless ripple profile (shown by solid lines in Fig. 5b) can be approximated by a similarity profile of "saw tooth" shape, given by

$$z^* = \begin{cases} C_1^{-1}[x^* \tan(\theta_1 - \beta) + \sin\beta], & 0 \le x^* \le C_1/\tan\theta_1 \\ C_1^{-1}(1 - x^*)\tan(\theta_2 + \beta), & C_1/\tan\theta_1 < x^* \le 1 \end{cases}, \tag{7}$$

where $z^* = z/h_0$, $x^* = x/\Lambda_0$, $C_1 \approx 0.22$ - averaged ripple steepness and θ_1 and θ_2 are "universal" on-shore θ_1 ($\approx 34^o$) and offshore θ_2 ($\approx 18^o$) ripple slope angles. It is interesting that when the slope of the tank bottom is subtracted from θ_1, the on-shore ripple angle becomes equal to the "avalanche" (repose) angle for the particular sand used for the experiments.

3.3 Ripple drift

At larger times, $t \gg t_1$, the morphology of the rippled bed continues to change and ripples demonstrate large time instabilities and drift slowly in the on-shore direction. Under symmetric oscillatory flow this drift is predominantly

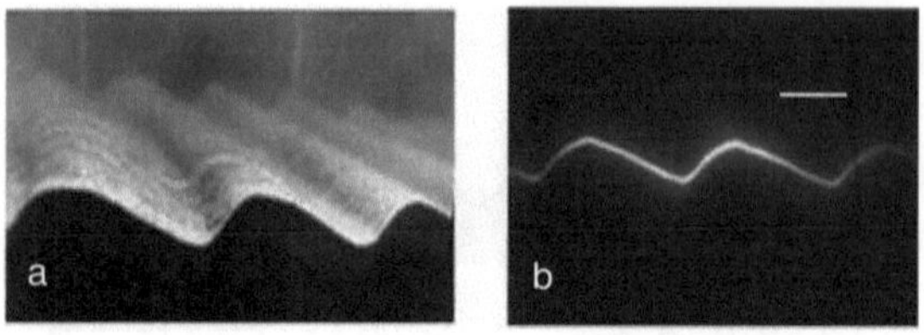

Fig. 4. Photograph (a) and a structured light profile (b) of asymmetric ripples under shoaling waves. Onshore direction is to the left and white line in (b) gives the scale (10 cm).

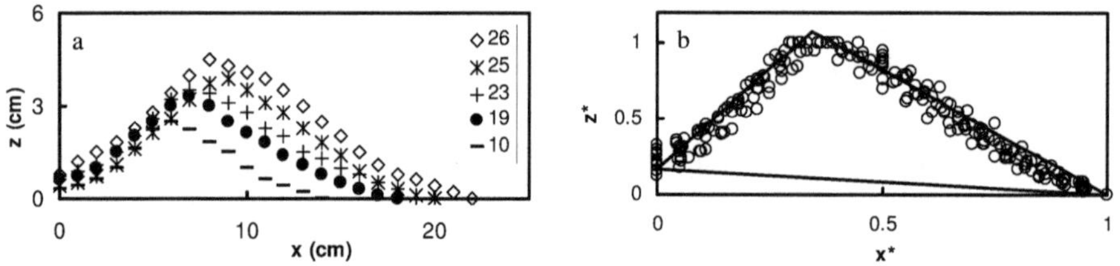

Fig. 5. Typical dimensional (a) and dimensionless (b) equilibrium ripple profiles taken at different sections (see legend) along the slope. Symbols - measurements, solid line in (b) - parameterization (7).

"stochastic" (with zero mean ripple displacement over large time interval, Voropayev et al., 1999), but under nonlinear shoaling waves the dominant is the "unidirectional" drift when all ripples slowly drift in the onshore direction with typical local velocity

$$u = U_+ C_2 A^{3/4} \Psi^{1/2}. \tag{8}$$

Here $C_2 \approx 10^{-4}$ - empirical coefficient, A - wave asymmetry parameter, $A = (U_+/U_-)^2$, and U_+ and U_- - maximum positive (onshore) and negative (offshore) near-bed water velocities at 10 cm above the bed. Note, that at large times (several hours) a sand bar was formed near the breaking point (Voropayev et al., 2003b) that strongly modified the flow. All quantitative data on ripple dynamics and scour/burial were taken before a noticeable sand bar was formed.

3.4 Effect of variable wave forcing

In the ocean, waves frequently change their intensity. To model the ripple dynamics under these conditions, a series of experiments was conducted (Testik et al., 2005a) with variable wave forcing. In these experiments waves with relatively large (L), moderate (M) and small (S) intensities were generated, and three basic cases of cyclic variation of wave forcing, namely L-M-L, M-L-M, and L-S-L were studied. Observations show that upon the change of wave forcing the ripples adjust themselves to the new forcing through proper adjustment processes. Depending on the forcing transitions (L-M, M-L or L-S), three basic adjustment processes were documented as ripple splitting, ripple re-growth and ripple flattening.

Typical example showing the process of ripple splitting and re-growth for L-M transition is shown in Fig. 6. Initially (Fig. 6a) ripples are at an equilibrium state and they are formed under forcing (L). When forcing is changed from (L) to (M), the size of initially large ripples decreases, and the adjustment takes place in two steps. First, the ripple splitting occurs and their size is halved (Fig. 6b), thus doubling the number of ripples. This system of transient ripples, with length and height half of the initial ones, is unstable under forcing (M). With time the transient ripples (Fig. 6b) increase their size to the equilibrium value, which are predetermined by the forcing (M) (Fig. 6c). To explain the results of observations, a model was advanced and Fig. 7 gives an example of how accurately this model describes the results obtained in the run L-M-L.

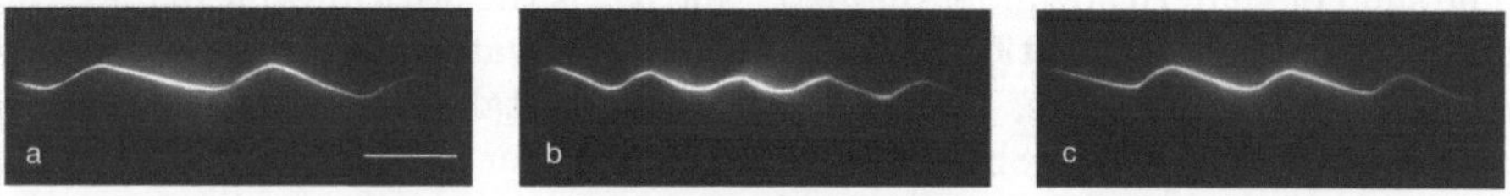

Fig. 6. Initial ripples under (L) forcing (a). After wave forcing was changed from (L) to (M), ripples first split into two smaller ripples (b) and then re-grow (c) to their equilibrium sizes for (M) forcing. White line in (a) gives the scale (10 cm).

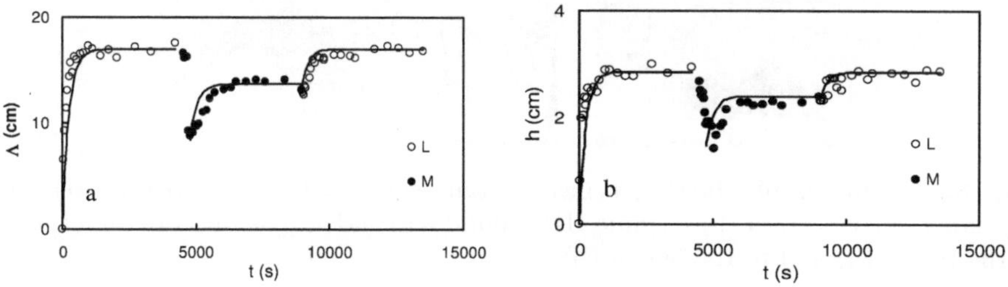

Fig. 7. The ripple length (a) and height (b) adjustment with time for run L-M-L. Symbols - measurements, solid lines - model predictions.

4 Scour/burial of short cylinder

4.1 Scour/burial regimes

The experiments on object burial delved into several issues, in particular, burial mechanisms and scour around object (Voropayev et al., 2003a). In this context, a scour/burial regime diagram was developed for the case of constant wave forcing, wherein four main scour/burial regimes and a sheet flow regime were identified as follows:

Regime I - No scour ($KC < KC^* \approx 2$ or $\theta < \theta^* \approx 0.018$; here, θ^* is the critical θ for the initiation of sediment motion in the proximity of a cylindrical mine and KC^* is the critical KC for the generation of vortices for an impulsively started flow around a cylinder).

Regime II - Initial scour ($KC > KC^*$ and $\theta > \theta^*$; scour usually occurs on the onshore side of the cylinder, Fig. 8a, due to wave asymmetry).

Regime III - Expanded scour ($KC^* < KC < KC^{**} \approx 14$, $\theta > \theta^{**} \approx 0.054$; here, KC^{**} is the critical value of KC, below which ripple height is expected to be smaller than the cylinder diameter and θ^{**} is the critical value of θ for the initiation of ripple formation. When background water motion is sufficiently energetic to form ripples, regime II becomes unstable and transforms into expanded scour, see Fig. 8b. If the cylinder diameter is larger than the ripple height, the flow disturbances due to the cylinder are dominant, the expanded scour pattern is stable and it persists for many hours.

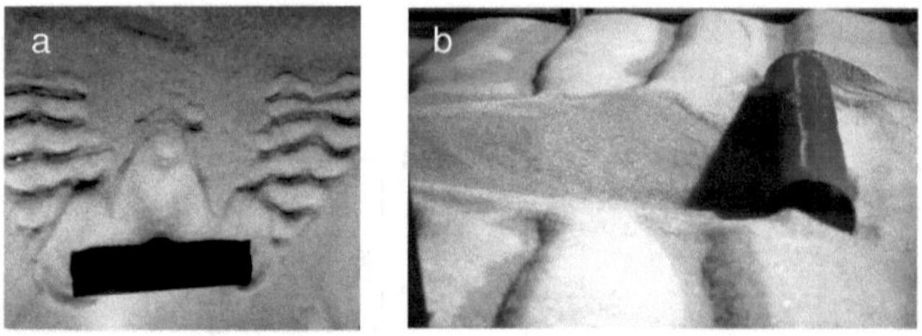

Fig. 8. A photograph showing typical initial scour pattern (a) (regime II) and expanded scour pattern (b) (regime III) around a cylinder. The onshore direction is to the top in (a) and to the left in (b).

Regime IV - Ripple dominated scour, with possible periodic burial ($KC > KC^{**} \approx 14$, $\theta > \theta_{cr}* \approx 0.054$; If the ripple height is larger than the cylinder diameter, the expanded scour pattern transforms into regime IV, wherein the cylinder is buried periodically under drifting ripples).

Regime V - Sheet flow ($\theta > \theta_0 \approx 0.83$; here θ_0 is the critical value of θ, above which sheet flow occurs. At such high values of Shields parameter, bottom features are “washed” away to form a planar bathymetry).

Collectively, the above limiting asymptotic estimates for KC and θ values, that demarcate different regimes, are depicted on the scour/burial regime diagram shown in Fig. 9 by dashed lines. Empirical parameterizations for transitional boundaries, that best fit the experimental data while being consistent with the limiting asymptotic behaviors between basic regimes discussed above, were identified and parameterized as follows:

$$KC = KC^* + (\theta^{**} - \theta)/(\theta - \theta^*), \tag{9}$$

$$KC = KC^* + (\theta_0 - \theta)^5/(\theta - \theta^{**}), \tag{10}$$

$$KC = KC^{**} + 1/(\theta - \theta^{**})^{2/3}. \tag{11}$$

These transitional boundaries are shown in Fig. 9 by solid lines.

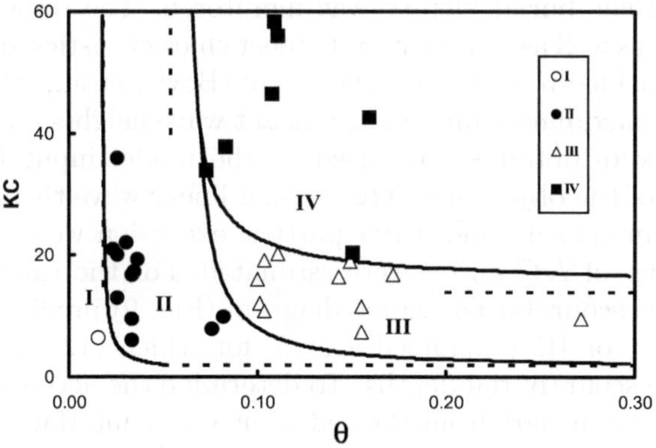

Fig. 9. Scour/burial regime diagram showing four regimes (I-IV) depending on the values of KC and θ parameters. Dashed lines show limiting asymptotes, solid lines show transition boundaries between different regimes in accordance with (9)-(11).

4.2 Scour depth and length

Scour measurements (Voropayev et al., 2003a) near the cylinder showed fast scour at the beginning, which slowed down with time similar to (4)

$$S(t)/S_0 = 1 - \exp(-t/t_3), \tag{12}$$

where $t_3 = C_3/\omega$ is the scour characteristic time, $S(t)$ the maximum depth of scour at time t and $C_3 \approx 3300$. Thus the typical time at which the system reaches an equilibrium state ($3t_3$, time for $S \approx 0.95\ S_0$) is about 1600 ($\approx 9900/2\pi$) wave periods. The steady state maximum scour depth S_0 was found to depend strongly on both KC and θ as

$$S_0/D = C_4[1 - \exp(-m(KC - KC^*))][1 - \exp(-n(\theta - \theta^*))], \tag{13}$$

whereas the dimensionless scour length L_0/D, depended only on KC as

$$L_0/D = C_5 KC. \tag{14}$$

The best-fit empirical constants in (13), (14) are $C_4 \approx 1.3$, $m \approx 0.06$, $n \approx 40$ and $C_5 \approx 0.35$. Note that the strong dependence of scour on KC is consistent with numerous recent data on scour around different two- and three-dimensional objects and should not be ignored (Sumer and Fredsoe, 2002).

5 Comparison with field observations

The results of laboratory based modeling described above collectively form the basis of a object burial predictive model recently proposed by Testik et al. (2006b). To verify our model, we compare below the model predictions with available observations of a recent field experiment, in which real size bottom objects of interest (four acoustic instrumented mines, AIMs), were deployed and their burial history was monitored. The data on background flow conditions as well as sediment and object characteristics and burial depths were kindly provided to us by the authors of (Bower et al., 2004, 2006). From these data the maximum values of significant wave heights and corresponding wave periods were obtained and used as the model input for a particular (analysis) day of the object site. Then, using linear wave theory, the values of the near bottom velocity and water particle excursion were calculated, from which the values of KC and θ were estimated. For the range of KC and θ so obtained, the scour/burial regime diagram (Fig. 9) predicts either regimes II (initial scour) or III (expanded scour), for which (12), (13) can be used to estimate the scour/burial depths, To determine the actual burial (distance of the AIMs bottom part from the sediment-water interface), the heights of sand ripples formed around the scour pit were also estimated from (5), (6) and these values were added to the burial depth prediction due to scour alone. The burial predictions of ASU model are compared with field measurements in Fig. 10. As can be seen, the agreement is satisfactory. Note, that the ratios of the ripple height to the object diameter in laboratory ($h_0/D \approx 0.13$) and field (≈ 0.2) are comparable and the effect of ripples on the burial depths is approximately the same in both cases.

6 Conclusions

Many extensions are desirable for the ASU model of Testik et al. (2006b). First, although only wave-induced scour is considered as the main burial mechanism (fortuitously this mechanism was dominant in the field experiment), there are other major burial mechanisms such as tide-induced scour, gravity sinking, liquefaction and impact/initial burial that can become important in field conditions. Such mechanisms need to be implemented in the model based on existing and future work. Second, in laboratory modeling of object scour/burial, the focus was on fixed wave conditions, but natural conditions are far from steady, requiring the consideration of time evolving wave

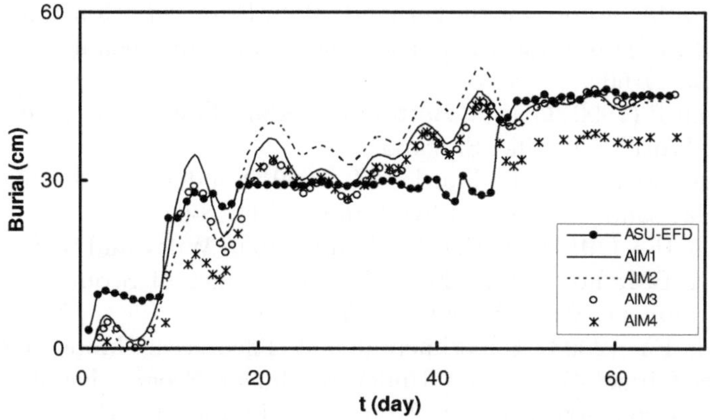

Fig. 10. Comparison of mine burial predictions of ASU model with the field observations for four AIMs (see legend).

fields, similar to that considered in our modeling of ripple dynamics. The local (temporal) scaling developed in our study will help to gain a quantitative understanding of sediment transport under such evolving wave fields. Finally, in our studies we used homogeneous sand. In the ocean, sand is usually heterogeneous and sand segregation on ripples and surrounding the objects is frequently observed in the field. To clarify this effect, which is important in acoustic detection of objects, additional laboratory modeling with heterogeneous (e.g., bi-modal, as a first step) sediments is required.

Acknowledgment

This research was supported by the Office of Naval Research.

References

[1] Andersen KH (2001) A particle model of rolling grain ripples under waves. Phys Fluids 13(1):58-64.

[2] Blondeaux P (1990) Sand ripples under sea waves. Part 1. Ripple formation. J Fluid Mech 218:1-17.

[3] Bower GR, Richardson MD, Briggs KB, Elmore PA, Braithwaite EF, Bradley J, Griffin S, Wever TF, Luhder R (2006) Measured and predicted burial of cylinders during the Indian Rocks Beach experiment, January - March 2003. IEEE J Oceanic Eng:in press.

[4] Bower GR, Richardson MD, Briggs KB, Vaughan WC, Kennedy CS, Braithwaite EF, Wever T, Luhder R (2004) Indian rocks beach experiment, January-March 2003. Naval Research Laboratory, Memorandum.

[5] Faraci C, Foti E, Baglio S (2000) Measurements of sandy bed scour processes in an oscillating flow by using structured light. Measurements 28:159-174.

[6] Luccio PA, Voropayev SI, Fernando HJS, Boyer DL, Houston WN (1998) Motion of cobbles in the swash zone on an impermeable slope. Coastal Eng 33:41-60.
[7] Nielsen P (1992) Coastal Bottom Boundary Layers and Sediment Transport. World Scientific, Singapore.
[8] Scherer MA, Melo F, Marder M (1999) Sand ripples in an oscillating annular sand-water cell, Phys Fluids 11(1):58-67.
[9] Sleath JFA (1984) Sea Bed Mechanics. John Wiley and Sons, New York.
[10] Sumer BM, Fredsoe J (2002) The Mechanics of Scour in the Marine Environment. World Scientific, Singapore.
[11] Testik FY (2003) Experimental and Theoretical Modelling of Sand-Water-Object Interaction Under Nonlinear Progressive Waves. Ph.D. Dissertation, Arizona State University, Tempe, AZ, USA.
[12] Testik FY, Voropayev SI, Fernando HJS (2005a) Adjustment of sand ripples under changing water waves. Phys Fluids 17(7):072104.
[13] Testik FY, Voropayev SI, Fernando HJS (2005b) Flow around a short horizontal bottom cylinder under steady and oscillatory flows. Phys Fluids 17(4):047103.
[14] Testik FY, Voropayev SI, Balasubramanian S, Fernando HJS (2006a) Self-similarity of asymmetric sand ripple profiles formed under nonlinear shoaling waves. Phys Fluids 18(10):027610.
[15] Testik FY, Voropayev SI, Fernando HJS, Balasubramanian S (2006b) Mines burial in the shoaling zone: scaling of laboratory results to oceanic situation. IEEE J Oceanic Eng: in press.
[16] Vittory G, Blondeaux P (1990) Sand ripples under sea waves. Part 2. Finite-amplitude development. J Fluid Mech 218:19-39.
[17] Voropayev SI, Cense A, McEachern GB, Boyer DL, Fernando HJS (2001) Dynamics of cobbles in the shoaling region of a surf zone. Ocean Eng 28(7):763-788.
[18] Voropayev SI, McEachern GB, Boyer DL, Fernando HJS (1999) Dynamics of sand ripples and burial/scouring of cobbles in oscillatory flow. Appl Ocean Res 21(5):249-261.
[19] Voropayev SI, Roney J, Fernando HJS, Boyer DL, Houston WN (1998) The motion of large bottom particles (cobble) in a wave induced oscillatory flow. Coastal Eng 34(3-4):197-219.
[20] Voropayev SI, Testik FY, Fernando HJS, Boyer DL (2003a) Burial and scour around short cylinder under progressive shoaling waves. Ocean Eng 30(13):1647-1667.
[21] Voropayev SI, Testik FY, Fernando HJS, Boyer DL (2003b) Morphodynamics and cobbles behaviour in and near the surf zone. Ocean Eng 30(14):1741-1764.
[22] Quinn R (2006) The role of scour in shipwreck site formation pocesses and the preservation of wreck associated scour signatures in the sedimentary record - evidence from seabed and sub-surface data. J. Arch. Science 33:1419-1432.

[23] Zhao Z, Fernando HJS (2006) Numerical simulation of scour around pipelines using an Euler-Euler coupled two phase model. Theor Comp Fluid Dyn:under revision.

On the influence of suspended sediment transport on the generation of offshore sand waves

Fenneke van der Meer[1], Suzanne J.M.H. Hulscher[1] and Joris van den Berg[2]

[1] Water Engineering and Management, University of Twente, Enschede, The Netherlands F.M.vanderMeer@utwente.nl
[2] Numerical Analysis and Computational Mechanics, University of Twente, Enschede, The Netherlands.

Summary. Sand waves are bed-forms occurring in shallow seas. Although their characteristics are mainly affected by bed load transport, during rough weather suspended sediment transport can influence their characteristics. As a first step to model these influences, we added suspended sediment transport to a numerical 2DV model that was specifically developed for simulating sand waves. In this paper, results are presented for initial, small amplitude, sand waves. Incorporating suspended sediment transport increases the growth rate of sand waves significantly while their wave length is more robust. Furthermore, we found that the results are sensitive to flow conditions, as expected, and sediment diffusivity, which needs a more advanced description.

1 Introduction

The sea bed of a shallow sea is rarely flat. Various bed-forms occur, varying from small scale ripples and mega-ripples to large scale sand banks. Sand waves are bed-forms with a scale between these two extremes. The wavelengths of these bed-forms vary between 100 and 800 meters, and heights can reach up to one third of the water depth (i.e. a maximum of around 10 meters in 30 meters of water). These characteristics, together with the fact that sand waves can migrate several meters per year and that they cover the majority of the Southern North Sea ([16], Figure 1, mean that they affect human activities in shallow seas. Therefore, we aim to model and so better understand the dynamics of these sand waves and the influence of both the tidal motion and weather conditions.
Observations indicate that sand waves change due to suspended sediment transport, especially during rough weather conditions. For example, [11] investigated sand waves in the Dutch coastal area, together with their physical environment. He concluded that sand waves occur under sufficiently high cur-

Bernard J. Geurts et al. (eds), Particle Laden Flow: From Geophysical to Kolmogorov Scales, 29–41.

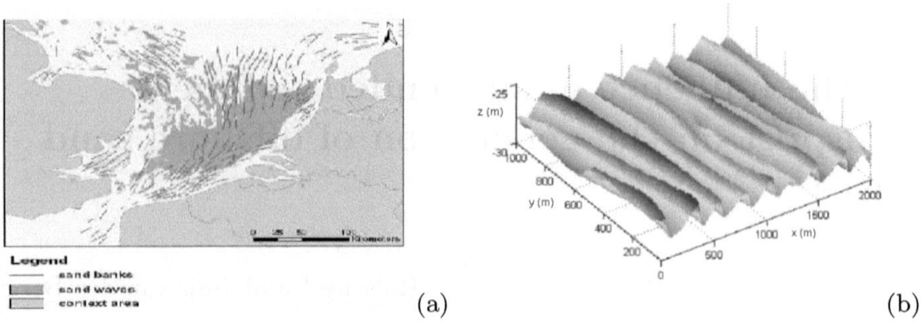

Fig. 1. (a) Sand wave and sand bank occurrence in the Southern part of the North Sea; (b) measured field data of a sand wave field in the North Sea.

rent velocities, low to moderate wave activity and an asymmetry in the tidal ellipse. Observations indicated that sand waves were lower on locations with more suspended sediment transport. Recently, [13] investigated weather influences on compound sand waves and mega-ripples. Although mega-ripples were found to be directly influenced by an individual storm, [13] concluded that sand wave morphology is a result of the general wind-wave climate. This, together with the local setting in which sand waves occur, was reasoned to lead to variation in sand wave shapes.

Offshore in shallow seas, bed load transport is expected to be the main sediment transport mechanism. As the water depth is in the order of tens of meters, under normal conditions short surface waves rarely interact with the sea bed. As grain sizes in sand wave areas are typically around the 200-300μm, vertical velocity due to the tidal current is lower than the fall velocity for the sediment for most of the tide, so the suspension of sediment only occurs during part of the tidal cycle. However, under storm conditions, suspended sediment can play an important role, especially in relatively shallow water. [2] found that under normal conditions and a flat bed, suspended sediment transport could be 30% of the bed load transport. [4] found seasonal dependency of sand wave height and migration in the Marsdiep. In this long term data set (1998-2005), sand waves were on average 30% higher after calm summer periods and lower after winter seasons. This variability was greater in locations where, due to finer sediment and stronger tidal currents, suspended sediment is more abundant.

Though techniques to measure sand wave characteristics as height and migration are improved over the past decades, still measurements are expensive and little detailed data is available. Modeling sand wave characteristics improves our knowledge of sand wave behavior and the processes underlying this behavior. In this way modeling can help where data is unavailable or insufficient.

The described studies indicate that suspended sediment transport influences sand waves, yet the effects have not yet been thoroughly investigated. Though the occurrence of suspended sediment transport increases the total sediment transport, it is not clear beforehand, whether suspended sediment will increase the sand wave growth, or will repress this growth. This depends on the distribution of erosion and deposition of the suspended sediment over the sand wave.

In this paper, we aim to contribute to an understanding of the effects of suspended sediment transport on the initial stage of sand wave formation. Hereto, we investigated the effect of implementing suspended sediment transport in the 2DV numerical sand wave code and the sensitivity of several parameters. After a short state of the art overview (Section 2.1), we will discuss the used model and the equations implemented to describe suspended sediment (Section 3). In Section 4, the simulations and their results are described, after which in Section 5 the results are discussed and conclusions are drawn.

2 Sand wave modeling

2.1 State of the art

[8] described sand waves as free instabilities of the coupled system of a sandy sea bed and a tidal flow. In such a system, vertical vortices play a crucial role. Small perturbations of the sea floor cause small perturbations in the flow field and vice versa. The bed can be either stable, which means that disturbances will be damped, or unstable, which means that bed perturbations will grow and the sea bed is changed. When perturbations are unstable (i.e. triggering growth) the flow field is changed such that, averaged over the tidal cycle, small vertical rest circulation cells occur (Figure 2). These cells cause small net transport to the crests of the perturbation, thereby causing growth. This process can be described using a linear stability analysis. A linear analysis is valid only for small, formally only infinitesimal small, amplitude perturbations. To overcome this limitation, numerical tools have been developed that allow the simulation of fully developed sand waves ([12],[15]). [12] showed that non-linear sand waves can be simulated with only bed load transport and an unidirectional steady current. [1] investigated the effects of tidal waves and asymmetry, explaining migration of small linear sand waves. [15] extended to fields of sand waves and showed sand waves to develop from random small bottom disturbances. Recently [3] showed effects of wind and waves on a sand pit situation, in which waves had a quantitative but no qualitative effect on the sand pit. In these studies, the effects of suspended sediment on sand wave formation, i.e. the aim of this paper, have not yet been studied.

3 Sand wave model

The simulation method used in this paper is based on the model described by [8], who applied a linear stability analysis. The linear stability analysis predicts

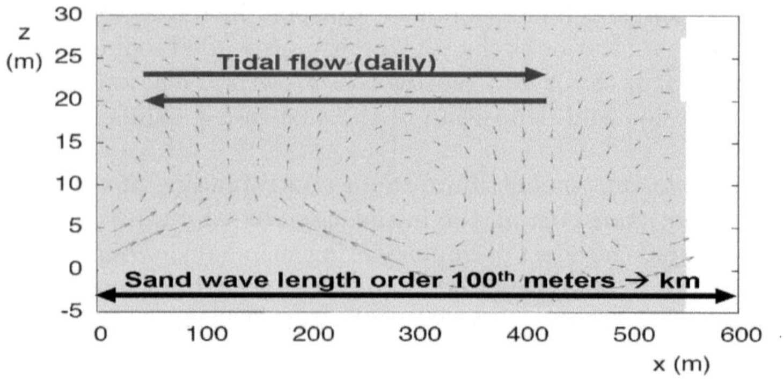

Fig. 2. Averaged over a tidal cycle, vertical circulations cells can occur due to small perturbations in the sea floor, causing growth or decay of these perturbations.

sand waves to grow exponentially in their initial stage. Starting with a small sinusoidal perturbation of the sea floor, characterized by a certain wavelength, growth or decay of the particular perturbation can be predicted numerical as well. For an initial bed with the form $h = A\sin(\frac{2\pi}{L}x)$, with A being the amplitude and L the wavelength of the sand wave, the growth rate can be determined by the change in the sand wave height using $\omega = \frac{1}{\partial t}\log(\frac{h_{new}}{h_{old}})$. The growth rate is expressed in 1/s. When the growth in height corresponding to various wave lengths is known, the wave length that induces the fastest growth in height can be found (fastest growing mode, FGM). The FGM indicates the dominant sand wavelength that is expected to be the most likely to occur in reality as it grows fastest in height.

We start the simulation by prescribing sinusoidal, small amplitude, bed waves. The tidal flow is modeled as a symmetrical sinusoidal current prescribed by means of a forcing. Using the bathymetry, a tidal flow is calculated. Since the flow changes over a time-scale of hours and the morphology over a time-scale of years, the bathymetry is expected to be invariant over a single tidal cycle. Once the tidal flow is known, the bed changes are calculated over this typical tide, using sediment transport equations. This is repeated until the bed evolution exceeds a certain value, after which a new tidal flow is calculated. This, in turn, affects the bed and so the process is iterative. In this way, we are able to simulate the morphological time scale accurately, while avoiding long computation times.

The model consists of the hydrostatic flow equations for 2DV flow (Equations 1 and 2). In the horizontal direction, periodic boundary conditions are used. For model details, we refer to [15].

$$\frac{\partial u}{\partial x} + \frac{\partial w}{\partial z} = 0 \tag{1}$$

$$\frac{\partial u}{\partial t} + u\frac{\partial u}{\partial x} + w\frac{\partial w}{\partial z} = -g\frac{\partial \zeta}{\partial x} + \frac{\partial}{\partial z}\left(A_v \frac{\partial u}{\partial z}\right) \quad (2)$$

In these equations x, z represent the horizontal and vertical directions and u and w the horizontal and vertical flow velocities. The variable t denotes time, ζ is the water surface elevation, g is the constant of gravity and A_v is the constant eddy viscosity.
Boundary conditions at the bed disallow flow through the bottom (equation 3). Further, a partial slip condition compensates for the constant eddy viscosity, which overestimates the eddy viscosity near the bed (equation 3). The parameter S denotes the amount of slip, with $S = 0$ indicating perfect slip and $S = \infty$ indicating no slip. At the water surface, there is no friction and no flow through the surface (equations 4).

$$w - u\frac{\partial h}{\partial x} = 0|_{seabed} \;\; ; \;\; A_v\frac{\partial u}{\partial z} = Su|_{seabed} \quad (3)$$

$$\frac{\partial u}{\partial z} = 0|_{surface} \;\; ; \;\; w = \frac{\partial \zeta}{\partial t} + u\frac{\partial \zeta}{\partial x}|_{surface} \quad (4)$$

The flow and the sea bed are coupled through the continuity of sediment (equation 5). Sediment is transported in two ways: as bed load transport (q_b) and as suspended load transport (q_s), which are modeled separately. Here we use a bed load formulation after [9] (equation 6).

$$\frac{\partial h}{\partial t} = -\left(\frac{\partial q_b}{\partial x} + \frac{\partial q_s}{\partial x}\right) \quad (5)$$

$$q_b = \alpha|\tau_b|^b\left[\frac{\tau_b}{|\tau_b|} - \lambda\frac{\partial h}{\partial x}\right] \quad (6)$$

Grain size and porosity are included in the proportionality constant α, τ_b is the shear stress at the bottom, h is the bottom elevation with respect to the spatially mean depth H and the constant λ compensates for the effects of slope on the sediment transport. For more details, we refer to [9] or [18].
In order to model suspended sediment transport q_s, we describe sediment concentration c throughout the water column, i.e. a 2DV model. Horizontal diffusion is assumed to be negligible in comparison with the other horizontal influences. The vertical flow velocity, w, is smaller than the fall velocity for sediment, w_s, and can be neglected in this equation, leading to equation (7). This means that the sediment is suspended only by diffusion from the bed boundary condition (equation 12). As the flow velocity profile is already calculated throughout the vertical direction, suspended sediment transport q_s can be calculated using equation (8).

$$\frac{\partial c}{\partial t} + u\frac{\partial c}{\partial x} = w_s\frac{\partial c}{\partial z} + \frac{\partial}{\partial z}\left(\epsilon_s\frac{\partial c}{\partial z}\right) \quad (7)$$

$$q_s = \int_a^H u(z)c(z)dz \tag{8}$$

$$w_s = \frac{\nu D_*^3}{18 D_{50}} \tag{9}$$

$$D_* \equiv \left(\frac{g(s-1)}{\nu^2}\right)^{1/3} D_{50} \tag{10}$$

$$\epsilon_s = A_v \tag{11}$$

The parameter ϵ_s denotes the vertical diffusion coefficient (here taken equal to A_v), a is a reference level above the bed above which suspended sediment occurs, D is the grain size. The dimensionless grain size is denoted by D_*, $(s-1)$ is the relative density of sediment in water $(\frac{\rho_s - \rho_w}{\rho_w})$, with ρ_w the density of water and ρ_s the density of the sediment and ν is the kinematic viscosity. Equations (9-11) are due to [18].
Suspended load is defined as sediment which has been entrained into the flow. By definition, it can only occur above a certain level above the sea bed. At this reference height, a reference concentration can be imposed as a boundary condition. Various reference levels and concentrations exist for rivers, near-shore and laboratory conditions. Those often applied are [17, 14, 5, 21]. For offshore sand waves, the choice of a reference height is more difficult than it is for the shallower (laboratory) test cases. In this case, the reference equation of [17] (equation 12) is used, with a reference height of 1 percent of the water depth, corresponding with the minimum reference height proposed in [17].

$$c_a = 0.015 \frac{D}{0.01 H D_*^{0.3}} \left(\frac{|\tau| - \tau_{cr}}{\tau_{cr}}\right)^{1.5} \tag{12}$$

The reference concentration at height a above the bed is given by c_a and τ_{cr} is the critical shear stress necessary to move sediment.
Both the gradient and the quantity of suspended sediment are largest close to the reference height. Therefore, concentration values are calculated on a grid with a quadratic point distribution on the vertical axis, such that more points are located closer to the reference height and fewer points are present higher in the water column. To complete the set of boundary conditions for sediment concentration, we disallow flux through the water surface.

4 Model results

In this paper, we concentrate fully on the influence of suspended sediment on the initial state of sand waves. We started each simulation with a sinusoidal bed-form with an amplitude of 0.1m.

Next, we investigated the (initial) growth rate and the fastest growing sand wavelength (FGM). Table 1 shows some basic values used in the simulations and the characteristics of the simulations are given in Table 2. Where possible, typical values for sand waves in the North Sea are used. Note that $\bar{u}$ is defined as the depth-averaged maximum flow velocity.

Table 1. Parameter values for the reference simulation

parameter	value	unit	parameter	value	unit
$\bar{u}$	1	m/s	ϵ_v	0.03	m^2/s
H	30	m	D	300	μm
A_v	0.03	m^2/s	w_s	0.025	m/s
S	0.01	m/s	a	0.3	m
α	0.3	-			

Table 2. Simulations

simulation	bed load	suspended load	varied parameter	simulation	bed load	suspended load	varied parameter
reference	√	-	-	3	√	√	ϵ_v
1	√	√	-	4	√	√	u
2	√	√	ref. height a	5	√	-	u

4.1 transport simulations

Figure 3(a) shows the growth rate for different sand waves lengths simulated in the reference simulation. Moreover, the figure shows that the FGM is approximately 640m. For simulation 1, we included suspended sediment in the reference computation. Figure 3(b) shows a comparison between the reference simulation and simulation 1. The growth rate is shown for a range of wavelengths. Most remarkable is the increase of the growth rate by a factor of approximately 10. This was unexpected as suspended sediment is assumed to be of minor importance in these circumstances. The FGM for simulation 1 is 560m, 80m less than in the reference simulation.

In figure 4, the concentration profile in the water column at a crest point over the tidal period is shown (upper figure), compared with the flow velocities (lower figure). The sediment is only entrained into the first few meters of the water column. The sediment concentration follows the flow without an apparent lag, as the flow velocity near the bed is small and slowly changes over time. However, these small variations in velocity are enough for the suspended sediment to be entrained and to settle again within one tidal cycle. Close to the reference height, the maximum sediment concentration is around $3{\cdot}10^{-4}$ m^3/m^3 ($0.8\ kg/m^3$).

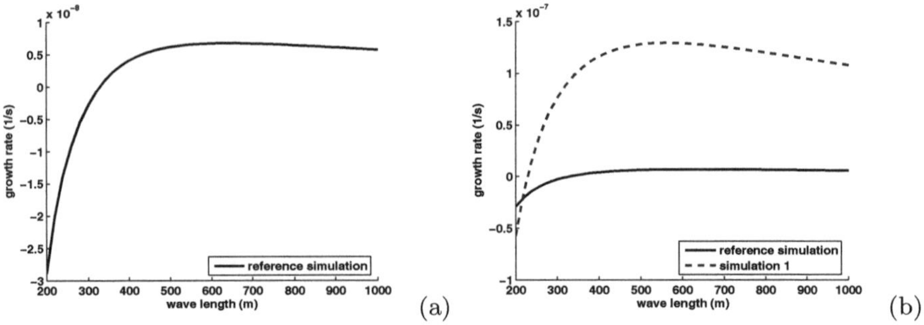

Fig. 3. (a) Growth rate – reference simulation; (b) growth rate – simulation 1 (solid), compared with reference simulation (dashed). Parameters in Table 1.

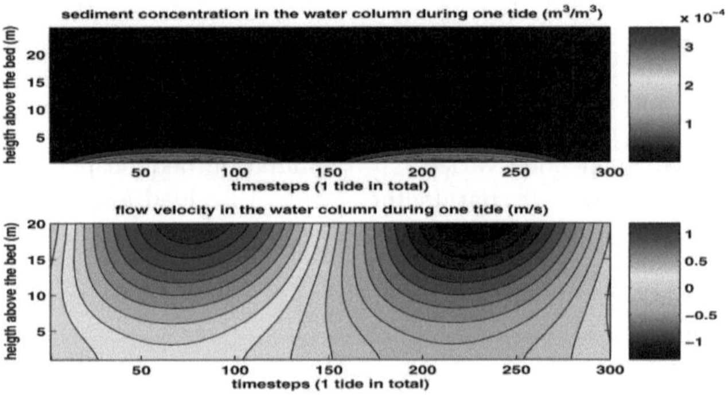

Fig. 4. Sediment concentration (upper) and flow velocity (lower) on one location over a tidal period, for simulation 1. More details see Fig 6 (upper).

4.2 sensitivity simulations

To study the influence of the reference height on the sediment entrainment and suspended transport, the reference height in simulation 2 equation (12) is

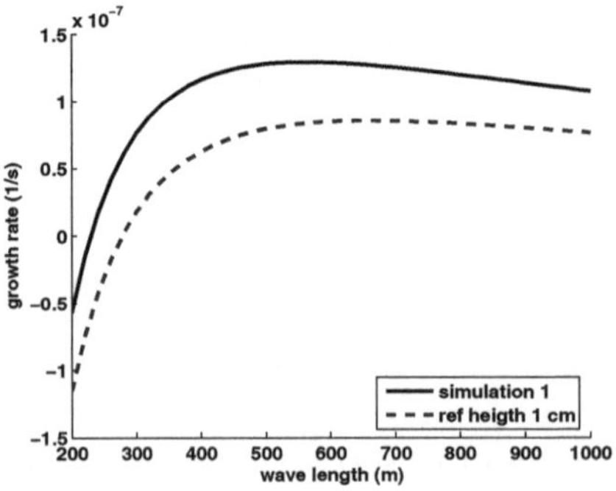

Fig. 5. Growth rate for simulation 2 (solid), compared to simulation 1 (dashed). For simulation characteristics, see Table 1.

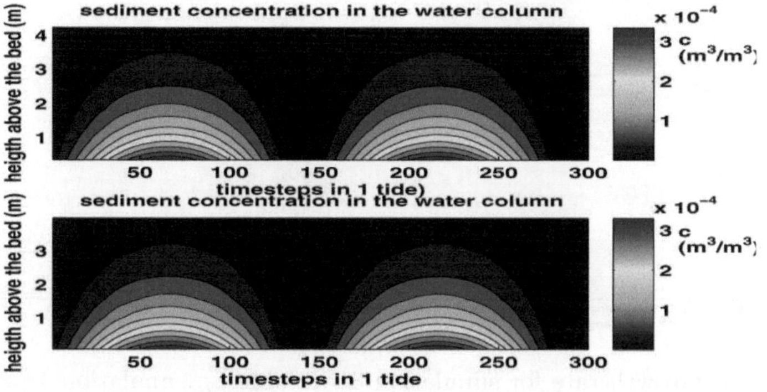

Fig. 6. Sediment concentration in the first 4 meters above a certain point of the sand wave during one tide. Comparison between simulation 1 (upper) and 2 (lower)

decreased to 0.01m above the bed. This height is used as the lowest measurable height for suspended sediment in shallow seas ([10, 6]). The results are shown in figures 5 and 6. It can be seen in figure 5 that the growth rate decreases for a lower reference height, whereas the FGM becomes 660m. Note that the growth rate, compared to the situation without suspended sediment, is still larger. In figure 6, it can be seen that, for the first 4 meters above the reference height, no change occurs, except that the sediment is entrained about 0.30m higher in the reference simulation. This difference is a direct result of the change in reference height itself (from 0.30m to 0.01m). Therefore the difference in growth rate is solely due to the contribution of these 0.29m to the integration of $u \cdot c$ over the water column.

Table 3. Simulation results, for varied values the first (second) value is for the +50% (-50%) simulation

simulation	FGM (m)	growth rate for FGM (1/s)	simulation	FGM (m)	growth rate for FGM (1/s)
reference	640	6.75e-9	3	860 - 350	1.24e-7 - 1.12e-7
1	560	1.29e-7	4	810 - 340	2.40e-7 - 3.87e-8
2	660	8.55e-8	5	670 - 610	1.23e-8 - 2.20e-9

In simulations 3 and 4, a sensitivity analysis was carried out for the diffusion coefficient and the flow velocity. The value of sediment diffusivity, ϵ_v, in the reference situation was assumed to be equal to the eddy viscosity Av, though its value is not established. Both ϵ_v and $\bar{u}$ were varied by $\pm$ 50% of their reference values. Their influence on the growth rate ω and the FGM are shown in figures 7(a) and 7(b). It can be seen that the FGM increases significantly for increasing ϵ_v (FGM becomes 860m), and decreases for decreasing ϵ_v (FGM

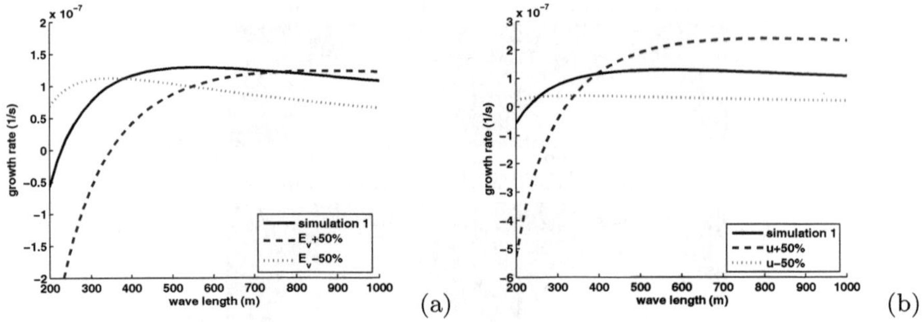

Fig. 7. (a) Growth rate for simulation 3, variable ϵ_v; simulation 1 (solid), ϵ_v+50% (dashed), ϵ_v-50% (dotted). (b)Growth rate for simulation 4, variable $\bar{u}$; simulation 1 (solid), $\bar{u}$+50% (dashed), $\bar{u}$-50% (dotted).

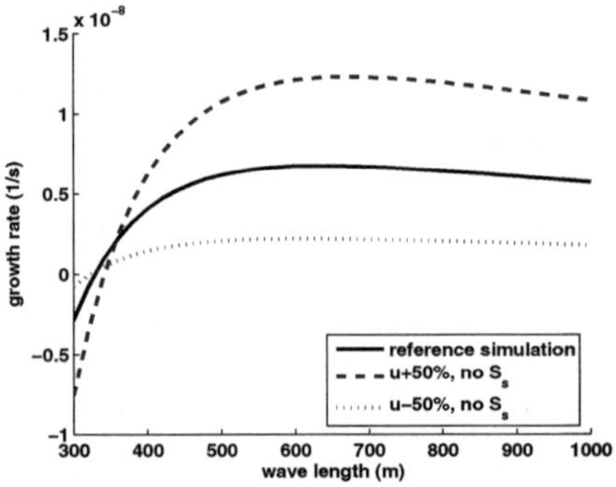

Fig. 8. Growth rate for simulation 5, variable $\bar{u}$ and no q_s; reference (solid), $\bar{u}$+50% (dashed), $\bar{u}$-50% (dotted)

becomes 350m). The growth rate of the FGM remains of the same order of magnitude. However, smaller wavelengths are damped more severely for increasing sediment diffusivity.

For the flow velocity $\bar{u}$, the FGM again tends to increase with increasing $\bar{u}$ and vice-versa (for values, see Table 3), and smaller wavelengths are damped more for higher values of $\bar{u}$. For the growth rate, we now see a different effect. As expected from the nonlinear $\bar{u}$ in the sediment transport equation, the growth rate is highly affected by $\bar{u}$. The higher the value of $\bar{u}$, the higher the initial growth rate for the FGM.

As shown in figure 7(b), suspended sediment transport increases the effect of variation in $\bar{u}$. If we compare this influence to the influence of varying $\bar{u}$ without suspended sediment transport (figure 8) it is clear that suspended sediment increases the effect of changing velocities on the FGM (45% change instead of 5% change in sand wavelength, for varying $\bar{u}\pm$50%). For the growth rate of the FGM, this influence is less pronounced; the decrease (increase) of growth rate with higher(lower) $\bar{u}$ is 82% (67%) for the case without suspended sediment and 86% (70%) for the case with suspended sediment.

5 Discussion and conclusions

In the reference simulation, ϵ_v is assumed to be equal to the value of A_v. Various coupling equations exist to relate ϵ_v to A_v, varying from ϵ_v being larger than to being smaller than A_v. [2] therefore assumed ϵ_v equal to A_v, as no generally accepted method is available. Figure 7(a) shows that varying the value of ϵ_v influences the FGM significantly, though the growth rate itself is hardly influenced. Possibly the large difference in growth rates between the case with and without suspended load transport (reference simulation and simulation 1) is caused, not by the value of ϵ_v, but by the constant value of both the eddy viscosity and sediment diffusivity. Due to these constant values, A_v might be overestimated near the bed, which is corrected for by the partial slip boundary condition. Such a correction is not used for the ϵ_v, possibly leading to an increase of suspended sediment. Due to the constant ϵ_v this sediment can also be entrained higher into the water column.
Unfortunately, little field data for offshore sediment transport is available at the moment, hindering a direct comparison with the results. [6] measured suspended sediment offshore in the North Sea at a water depth of 13 meters. Only during minor storms suspended sediment was detected. Maximal values were around 2.3 kg/m^3 for 0.3m above the bed and 0.2 kg/m^3 for 1m above the bed. For simulation 1, these values were 8 kg/m^3 and 0.34 kg/m^3. [7] measured sediment concentrations during a severe storm in the North Sea close to the coast of the UK. They found, even under conditions of storm, finer sediment ($\sim$100μm) and a 25m water depth, that the sediment concentration had decreased by about three orders of magnitude after 1 meter ($\pm$ 40 kg/m^3 to 0.03 kg/m^3). However, in the simulations this decrease was slower, leading to higher concentrations higher in the water column ($\pm$ 8 kg/m^3 close to the reference height to 0.03 kg/m^3 at 3 meter above the bed). Although the sediment concentration predicted in the model seems to be in a comparable order of magnitude, transport rates are too large. The most likely cause is the high entrainment of sediment into the water column. Further study on this topic, and the effect of a depth dependent ϵ_v is currently investigated.
As w turned out to be around an order of magnitude smaller then w_s during most of the tide, this term was neglected in the sediment continuity equation (equation 7). However, for higher flow velocities or smaller grain sizes this term will become more important. In that case w should be incorporated and might increase the amount of suspended sediment during a part of the tidal cycle on certain locations on the sand waves, leading to further growth or decay of the sand waves. The effect depends on the specific locations (i.e. crests or troughs) were suspended sediment will erode or deposit.

[17] proposed a reference height for suspension with a minimum value of 1% of the water depth. However, [19] stated that this leads to unrealistically high reference levels in water depths of tens of meters. [19] therefore proposed to use a reference height of 0.01m instead. [10] and [6] also used this height as the lowest measurable height for suspended sediment in shallow seas. Both heights are tested in simulations 1 and 2. They turn out to differ only in the lowest part of the water column, which was excluded from the 1% (i.e. 0.3m) reference height and included in the 0.01m alternative. Thus, the reference height does not change the processes, but only includes or excludes the sediment in the first view centimeters above the bed.
Based on grain sizes, [11] expected suspended transport for grains smaller than 230-300μm. Grains smaller than 170μm would be transported in suspension only, in this case sand waves are rarely found. Recently, [20] showed that a mixture of grain sizes leads to grain size sorting over sand waves, but hardly affects the sand wave form and growth rate in the numerical code. Therefore, in this paper we assumed grains of only one grain size, corresponding with the medium grain size typically found on sand wave fields.

Concluding, the inclusion of suspended sediment transport in a sand wave model demonstrates significant influences of suspended load on the initial growth of sand waves. The influence of various parameters was investigated, showing that the reference height for suspended sediment is of minor importance, while the sediment diffusivity, ϵ_v, and especially the depth averaged maximum flow velocity, $\bar{u}$, largely influence both the FGM and the initial growth rate. Further research will focus on fully developed sand waves and the effects of wind and storm conditions, validated against field data.

Acknowledgment

This research is supported by the Technology Foundation STW, applied science division of NWO and the technology program of the Ministry of Economic Affairs. The authors are indebted to Jan Ribberink for his suggestions.

References

[1] Besio, G., Blondeaux, P., and Frisina, P. (2003). A note on tidally generated sand waves. J. Fluid Dynamics, 485, 171-190
[2] Blondeaux, P. and Vittori, G. (2005a). Flow and sediment transport induced by tide propagation: 1 the flat bottom case. J. Geoph. Res. - Oceans, 110 (C07020, doi:10.1029/2004JC002532)
[3] Blondeaux, P. and Vittori, G. (2005b). Flow and sediment transport induces by tide propagation:2 the wavy bottom case. J. Geoph. Res. - Oceans, 110 (C08003, doi:10.1029/2004JC002545)
[4] Buijsman,M. C. and Ridderinkhof, H. (2006). The relation between currents and seasonal sand wave variability as observed with ferry-mounted adcp. In: PECS 2006, Astoria, OR-USA

[5] Garcia, M. and Parker, G. (1991). Entrainment of bed sediment into suspension. J. Hydraulic Engg, 117 , 414-435
[6] Grasmeijer, B. T., Dolphin, T., Vincent, C., and Kleinhans, M. G. (2005). Suspended sand concentrations and transports in tidal flow with and without waves. In: Sandpit, Sand transport and morphology of offshore sand mining pits, Van Rijn, L. C., Soulsby, R. L., Hoekstra, P., and Davies, A. G.(eds.), U1-U13. Aqua Publications
[7] Green, M. O., Vincent, C. E., McCave, I. N., Dickson, R. R., Rees, J. M., and Pearson, N. D. (1995). Storm sediment transport: observations from the British North Sea shelf. Continental Shelf Res., 15 (8), 889- 912
[8] Hulscher, S. J. M. H. (1996). Tidal-induced large-scale regular bed form patterns in a three-dimensional shallow water model. J. Geoph. Res., 101, 727-744
[9] Komarova, N. L. and Hulscher, S. J. M. H. (2000). Linear instability mechanisms for sand wave formation. J. Fluid Mech., 413, 219-246
[10] Lee, G. and Dade, W. B. (2004). Examination of reference concentration under waves and currents on the inner shelf. J. Geoph. Res., 109 (C02021, doi:10.1029/2002JC001707)
[11] McCave, I. N. (1971). Sand waves in the North Sea off the coast of Holland. Marine Geology, 10 (3), 199-225
[12] Nemeth, A. A., Hulscher, S. J. M. H., and Van Damme, R. M. J. (2006). Simulating offshore sand waves. Coastal Engineering, 53, 265-275
[13] Passchier, S. and Kleinhans, M. G. (2005). Observations of sand waves, megaripples, and hummocks in the Dutch coastal area and their relation to currents and combined flow conditions. J. Geoph. Res. - Earth Surface, 110 (F04S15, doi:10.1029/2004JF000215)
[14] Smith, J. D. and McLean, S. R. (1977). Spatially averaged flow over a wavy surface. J. Geoph. Res., 12 , 1735-1746
[15] Van den Berg, J. and van Damme, D. (2006). Sand wave simulations on large domains. In: River, Coastal and Estuarine Morphodynamics: RCEM2005 , Parker and Garcia(eds.)
[16] Van der Veen, H. H., Hulscher, S. J. M. H., and Knaapen, M. A. F. (2005). Grain size dependency in the occurence of sand waves. Ocean Dynamics, (DOI 10.1007/s10236-005-0049-7)
[17] Van Rijn, L. C. (1984). Sediment transport, part ii: Suspended load transport. J. Hydraulic Engineering, 11, 1613-1641
[18] Van Rijn, L. C. (1993). Principles of sediment transport in rivers, estuaries and coastal seas, vol. I11. Aqua Publications, Amsterdam
[19] Van Rijn, L. C. and Walstra, D. J. R. (2003). Modelling of sand transport in DELFT3D-ONLINE. WL—Delft Hydraulics, Delft
[20] Wientjes, I. G. M. (2006). Grain size sorting over sand waves. CE&M research report 2006R-004/WEM-005
[21] Zyserman, J. A. and Fredsoe, J. (1994). Data-analysis of bed concentration of suspended sediment. J. Hydraulic Engg, 120, 1021-1042.

Sediment transport by coherent structures in a turbulent open channel flow experiment

W.A. Breugem and W.S.J. Uijttewaal

Delft University of Technology, Faculty of Civil Engineering and Geosciences, Environmental Fluid Mechanics Section, P.O. Box 5048, 2600 GA Delft, The Netherlands, `w.a.breugem@tudelft.nl`, `w.s.j.uijttewaal@tudelft.nl`

Summary. In order to obtain more insight into the vertical transport of suspended sediment, an experiment was performed using a combination of PIV and PTV for the measurement of the fluid and particle velocity respectively. In this experiment, the particles were fed to the flow at 16 and 75 water depths from the measurement section with an injector located at the centerline of the channel near the free surface. At 16 water depths from the sediment injection, most sediment is still near the free surface, and the sediment is transported downwards in sweeps, thus leading to a mean particle velocity that is faster than the mean fluid velocity. It appears that in this situation, downward going particles are indeed found in sweeps (Q4), whereas upward going particles are preferentially concentrated in both Q1 and Q2 events. In the fully developed situation on the other hand, upward going particles are preferentially concentrated in ejections, while downward going ones are found in both Q3 and Q4 events, with a relatively increased frequency in Q3, and a decreased one in Q4. The increased number of particles in Q2 and Q3, which have low fluid velocities, leads to a mean particle velocity lower than the mean fluid velocity.

1 Introduction

The transport of suspended particles in turbulent flows is important in many environmental flows. Therefore, much research already has been done. Nevertheless, modeling this highly complex phenomenon remains difficult.

The current state of the art in modeling sediment transport is by using a two-fluid model [e.g. 18]. A vertical momentum balance for the dispersed phase shows in the equilibrium situation (where the vertical accelerations and gradients of the vertical particle velocity are negligible) the following relation for the mean vertical particle velocity $\langle u_{p,y} \rangle$:

$$\langle u_{p,y} \rangle_p = \langle u_{f,y} \rangle_f + \langle u'_{f,y} \rangle_p + u_{y,T} \tag{1}$$

Here, $\langle u_{f,y} \rangle_f$ is the fluid velocity ensemble averaged over the fluid phase, $u_{y,T}$ the still water terminal velocity, and $\langle u'_{f,y} \rangle_p$, the drift velocity, i.e. *the*

Bernard J. Geurts et al. (eds), Particle Laden Flow: From Geophysical to Kolmogorov Scales, 43–55.

deviation of the mean fluid velocity as seen by the particles. The drift velocity results from averaging the relative velocity in the Stokes drag term of this equation. Turbophoresis is neglected, because for almost neutrally buoyant particles, it is counterbalanced by the pressure gradient working on the particles. This agrees with our intuition that fluid particles in a fluid should not concentrate near the wall. Equation 1 physically means that the mean particle velocity (i.e. the flux per unit of concentration) is equal to the settling velocity added to mean vertical fluid velocity and the extra drift term. Simonin et al. [18] used a gradient diffusion hypothesis for the closure of this term. In fact, the drift term is comparable to the $\langle c'v' \rangle$ term in conventional advection-diffusion models.

The importance of the drift velocity in this modeling approach implies that in order to understand dispersion, we need to know in which flow structures particles are located. It is already widely known that particles are not necessarily distributed homogeneously in a turbulent flow. Preferential sweeping [8], does not seem to be important for the situation we consider with a relative density ratio ρ_p/ρ_f just above one, as the sweeping of particle out of vortices by their inertia is compensated by the inward pressure gradient into the vortex. Nevertheless, some DNS results show preferential concentration for similar particles [19], but this seems to be mainly due to the initial conditions [20]. Particles subjected to gravity but without inertia moving in cellular flow fields were shown to have a complex behavior [12]. These particles can either get trapped inside the vortex, leading to an upward drift velocity, or remain outside the vortex, leading to a downward drift that enhances their apparent settling velocity, but that does not change their slip velocity. The combination of these two situations leads to a zero drift velocity for these kind of particles in homogeneous isotropic turbulence. From the previous, it appears that even particles without inertia can see a velocity field that is different from the overall velocity field, although not strictly due to preferential concentration.

The objective of this study is to provide more insight in the behavior of small, particles that are slightly heavier than the fluid and to find the flow structures that cause the vertical transport of these particles. In order to capture these flow structures, we perform an experiment, measuring simultaneously the fluid and particle velocities with Particle Image Velocimetry (PIV) and Particle Tracking Velocimetry (PTV) respectively. We inject polystyrene particles near the free surface and perform measurements at either 16 or 75 water depths from the injection point. In the first situation, the highest concentration is found near the free surface and the particles are on average moving downwards, whereas in the latter situation, a fully developed situation exists, in the wall normal direction. From now on, we will call the case with the sediment inlet at 16 water depth from the measurement section the "settling situation" and the one with the inlet at 75 water depths the "fully developed situation". The complete experimental setup is described in the next section. In section 3, we show the mean profiles and probability density functions of the drift velocity and we compare them with the statistics at randomly generated particle locations. In section 4, we determine the conditionally averaged drift velocity in the vicinity of a vortex head, followed by our conclusions in section 5.

2 Experimental setup

The experiments were performed in an open channel, with a length of 23.5 m a width of 0.495 m and a height of 0.50 m (fig. 1). The walls and bottom were made of glass in order to have a hydraulically smooth boundary. In order to perform the fluid velocity measurements using the PIV technique, the water was seeded with 10 μm hollow glass spheres ($\rho = 1100\ kg/m^3$).

As pseudo-sediment, polystyrene particles with a mean diameter of 347 μm were used, which had a density ρ_p of 1035 kg/m^3. The terminal velocity was determined in still water as $v_T = 2.2\ mm/s$, which compares well with the expected value of 2.1 mm/s ($Re_p = v_T d_p/\nu_f = 0.71$).

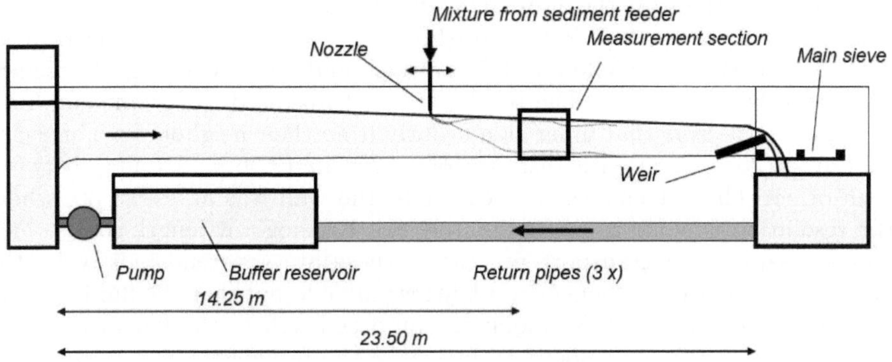

Fig. 1. Experimental setup

The experiments were performed at $Re = 10,000$ ($Re_* = 508$), which was obtained by setting the centerline velocity to 0.20 m/s and the water depth h to 5.0 cm. This velocity was chosen to ensure a sufficient amount of sediment was suspended ($u_*/v_T \approx 5$). The particles were fed to the flow mixed with water through a nozzle with an inner diameter of 1 cm at the channel's centerline and its center located at 0.7 cm below the free surface. The inflow velocity was manually adjusted to the channel velocity. The position of the nozzle was varied from 80 cm to 375 cm from the measurement section, i.e. at $x_{in}/h = 16$ and $x_{in}/h = 75$. In the latter situations, the vertical particle velocities were zero up to experimental accuracy, which means that a fully developed situation exists. Apart from that, the statistics of that situation did not differ much from a preliminary test where the sediment was injected at 160 water depths from the measurement section.

The volumetric sediment concentration that was introduced was $1.2\ 10^{-2}$. For each set, a sequence of 15×300 double images was recorded at 2 Hz. It was checked that the flow remained stationary for the time of the experiment. For comparison, also four sets of 300 double images were recorded at a frame rate of 2 Hz for the flow in the channel, without the nozzle and any sediment input, which we will call the clear water flow (CWF) data.

A 45 mm x 45 mm measurement section was located at a distance of 14.25 m from the channel entrance. At this location, a combination of both PIV and PTV was used to measure the streamwise and wall normal velocities of the polystyrene particles and the ambient fluid.

The data were processed with a modified version of the method of Kiger and Pan [9] to discriminate sediment from tracer particles. In this algorithm, a median filter with a size of 7 pixels is applied to remove the image of the tracer particles from the recorded images, resulting in an image of only the sediment particles. A PTV algorithm [21], which uses the displacement of the centroid of the particles to determine the particle velocities, was applied to the image with only sediment particles. By subtracting the image of the sediment particles from the original image, an image containing only the tracer particles was obtained. A PIV algorithm [11] is applied to this tracer image with first a 64 x 64 window (50 % overlap) and then two 32 x 32 window (75 % overlap) iterations. The results are postprosessed with a median filter to eliminate vectors that differ significantly from their neighbours. This leads to 126 x 126 vectors with distance of 0.37 mm (3.76 wall units) between each other. The velocity vector nearest to the wall was at $y+ > 30$, where the resolution is equal to approximately two Kolmogorov length scales. This seems adequate for transport process as these are goverend mostly by the large scale structure. Breugem and Uijttewaal [5] found that the fluid velocity profiles measured with this resolution compared well to the law of the wall, when using the friction velocity from a fit of the Reynolds stress profile. The fluid Reynolds stress profile was found to be linear and the mean vertical fluid velocity was found to be zero up to experimental uncertainty, which together indicate that secondary currents are negligible as might be expected with the present B/h ratio of 10.

3 One-point drift velocity statistics

The sediment concentration profile (fig. 2) shows a high concentration near the free surface in the $x_{in}/h = 16$ case, whereas it resembles the Rouse profile with most sediment near the bottom in the $x_{in}/h = 75$ case 2. In the same figure, the drift velocity profiles are also shown for both the settling and the fully developed case. These are defined by performing a bicubic interpolation of the PIV fluid velocities at the particle locations. It is clear that in the predominantly settling regime, the streamwise fluid velocity seen by the particles is higher than the average streamwise fluid velocity, whereas the opposite is true in the fully developed case. The latter result has been found before by for example Kiger and Pan [10]. The wall normal drift velocity is at first negative, meaning that the sediment particles see on average a downward velocity, which brings them down even more rapidly than gravity does (as the still water settling velocity is about $0.2u_*$). In the fully developed situation, the particles see on average an upward moving fluid velocity. From theory, it is expected that the vertical drift is equal to the settling velocity, but in the data, the drift is smaller (a maximum of $0.1u_*$ rather than the expected $0.2u_*$). There are two reasons for this discrepancy. First of all, there is a bias in the measured fluid velocity toward the particle velocity as a result of leftovers of the particle images after median filtering the images, which decrease the measured relative velocity. Furthermore, because the grid spacing of the PIV has about the size of a particle diameter, the fluid velocity that is used for determining the drift velocity by interpolation is not the undisturbed velocity, but rather the one that is affected by the disturbance field of the particle itself. In case of a Stokes flow, which is not completely valid here as Re_p of a freely settling particle is 0.71, the velocity disturbance around a moving particle needs about five particle diameters to decay [e.g. 4].

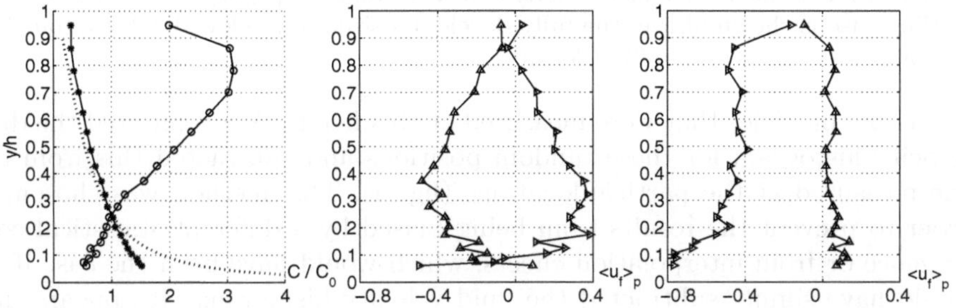

Fig. 2. Left: mean sediment concentration profiles. $-\circ-$: $x_{in}/h = 16$; $-*-$: $x_{in}/h = 75$; Dashed line: Rouse profile. Middle: Drift velocity profile $x_{in}/h = 16$; Right: $x_{in}/h = 75$; $-\triangleright-$: streamwise direction; $-\triangle-$: wall normal direction;

In order to determine which coherent structures are causing the drift velocity, we computed the probability density functions (pdf) at $y/h = 0.55$ (fig. 3). From these figures, it is first of all clear that in both cases the particle velocity and the drift velocity do not differ much. There is an asymmetry in the data, showing stronger sweeps than ejections and showing significantly less Q1 and Q3 events than Q2 and Q4. Yet, there exists a clear difference between both cases, with the peak of the histogram in the settling case in the Q4 quadrant, whereas it is at the center in the fully developed situation.

To determine in which flow structures, the particles are concentrated, we first picked random locations in the measured flow fields using the same num-

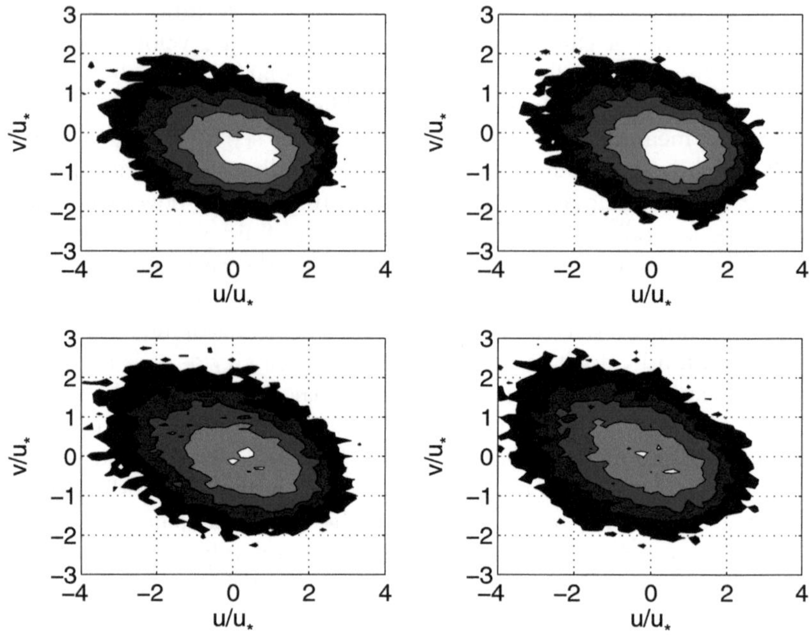

Fig. 3. PDF of particle velocity (left) and drift velocity (right) at $y/h = 0.55$. The settling case is shown above, the fully developed situation below. Each contour line indicates a doubled probability density.

ber as the particles that were measured at this height. We computed the drift velocity histogram for these random positions and subtracted this from the one measured at the particle locations (fig. 4). This method was chosen, in order to prevent the results from being biased by a different statistical convergence or from interpolation effects, which would have been the case if we would have simply subtracted the fluid velocity histograms. It appears that in the settling case, the upward moving particles are found in all upward flow structures (Q1 and Q2), whereas downward moving particles, which are much more common than upward moving ones, are preferentially concentrated in sweeps (Q4). This appears to happen over the complete water depth (not shown), except near the free surface ($y/h > 0.8$) , where downward moving particles are preferentially concentrated in both Q3 and Q4 events.

In the fully developed situation on the other hand, upward moving particles are preferentially concentrated in ejections (Q2), whereas downward moving particles are concentrated preferentially in inward interactions (Q3). Note that, because Q4 events are much more common than Q3 events in an open channel flow, particles end up about as many times in Q3 as in Q4 events due to the preferential concentration (fig. 3). In this situation, the number of upward and downward moving particles is equal at every flow depth except near the bottom [see 5], just as was found by Kiger and Pan [10].

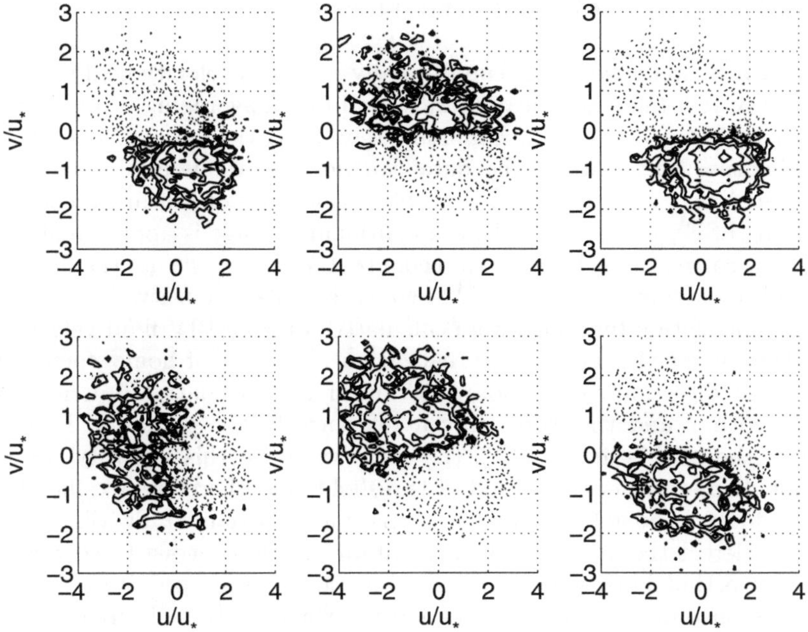

Fig. 4. PDF of the increase of the PDF at $y/h = 0.55$ with to respect to random sampled particles for all particles (left), upgoing particles (middle), and downgoing particles (right). The settling case is shown in the upper row, the fully developed situation in the lower row. Each contour line indicates a doubled intensity and negative values are indicated with dashed lines.

The results for the fully developed case agree with the data from Kiger and Pan [10] at $y/h = 0.6$ ($y+ = 340$). Both the preferential concentration of upgoing particles in ejections and of downgoing particles in inward interactions (Q3) can clearly be seen in their data. A decreased number of particles in sweeps and an increased one in ejections also agrees with the findings of Cellino and Lemmin [6] and Nikora and Goring [14], who find larger than average upward sediment fluxes ($\langle c'v' \rangle$) in these quadrants, noting that an increased upward flux in a sweep (i.e. in a downward flow structure) can only come from a decreased concentration (because Cellino and Lemmin [6] do not find the Reynolds stress contributions of the different quadrants to change with respect to clear water flows). Cellino and Lemmin [6] do not find the increased importance of Q3 events in their measurements. This may be attributed to the larger concentration in their measurements, as Nikora and Goring [14] report a large concentration dependence on the quadrant distribution with increased contributions for Q1 and Q3 events and increased concentration fluctuations in their low concentration cases.

4 Spatial drift velocity structure

Here, we are interested in the conditionally averaged velocity field when there is a vortex at a reference location $\mathbf{x}_0$. Conditional averaging is a widely used tool in turbulence research. Unfortunately, statistical convergence is quite slow, because only part of the data can be used to determine conditional averages. A way to overcome this problem is the use of Linear Stochastic Estimation (LSE) [e.g. 1]. In LSE, a conditional average is approximated from two-point correlations. In order to recognize the vortex, we use the fluctuating part of the swirling strength λ_s' [22], which is scalar quantity.

The correlation functions were calculated for each PIV fluid velocity field and then averaged over all 4500 realizations. Because of homogeneity in the streamwise direction, the standard deviation and correlation function do not change with x and therefore, only a reference height y_0 has to be chosen. Because the LSE is linear in λ_s', the exact value for this quantity is not important. It merely acts as a kind of threshold, and therefore a value of $1\ /s^2$ was used as was done before by Christensen and Adrian [7]. Swirling strength does not detect the direction of the rotation, this in contrast to vorticity. Yet it is known that both vortices in the direction of and opposite to the mean shear are encountered in boundary layer turbulence [17]. Therefore, we calculate the statistics conditioned on only those values of the swirling strength for which $\omega_z < 0$, i.e. only for vortices that rotate with the mean shear.

The conditionally averaged fluid flow results for $y_0 = 0.5h$ are given in figure 5. We use a streamline plot, rather than the normalized vector map Christensen and Adrian [7] use to visualize the flow direction clearly even at large distances from the hairpin vortex. In combination, we use a vector plot without renormalization, which gives a clear impression of the size of the structures. The swirling flow pattern is clearly visible at this location, and it is also clear that a strong Q2 event can be found below the vortex head, which extends over the complete water depth. This means that the whole flow structure (vortex and the induced flow) can be classified as an attached eddy. It is also interesting to note the absence of strong Q4 events in this flow structure. A small Q3 event is visible upstream and below the vortex head.

The conditionally averaged drift velocity is shown in figure 6. Here, there is a significant smaller number of vectors than in the fluid velocity LSE, because larger bins were used in order to get converged statistics. Note that the drift velocity, is not a zero-mean quantity (fig. 2). It appears that in the settling case, the particles see large scale sweeps upstream and above the vortex head. Around the vortex, it can be seen that the particles see an even intenser drift at the downstream side of the vortex pulling the particles down around it. In the fully developed case, the drift velocity again looks very similar to the fluid velocity structure. Note that some care should be taken in interpreting these results, because it does not show the amount of particles at a location and some results therefore might be coming from a rather small number of sediment particles and at the same time contribute little to the actual transport, as there are no particles at those locations. In the settling case, the conditionally drift velocity becomes clearly smaller for a vortex higher in the water column (not shown), with the most significant contribution above the vortex head. In the fully developed situation on the other hand, the contribution to the drift velocity becomes higher for vortices higher in the water column (when its intensity is not changed), and it is located below the vortex

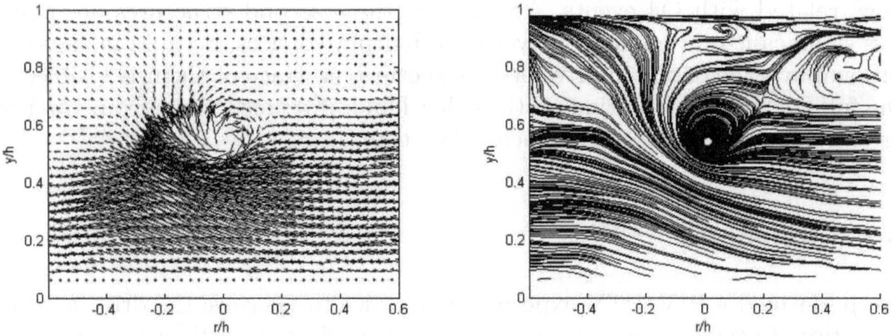

Fig. 5. Conditionally averaged *fluid flow* structure at $y_0 = 0.5h$, calculated with LSE. Only every other vector is shown.

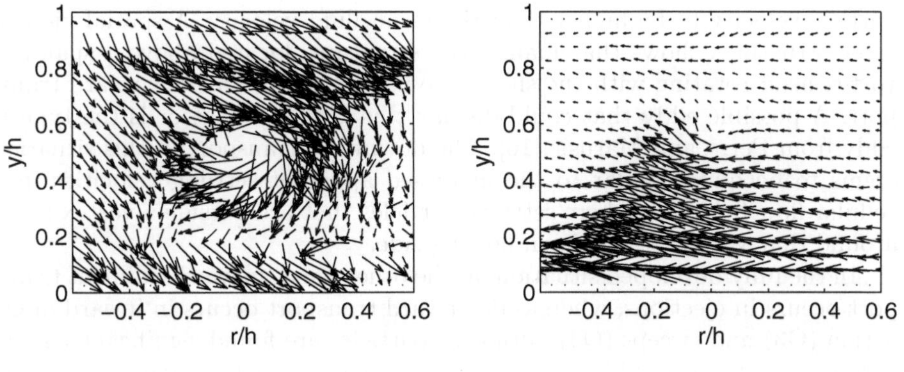

Fig. 6. Conditionally averaged *drift velocity* structure at $y_0 = 0.5h$. Left for the settling case, right for the fully developed situation.

The clear difference between the conditional average of the drift velocity in the settling case and the fluid velocity must mean that apparently only some vortices are important for transporting the particles down in this case. I.e., although the downward and upward drifts are both related with a spanwise vortex, the complete vortical structure transporting them is presumably very different. It appears that the concerned structure consists of a vortex head with a sweep upstream and above of it. A model for this kind of structure could be the type B eddies of Perry and Marusic [16], which consists of an spanwise oscillating vortex tube, inclined 45 degrees in the streamwise direction and rotating with the mean shear. They proposed this structures in order to obtain a better comparison with experimental data without claiming their existence. Interestingly, a similar structure was found in conditionally averaged structures by Adrian and Moin [2] in a DNS of a homogeneous shear flow, related with Q4 events. Apparently, these second structures are much less significant in a boundary layer than hairpin vortices, because in the conditional average of the fluid velocity structure, no trace of them is visible. In the fully developed situation on the other hand, it seems that hairpin vortices are responsible for the drift velocity structure.

5 Conclusion

We performed a PIV/PTV experiment in order to measure the drift velocity statistics of pseudo sediment particles in a turbulent flow. We varied the distance between the introduction of sediment and the measurement location. At 16 water depths from the measurement section, we found that most of the sediment was still near the free surface and moving downwards, preferentially concentrated in sweeps (Q4), whereas a smaller number of upward moving particles are found both in Q1 and Q2 structures, thus causing the mean particle velocity to be higher than the mean fluid velocity. A spatial view of these structures shows that mainly the structures located above a spanwise vortex head rotating with the shear, are important for this downward transport. A possible eddy that could display this kind of behavior is the type B eddy from Perry and Marusic [16]. The downward transport in this situation seems to be quite similar to the increased apparent settling velocities in a cellular flow field [12], where settling particles that are outside a vortex move around that vortex at the down flowing side.

In the fully developed situation on the other hand, upward sediment transport occurs in ejections, whereas downward transport occurs in inward interaction (Q3) and sweeps (Q4), although particles are found significantly more in Q3 events than could be expected from random sampling, and significantly less in Q4 events. The streamwise velocity that is lower than the mean in Q2 and Q3 events causes the mean streamwise particle velocity to be slower than the mean fluid velocity. The ejections are clearly related to hairpin vortex structures. In this situation, the number of upward and downward moving particles is approximately equal.

The physical mechanism for the increased concentration in Q3 events in the fully developed situation is shown in figure 7. Particles from the near the bottom, where the largest concentration exists, are transport upwards by an ejection related to a hairpin vortex. These hairpin vortices travel in

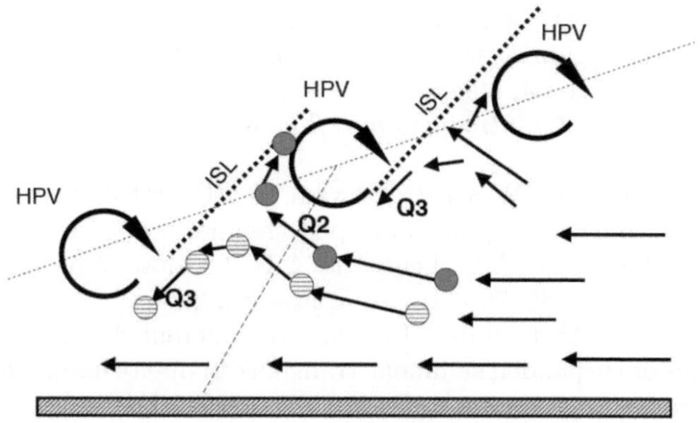

Fig. 7. Physical mechanism of particle transport in the fully developed situation. The image shows a hairpin vortex packet and two typical particles in a frame moving with the hairpin vortex packet convective velocity. ISL: Internal shear layer, HPV: Hairpin vortex

packets [3] and therefore a Q3 event related to an upstream hairpin vortex can transport a part of the particles downwards (dotted in fig. 7). Another part of the particles (filled in fig. 7) is transported upwards of the internal shear layer that connects the two vortices. From there, it might remain at the same vertical location [15], be transported further upwards by another hairpin vortex packet or be transported down by a sweep (similar as what happens in the settling case). Note that these light particles do not seem to settle down by gravity [also found by 13], but are transported downwards by coherent structures. Yet, the influence of gravity on the concentration profile in the fully developed situation is evident.

Acknowledgment

This research is supported by the Dutch Technology Foundation STW, applied science division of NWO and the Technology program of the Ministry of Economic affairs. The financial support from WL|Delft Hydraulics and KIWA Water Research is highly appreciated.

References

[1] R.J. Adrian. Stochastic estimation of conditional structure: a review. *Applied Scientific Research*, 53:291–303, 1994.

[2] R.J. Adrian and P. Moin. Stochastic estimation of organized turbulent structure: homogeneous shear flow. *Journal of Fluid Mechanics*, 190: 531–559, 1988.

[3] R.J. Adrian, C.D. Meinhart, and C.D. Tomkins. Vortex organization in the outer region of the turbulent boundary layer. *Journal of Fluid Mechanics*, 422:1–54, 2000.

[4] G.K. Batchelor. *An introduction to fluid dynamics.* Cambridge university press, 1967.

[5] W.A. Breugem and W.S.J. Uijttewaal. A PIV/PTV experiment on sediment transport in a horizontal open channel flow. In R.M.L. Ferreira, E.C.L.T. Alves, J.G.A.B. Leal, and A.H. Cardoso, editors, *River Flow*, volume 1, pages 789–798. Taylor & Francis, 2006.

[6] M. Cellino and U. Lemmin. Influence of coherent flow structures on the dynamics of suspended sediment transport in open-channel flow. *Journal of Hydraulic Engineering*, 130(11):1077–1088, 2004.

[7] K.T. Christensen and R.J. Adrian. Statistical evidence of hairpin vortex packets in wall turbulence. *Journal of Fluid Mechanics*, 431:433–443, 2001.

[8] J.K. Eaton and J.R. Fessler. Preferential concentration of particles by turbulence. *International Journal of Multiphase Flow*, 20(Suppl):169–209, 1994.

[9] K.T. Kiger and C. Pan. Piv technique for the simultaneous measurement of dilute two-phase flows. *Journal of Fluids Engineering*, 122:811–818, 2000.

[10] K.T. Kiger and C. Pan. Suspension and turbulence modification effects of solid particulates on a horizontal turbulent channel flow. *Journal of Turbulence*, 3(019):1–21, 2002.

[11] LaVision. FlowMaster Manual. PIV Hardware Manual for Davis 6.2. Technical report, LaVision GmbH, February 2002.

[12] M.R. Maxey and S. Corrsin. Gravitational settling of aerosol particles in randomly oriented cellular flow fields. *Journal of the Atmospheric Sciences*, 43(11):1112–1134, 1986.

[13] B. Mutlu Sumer and R. Deigaard. Particle motions near the bottom in turbulent flow in an open channel. part 2. *Journal of Fluid Mechanics*, 109:311–337, 1981.

[14] V.I. Nikora and D.G. Goring. Fluctuations of suspended sediment concentration and turbulent sediment fluxes in an open-channel flow. *Journal of Hydraulic Engineering*, 128(2):214–224, 2002.

[15] Y. Niño and M.H. García. Experiments on particle-turbulence interactions in the near-wall region of an open channel flow: implications for sediment transport. *Journal of Fluid Mechanics*, 326:285–319, 1996.

[16] A.E. Perry and I. Marusic. A wall-wake model for the turbulence structure of boundary layers. part 1. extension of the attached eddy hypothesis. *Journal of Fluid Mechanics*, 298:361–388, 1995.

[17] L.M. Portela. *Indentification and characterization of vortices in the turbulent boundary layer.* PhD thesis, Stanford University, 1997.

[18] O. Simonin, E. Deutsch, and J.P. Minier. Eulerian prediction of the fluid particle correlated motion in turbulent two-phase flows. *Applied Scientific Research*, 51(1-2):275–283, 1993.

[19] K.D. Squires and H. Yamazaki. Preferential concentration of marine particles in isotropic turbulence. *Deep-Sea Research Part I*, 42(11/12): 1989–2004, 1995.

[20] K.D. Squires and H. Yamazaki. Addendum to the paper "preferential concentration of marine particles in isotropic turbulence". *Deep-Sea Research Part I*, 43(11-12):1865–1866, 1996.

[21] G.A.J. van der Plas, M.L. Zoeteweij, R.J.M. Bastiaans, R.N. Kieft, and C.C.M. Rindt. *PIV, PTV and HPV User's Guide 1.1.* Eindhoven University of Technology, April 2003. Draft.

[22] J. Zhou, R.J. Adrian, S. Balachandar, and T.M. Kendall. Mechanisms for generating coherent packets of hairpin vortices in channel flow. *Journal of Fluid Mechanics*, 387:353–396, 1999.

Transport and mixing in the stratosphere: the role of Lagrangian studies

Bernard Legras and Francesco d'Ovidio

Laboratoire de Meteorologie Dynamique, Ecole Normale Supérieure and CNRS, 24 rue Lhomond, 75231 Paris Cedex 05, France legras@lmd.ens.fr, www.lmd.ens.fr/legras

1 Introduction

The stratosphere is an important component of the climate system which hosts 90% of the ozone protecting life from the ultra-violet radiations and, through the region called upper troposphere / lower stratosphere (UTLS) that encompasses the tropopause, has some control on the weather, the chemical composition of the atmosphere and the radiative budget. Because the temperature grows with altitude in the stratosphere, convection is inhibited by stratification, and the motion is mainly layer-wise on isentropic surfaces, with time scales of the order of weeks to months. The cross-isentropic adiabatic circulation is slow with time scales of the order of the season to several years. Below 30km, many chemical species, among which ozone, do not have significant sources or sinks and exhibit a chemical life-time of the order of several months to years. Such species can be treated as passive scalars transported by the flow. Their distribution is then dependent on the transport and mixing properties. Two useful quantities are the potential temperature $\theta = T(p_0/p)^{R/C_p}$ which is related to entropy by $S = C_p \ln \theta$ and the Ertel potential vorticity (or PV) $P = (\nabla \times \mathbf{u} \cdot \nabla \theta)/\rho$ which is a passive tracer under adiabatic and inviscid approximation. Owing to the separation between fast horizontal adiabatic motion and slow vertical diabatic motion, the potential temperature is often used as a vertical coordinate. PV is not practically measurable by in situ or remote instruments unlike many chemical tracers but can be easily calculated from model's output. It is most often used as a diagnostic of transport and dynamical activity.

Observations by in situ instruments and remote sensing show that the stratosphere exhibits well-mixed regions separated by transport or mixing barriers. Particularly, in the winter hemisphere, two dominating barriers are formed at the periphery of the polar vortex and in the sub-tropics that isolate the mid-latitudes from both the polar and tropical regions [1]. Since vertical, diabatically induced, motion is upward in the tropics and downward in the extra tropics, with the largest descent within the polar vortex, vertical tracer

Bernard J. Geurts et al. (eds), Particle Laden Flow: From Geophysical to Kolmogorov Scales, 57–69.

gradients are turned into step horizontal gradients on isentropes intersecting the barriers. In the UTLS, the barrier associated with the subtropical jets near 30N and 30S in latitude separates the upper troposphere from the lowermost stratosphere on isentropic surfaces crossing the tropopause. The layer of the stratosphere just above the tropopause undergoes exchanges with the mid-latitude troposphere mainly due to upper level frontogenesis, which is a consequence of baroclinic instability and/or intense convective events which are often themselves associated with frontogenesis. At higher levels, that is for $380K > \theta > 350K$, the exchanges across the tropopause occur between the lower stratosphere and the Tropical Tropopause Layer (TTL) which is an intermediate region between the tropical convective layer and the stratosphere. Figure 1 summarizes these processes.

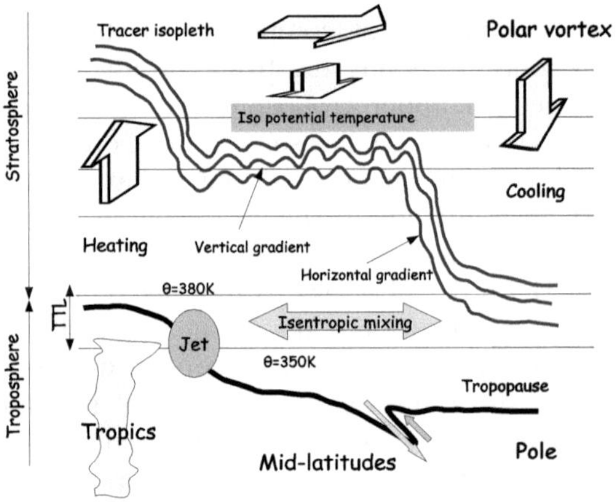

Fig. 1. Scheme summarizing the Brewer-Dobson meridional circulation in the stratosphere end exchange processes.

The scope of theory and modeling is to provide a qualitative and quantitative account of these observed properties. A number of progresses in this matter over the last ten years have been due to the extensive usage of Lagrangian calculations of parcel trajectories based on the analyzed winds provided by the operational meteorological centers. It is the goal of this presentation to review the ongoing work in this direction.

2 Isentropic stirring

The strong stratification of the stratosphere accompanied by weak net diabatic contribution constrains parcels to move on isentropic surfaces. Lagrangian isentropic motion differs from fully turbulent motion and is akin to two-dimensional turbulence or chaotic motion in a plane, where the flow is smooth and advection is dominated by the large structures. Such dynamics is known to produce transport barriers where tracer gradients intensify.

The effective diffusivity [2, 3, 4] has been used with success as a diagnostic of such effects in atmospheric flows. The method applies to a tracer, usually PV, with a mean latitudinal gradient such that the longitude-latitude coordinates can be replaced by the area of embedded tracer contours and an azimuthal coordinate along the contours. By a weighted averaging along the contours, the advection-diffusion equation $\partial c/\partial t + \mathbf{u}\boldsymbol{\nabla} c = \kappa\nabla^2 c$ is replaced by a purely diffusive equation

$$\frac{\partial C(A,t)}{\partial t} = \frac{\partial}{\partial A}\left(\kappa_{\text{eff}}(A,t)\frac{\partial C(A,t)}{\partial A}\right),$$

where A is the area of the contour $\gamma(C,t)$ and the effective diffusivity is

$$\kappa_{\text{eff}(A,t)} = \kappa_0 \frac{L_{\text{eq}}^2(A,t)}{A\left(4\pi - \frac{A}{r^2}\right)} \tag{1}$$

with

$$L_{\text{eq}}(A,t) = \oint_{\gamma(C,t)} |\nabla c| dl \oint_{\gamma(C,t)} \frac{1}{|\nabla c|} dl\,,$$

and r is the radius of the Earth. The device in (1) is to bind the complex stirring of the passive scalar in the averaging over the contour. It can be shown [5] that L_{eq} is always larger than the actual length of the tracer contour but that in practice the two quantities scale similarly. Hence, effective diffusivity is small where the contours are less deformed, that is where transverse gradients are less intensified, leading to less mixing.

Another measure of atmospheric stirring is provided by the local Lyapunov exponent [6, 7] which estimates the Lagrangian stretching as the separation rate of two close parcels over a given period of time or over a given growth. Around the Antarctic stratospheric polar vortex, a minimum in both the effective diffusivity and local Lyapunov exponent is observed along the streak lines at the center of the jet [7, 8]. This minimum is surrounded by a very active mixing region with large stretching where air is brought from and to the mid-latitudes within a few days. However, the very stable Antarctic polar vortex is rather an exception than the rule among atmospheric flows which usually exhibit much less stable patterns. Over most of the atmosphere, the Lyapunov exponents and effective diffusivity are rather anti-correlated than correlated, contrary to the simple expectation.

The main reason of this discrepancy is that atmospheric flow, like most quasi-2D flows, is dominated by extended shear regions that stretch material lines but contribute weakly to the growth of tracer gradients. Let us first consider the deformation of a small material circle surrounding a parcel at time t. As time runs forward or backward, the flow defines a pair of linear transformations T and T^{-1} which map the circle onto ellipses at future or past time $t+\tau$ or $t-\tau$. The local forward and backward Lyapunov exponents are $\lambda_f = 1/\tau \ln \sigma_f$ and $\lambda_b = 1/\tau \ln \sigma_b$ where σ_f and σ_b are the singular values of T and T^{-1}. The singular vectors of these transformations are the Lyapunov eigenvectors. The eigenvector pointing to the smallest axis of the ellipse in the transformation indicates the unstable manifold or stable material line for backward time and the stable manifold or unstable material line for forward time [9, 8]. The convergence of the eigenvectors is at least as fast and generally much faster than the Lyapunov exponent [10]. Consequently, the tracer gradient tends, at any time, to be perpendicular to the local unstable manifold [11, 12]. This leads to an estimate of the future growth of the tracer gradient as the product of the forward Lyapunov exponent with the sine of the angle between the local stable and unstable manifolds that we denote as $\lambda_\perp$, the *transverse Lyapunov exponent.* Since stable and unstable manifolds are parallel in the direction of the wind for a pure shear, the transverse Lyapunov exponent vanishes in this case. As a matter of fact, when the gradient is perpendicular to the shear, it does not intensify at all.

Fig. 2 shows, strikingly, that that the maximum in stretching over the subtropical jet in the northern hemisphere (near 30N) is turned into a minimum of the transverse Lyapunov exponent, in agreement with the fact revealed from effective diffusivity that the jet acts as a barrier to mixing.

Fig. 3 shows that the transverse Lyapunov exponent correlates linearly with $\ln \kappa_{\text{eff}}$ while no correlation emerges with the Lyapunov exponent.

A quantitative relation between effective diffusivity and the transverse Lyapunov exponent can be found by assuming that, owing to the chaotic stirring by the atmospheric flow, $L_{\text{eq}} \sim \exp(\lambda_\perp T_\kappa)$ where T_k is a characteristic time after which diffusion becomes dominant and balances the exponential growth. Using this expression in (1), we define the *Lyapunov diffusion* κ_λ as

$$\ln \frac{\kappa_\lambda}{\kappa_0} = A + 2\lambda_\perp T_\kappa \qquad (2)$$

where A is a constant. Both A and T_k can be obtained by a fit of κ_λ to κ_{eff}. Fig.4 shows that indeed κ_λ reproduces very well the variations of κ_{eff}.

Hence the transverse Lyapunov exponent provides a diagnostic of atmospheric stirring that matches very well the effective diffusivity but with a number of advantages with respect to this latter, in particular in resolving structures in longitude. This is visible in Fig. 2 where the belt of low values along the jet is interrupted in the eastern Pacific opening a gate to mixing across this region between high and low latitudes. During the strong El Niño event of 1998, Fig. 5 shows a much more continuous belt over the Pacific, suggesting that El Niño is associated with a closing of the mixing gate. This result is also supported by analysis of meridional fluxes [13] and the climatology of intrusions of stratospheric air in the tropical upper troposphere [14]. For further discussion, see [15].

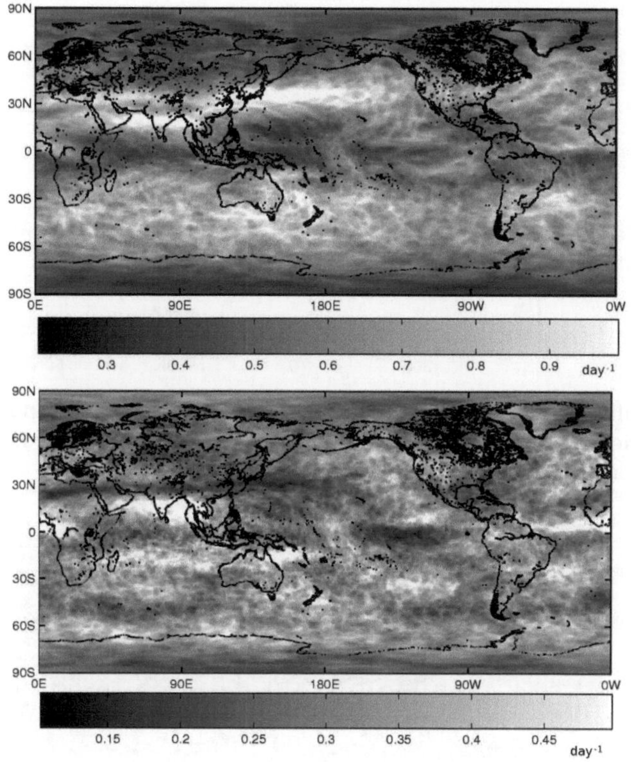

Fig. 2. Forward Lyapunov exponent (top), measuring stretching intensity, and transverse Lyapunov exponent (bottom) averaged over five January months (1997, 1999, 2000, 2001 and 2002) on the surface $\theta = 350$K which intersects the tropopause within the subtropical jet.

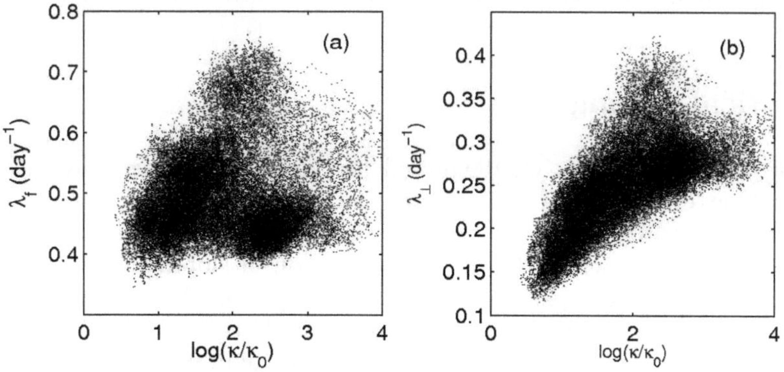

Fig. 3. Scatter plots of monthly-means of longitudinally averaged forward (a) and transverse (b) Lyapunov exponent against effective diffusivity over the period 1980-2000 on the θ =350K isentropic surface.

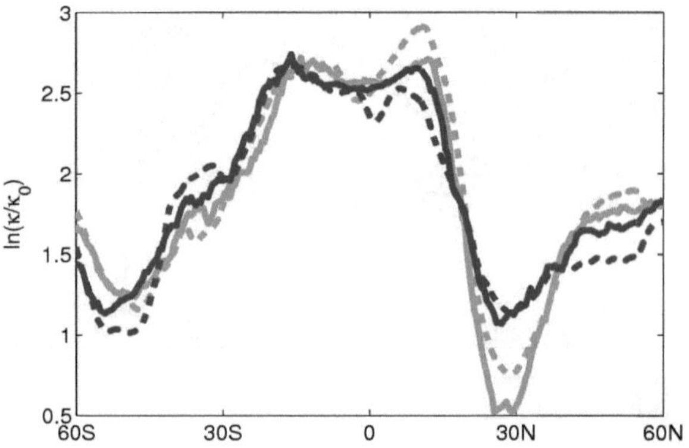

Fig. 4. Zonally averaged Lyapunov diffusivity (continuous line) and effective diffusivity (dashed line) for the El Niño winter 1997-1998 (gray) and La Niña winter 1998-1999 (black).

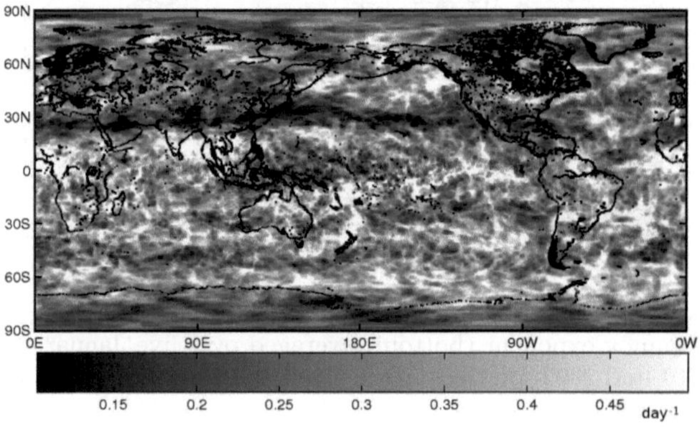

Fig. 5. Transverse Lyapunov exponent averaged over January 1998 on the surface θ =350K.

3 Vertical mixing

In the previous section, we diagnosed the stirring of tracers by layer-wise motion that elongates tracer contours and generates a large number of filamentary structures within an isentropic surface. It has been shown [16, 17] that filaments are merely the horizontal section of sloping sheets with large horizontal to vertical aspect ratio of the order of 200. This value is essentially the ratio of the vertical shear to the horizontal strain. Owing to this high aspect ratio, the dissipation of a tracer sheet is mainly a product of vertical mixing by small-scale turbulence. Present meteorological analysis provided by weather centers basically resolve the motion that induces stirring and generation of tracer sheets, but small-scale turbulence due to shear instability or gravity wave breaking is unresolved by any large-scale numerical atmospheric model. Local diffusion coefficient that may vary in space and time are often used to characterize the mixing produced by turbulence. Radar estimates of this quantity in active turbulent regions provide values of the order of 1-5 m^2s^{-1} but models [18, 17, 19] suggest that active regions are sparse in the atmosphere and that, on the average, much smaller turbulent diffusion is required to explain the observation of tracers.

Lagrangian reconstruction methods are often used to explain the spatial and temporal variations of atmospheric tracers. Such methods, sketched in Fig. 6, are based on the possibility to reconstruct small scales of the tracer distribution from the time series of the advecting wind and have been very successful in the lower stratosphere and the upper troposphere [20, 18, 21] showing that a large number of tracer structures seen in satellite images, aircraft transects and balloon profiles can be explained by advection. In most early studies, the reconstructed tracer was PV but recent works focus on reconstructed chemical tracers that can be compared more directly with observations. Diffusion can be introduced in such methods by dividing each parcel into a large number of particles which are advected backward adding a random velocity component in the vertical direction, such that over one time step

$$\delta z = w\delta t + \eta\delta t \,,$$

where the random component η fits a chosen vertical turbulent diffusivity by [19]

$$D = \frac{1}{2} < \eta^2 > \delta t \,.$$

The rationale of this approach is to integrate the adjoint equation for the Green function of the advection-diffusion equation which is well-posed for backward times [19]. It has been applied, using a different formulation, to identify pollution sources from a network of sensors [22].

Comparing high resolution airborne tracer measurements with such reconstructions done with several values of D provides an estimate of the Lagrangian averaged diffusivity which matches best the observed fluctuations. Fig. 7 provides an example of such comparison done during a campaign in the Arctic, showing that turbulent diffusivity is, on the average, much smaller inside the polar vortex than outside, the largest of these estimates being one order of magnitude less than the radar estimate. When sharp transitions are identified in both the observed and the reconstructed data, it is possible to provide a local estimate of Lagrangian turbulent diffusivity. Fig .8 shows that D varies by at least one order of magnitude across a 80km wide filament.

As diffusivity can be estimated independently from the strain, a relation between both quantity, which is usually assumed in most parameterization of turbulence, can be tested. It has been shown [19] that on the average the two quantities are correlated but that this cannot explain the type of variability shown in fig. 8 which is perhaps due to burst of gravity wave breaking.

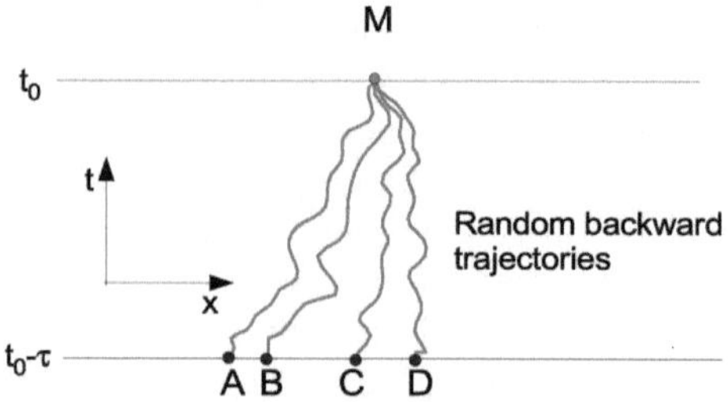

Fig. 6. In the standard reverse domain filling method [23] parcel trajectories starting at time t are integrated backward in time for a duration τ using available wind fields over the time interval. Provided a coarse resolution distribution of the tracer is known at time $t - \tau$, the value attributed to the parcel at time t is that of the coarse field at its location at time $t - \tau$. This procedure is able to reconstruct the small-scales at time t if transport is dominated by the resolved scales of motion. In the diffusive version [19], each parcel is a mixture of a cloud of particles that originate from a distribution of locations at time $t - \tau$ under backward advection plus diffusion. Then the value of the tracer at time t is an average of the values at $t - \tau$ for all the particles of the cloud. Unlike the standard procedure, the reconstruction is to a large extend independent of τ and the applied diffusion controls its smoothness.

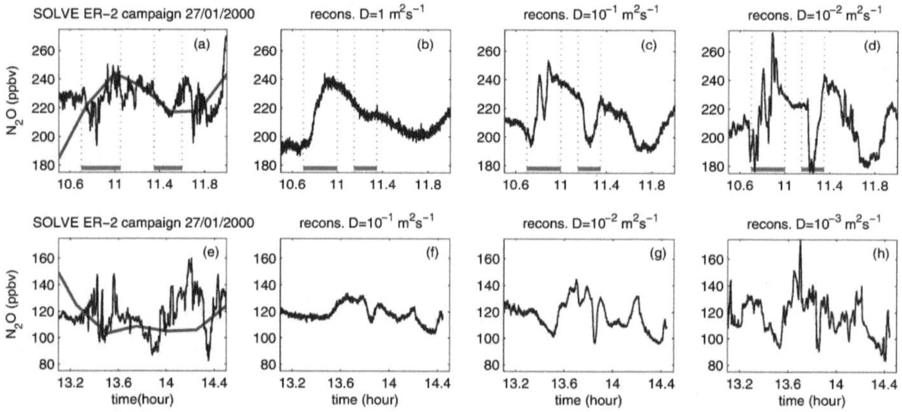

Fig. 7. (a): Sample of aircraft tracer measurement at 56 hPa in the surf zone outside the polar vortex during SOLVE campaign (winter 2000). The thick line shows the corresponding transect in the chemical transport model used to provide the coarse distribution of the tracer. (b-d): Advective-diffusive reconstructions for three values of the diffusivity $D = 1$, 0.1 and 0.01 $m^2\ s^{-1}$. The comparison is based on the roughness of the curves with some details identified near 11UT and 11:30UT. Clearly, the reconstruction is too smooth for $D = 1\ m^2\ s^{-1}$and too rough for $D = 0.1\ m^2\ s^{-1}$, while $D = 0.1\ m^2\ s^{-1}$seems of the right order. For a more quantitative assessment, see [19]. (e): Same as (a) inside the polar vortex. (f-h) Advective reconstructions inside the polar vortex where the comparison suggests that $0.01\ m^2\ s^{-1} > D > 0.001\ m^2\ s^{-1}$.

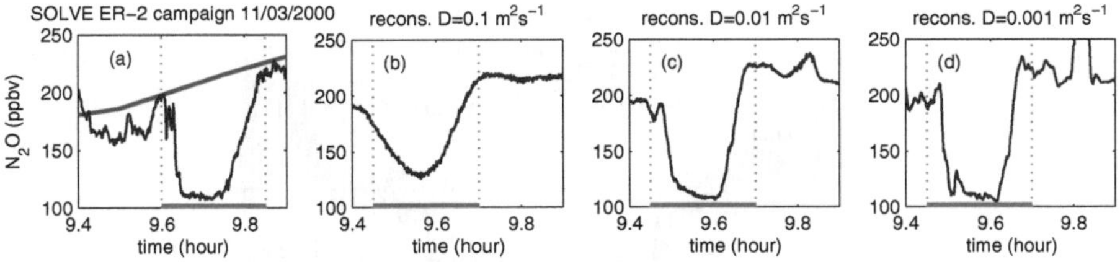

Fig. 8. (a) Tracer transect across a filament just outside the polar vortex during SOLVE campaign. The right edge of the filament fits very well an error function (solution of the advective diffusive equation with constant strain) with a width of 36km while the left edge is much steeper with a width of about 2.5km. (b-d) The reconstructions show that the two slopes cannot be reproduced with a single uniform diffusion indicating a large variation, by more than one order of magnitude, of turbulent diffusion across the width of the filament.

4 Meridional Brewer-Dobson circulation

Over time scales of several years the stratospheric circulation is characterized by an overturning from the tropics to the mid and polar latitudes (see Fig.1) known, since the pioneering work of Brewer [24], as the Brewer-Dobson circulation. It is an important requirement, for the distribution of long-lived chemical species that numerical models reproduce quantitatively this circulation. A large class of models of atmospheric chemistry, denoted as chemical-transport models, rely on the analyzed winds provided by the operational weather centers to advect the chemical compounds horizontally and vertically. Among those wind datasets, the ERA-40 reanalysis of the European Center for Medium Range Weather Forecast (ECMWF), already used in section 1, is particularly useful since it covers a 45-year period from 1957 to 2002 [25].

Most global weather forecast models, including that of ECMWF, use the hydrostatic approximation. This means that vertical velocities are calculated from the the continuity equation, that is basically from the vertical integration of the horizontal divergence. Such estimate is known to be noisy by nature as the divergent circulation is weak and badly constrained by observations. Moreover, the practice of weather centers is to archive instantaneous velocity fields at times separated by interval of several hours, hence strongly undersampling fluctuations, such as gravity waves, with time scales much shorter than the archiving interval. Most studies indeed rely, by tradition, on 6-hourly winds. Fig .9 shows that these winds induce a too strong meridional circulation (hence the age is too young in the extra-tropics). However, a 3-hourly dataset, also available from ECMWF reduces considerably the discrepancy with observations. A chemical-transport model using this dataset is able to reproduce with good accuracy the ozone column at mid-latitude (F. Lefèvre, 2005, personal communication). Since the 3-hourly dataset still contains a significant amount of spurious noise, an alternative is to move horizontally parcel on isentropic surfaces and to use diabatic heating rates, calculated from the

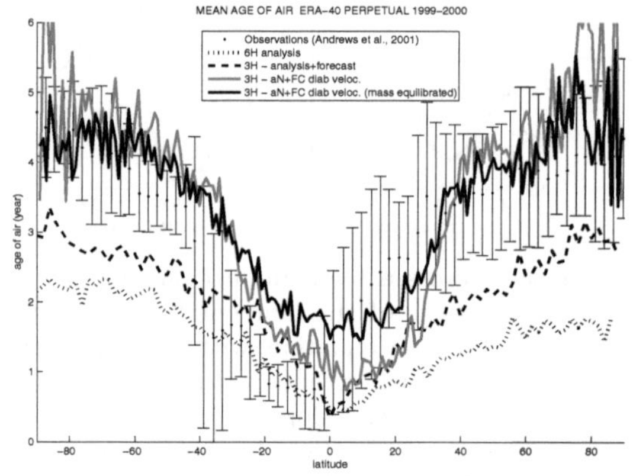

Fig. 9. Comparison of the age of air between observations and model calculations. The age of an air parcel is the time spent by this parcel in the stratosphere since it entered it at tropical latitudes [26]. Since an air parcel is a mixture, the mean age of air is the average over all particles within the parcel. The age of air can be measured using gases like SF_6, CH_4 and N_2O which are well mixed in the troposphere and are slowly destroyed in the stratosphere or CO_2 which is also well mixed in the troposphere and increases with time. The observation curve [27] is based an aircraft measurements at about 20km. Model calculations are performed using Lagrangian trajectories initialised almost uniformly at 20km integrated backward until they cross the tropopause. The age is averaged over latitude circles. The four curves are built using wind datasets over the cycle 1999-2000 and calculations are done over 15 years repeating this cycle. For parcels which have not left the stratosphere after 15 years, the age is extrapolated as in [28]. Dotted: reconstruction using the ERA-40 winds at 6-hour interval. Dash: reconstruction using the ERA-40 winds at 3-hour intervals (with forecasts interleaved with analysis as in [19]. Gray solid: reconstruction using the ERA-40 winds in the horizontal and heating rates for vertical motion. Black solid: same as previous with a correction on the horizontal isentropic divergence to balance the heating rate.

local radiative budget and averaged over 3-hour intervals, as vertical velocities. Fig 9 shows that the meridional circulation calculated from such data provides a good agreement with observations. Doing so, we have, however, introduced an inconsistency since mass conservation is no longer satisfied. This conservation is reestablished by forcing the horizontal divergence on isentropic surfaces to satisfy the equation

$$\partial\sigma/\partial t + \boldsymbol{\nabla}_\theta(\sigma\mathbf{u}) + \partial(\sigma\dot{\theta})/\partial\theta = 0$$

where $\sigma = -g\partial p/\partial\theta$. Fig. 9 shows that a further improvement in the agreement with observations is obtained in this way except near the equator. Hence, it is now clear that available analyses fulfill the constrain that the Brewer-Dobson circulation is fairly well reproduced, at least over the recent years where high quality satellite observations are available, but it is also clear that special care should be taken when using analyzed data for transport calculations over durations of months to years in order to avoid spurious diffusive transport due to noise and under-sampling. Another important factor impacting the quality of the analysis is the the model itself and the type of assimilation scheme [28, 30].

References

[1] L. Polvani, D. Waugh, R. Plumb, On the subtropical edge of the stratospheric surf zone, J. Atmos. Sci. 51 (11) (1995) 1401–1416.

[2] N. Nakamura, Two-dimensional mixing, edge formation, and permeability diagnosed in an area coordinate, J. Atmos. Sci. 53 (11) (1996) 1524–1537.

[3] P. Haynes, E. Shuckburgh, Effective diffusivity as a diagnostic of atmospheric transport. Part I: stratosphere, J. Geophys. Res. 105 (D18) (2000) 22,777–22,794.

[4] P. Haynes, E. Shuckburgh, Effective diffusivity as a diagnostic of atmospheric transport. Part II: troposphere and lower stratosphere, J. Geophys. Res. 105 (D18) (2000) 22,795–22,810.

[5] E. Shuckburgh, P. H. Haynes, Diagnosing transport and mixing using a tracer-based coordinate system, Phys. Fluids 15 (2003) 3342–3357.

[6] R. Pierrehumbert, Large-scale horizontal mixing in planetary atmospheres, Phys. Fluids A 3 (5) (1991) 1250–1260.

[7] B. Joseph, B. Legras, Relation between kinematic boundaries, stirring, and barriers for the Antarctic polar vortex, J. Atmos. Sci. 59 (2002) 1198–1212.

[8] T.-Y. Koh, B. Legras, Hyperbolic lines and the stratospheric polar vortex, Chaos 12 (2) (2002) 382–394.

[9] G. Haller, G. Yuan, Lagrangian coherent structures and mixing in two-dimensional turbulence, Physica D 147 (2000) 352–370.

[10] B. Legras, R. Vautard, A guide to Liapunov vectors, in: Predictability, Vol. I of Seminar Proceedings, ECMWF, Reading, UK, 1996, pp. 143–156.

[11] P. Klein, B. L. Hua, G. Lapeyre, Alignment of tracer gradient vectors in 2d turbulence, Physica D 146 (2000) 246–260.

[12] G. Lapeyre, B. Hua, P. Klein, Dynamics of the orientation of active and passive scalars in two-dimensional turbulence, Phys. Fluids 13 (2001) 251–264.

[13] M. Shapiro, H. Wernli, N. Bond, R. Langland, The influence of the 1997-1998 ENSO on extratropical baroclinic life cycles over the Eastern North Pacific, Quart. J. Roy. Met. Soc. 127 (2001) 331–342.

[14] D. Waugh, L. Polvani, Climatology of intrusions into the tropical upper troposphere, Geophys. Res. Lett. 27 (23) (2000) 3857–3860.

[15] F. d'Ovidio, B. Legras, E. Shuckburgh, Local diagnostic of mixing and barrier modulation at the tropopause, submitted to Geophys. Res. Lett., http://www.lmd.ens.fr/legras (2006).
[16] D. Waugh, R. Plumb, R. Atkinson, M. Schoeberl, L. Lait, P. Newman, M. Lowenstein, D. Toohey, L. Avallone, C. Webster, R. May, Transport of material out of the stratospheric Arctic vortex by Rossby wave breaking, J. Geophys. Res. 99 (D1) (1994) 1071–1088.
[17] M. Balluch, P. Haynes, Quantification of lower stratospheric mixing processes using aircraft data, J. Geophys. Res. 102 (D19) (1997) 23,487–23,504, (97JD00607).
[18] D. Waugh, R. Plumb, J. Elkins, D. Fahey, K. Boering, G. Dutton, C. Volk, E. Keim, R.-S. Gao, B. Daube, S. Wofsy, M. Lowenstein, J. Podolske, K. Chan, M. Proffitt, K. Kelly, P. Newman, L. Lait, Mixing of polar air into middle latitudes as revealed by tracer-tracer scatter plots, J. Geophys. Res. 102 (D11) (1997) 13,119–13,134, (96JD03715).
[19] B. Legras, I. Pisso, G. Berthet, F. Lefèvre, Variability of the Lagrangian turbulent diffusion in the lower stratosphere, Atmos. Chem. Phys. 5 (2005) 1–18.
[20] C. Appenzeller, H. Davies, W. Norton, Fragmentation of stratospheric intrusions, J. Geophys. Res. 101 (D1) (1996) 1453–1456.
[21] A. Mariotti, M. Moustaoui, B. Legras, H. Teitelbaum, Comparison between vertical ozone soundings and reconstructed potential vorticity maps by contour advection with surgery, J. Geophys. Res. 102 (D5) (1997) 6131–6142.
[22] J. Issartel, J. Baverel, Inverse transport for the verification of the Comprehensive Nuclea Test Ban Treaty, Atmos. Chem. Phys. 3 (2003) 475–486.
[23] R. Sutton, H. Maclean, R. Swinbank, A. O'Neill, F. Taylor, High-resolution stratospheric tracer fields estimated from satellite observations using Lagrangian trajectory calculations, J. Atmos. Sci. 51 (1994) 2995–3005.
[24] A. Brewer, Evidence for a world circulation provided by the measurements of helium and water vapour distribution in the stratopshere, Quart. J. Roy. Met. Soc. 75 (1949) 351–363.
[25] S. Uppala, *et al.*, The ERA-40 re-analysis, Quart. J. Roy. Met. Soc. 131 (2005) 2961–3012.
[26] D. Waugh, T. Hall, Age of stratospheric air: theory, observations, and models, Rev. Geophys. 40 (4).
[27] A. Andrews, *et al.*, Mean ages of stratospheric air derived from in situ observations of CO_2, CH_4, and N_2O, J. Geophys. Res. 106 (D23) (2001) 32,295–32,314.
[28] M. Scheele, P. Siegmund, P. van Velthoven, Stratospheric age of air computed with trajectories based on various 3D-Var and 4D-Var data sets, Atmos. Chem. Phys. 5 (2005) 1–7.
[29] M. Schoeberl, A. Douglass, Z. Zhu, S. Pawson, A comparison of the lower stratospheric age spectra derived from a general circulation

model and two assimilation systems, J. Geophys. Res. 108 (D3) (2003) doi:10.1029/2002JD002652.

[30] B. Monge-Sanz, M. Chipperfiels, A. Simmons, S. Uppala, Mean age of air and transport in a CTM: comparison of different ECMWF analyses, Geophys. Res. Lett., in press (2006).

Numerical modeling of heat and water vapor transport through the interfacial boundary layer into a turbulent atmosphere

A.S.M. Gieske

International Institute for Geo-Information Science and Earth Observation (ITC), Water Resources Division, PO Box 6, Enschede, The Netherlands gieske@itc.nl

Summary. A stochastic numerical model is developed to simulate heat and water vapor transfer from a rough surface through a boundary layer into the fully turbulent atmosphere. The so-called interfacial boundary layer is conceptualized as a semi-stagnant layer of air in the roughness cavities at the surface into which the smallest eddies penetrate to random approach distances and with random inter-arrival times, carrying away energy, molecules, or any other scalar admixture. The model makes use of the one-dimensional transient heat conduction equation where the boundary conditions are updated in time and space by random deviates from a general gamma distribution. The one-dimensional transfer equation is solved by the implicit finite difference method which allows conversion to a standard tridiagonal matrix equation.

The algorithm is simple to implement and allows generation of large ensembles for statistical analysis in short periods of time. The simulations were used to compare and contrast earlier results obtained for heat and mass transfer through Earth surface-air interfaces. It is shown that even small increases in boundary-layer thickness may significantly enlarge the inverse Stanton roughness number St_k^{-1} reducing heat transfer from the surface. Review of experimental work suggests an updated relation for the heat transfer coefficient from bare soils into the atmosphere. Work is under way to incorporate the results into the atmospheric and remote sensing research related to the determination of the Earth's sensible and latent heat fluxes.

1 Introduction

In the application of remote sensing techniques to determine evapotranspiration, use is normally made of the energy balance equation

$$E = R_n - G - H \tag{1}$$

where E is the evapotranspiration, R_n the net solar radiation, G the soil heat flux and H the sensible heat flux. The net solar radiation and soil heat flux can be determined reasonably well with remote sensing techniques in the visible and thermal infrared parts of the spectrum, while several parameterizations

Bernard J. Geurts et al. (eds), Particle Laden Flow: From Geophysical to Kolmogorov Scales, 71–83.

exist to determine the sensible heat flux H using thermal infrared imagery to determine surface temperatures. Therefore, even if the transport mechanisms for latent and sensible heat E and H are the same, it is more practical to determine the sensible heat flux first because of the possibility to obtain remotely sensed surface temperatures. For this reason the transport problem is formulated here in terms of heat transport, closely following [1] and [3].

The transport of heat and water vapor in a fully turbulent atmosphere can be described using Reynolds analogy for turbulent flow because it may be assumed that transport is predominantly linked to eddy air movement at scales where molecular diffusive transport can be neglected. Under these circumstances turbulent transport of momentum, heat, water vapor, carbon dioxide and dust particles are analogous. However, close to the surface Reynolds analogy ceases to be valid. Wind speed becomes zero at the roughness length for momentum transport (z_0) while temperature and water vapor concentration on the other hand approach finite values. Conduction and diffusion are the dominant transport mechanisms close to the surface even in fully rough regimes. Kays and Crawford [8] express this as the heat transfer by conduction through what may be a semi-stagnant fluid in the roughness cavities at the surface. This stagnant layer is here referred to as interfacial sub-layer following the usual micro-meteorological convention [3, 12].

The non-dimensional temperature difference δt_0^+ across the interfacial sub-layer is related to the roughness Stanton number St_k as

$$\delta t_0^+ = \frac{\Delta T \rho C_p u_*}{H} = St_k^{-1} \tag{2}$$

where ΔT is the temperature difference, ρ the air density, C_p the specific heat and u_* the friction velocity. The roughness Stanton number St_k must be determined by experiment [8] and is a function of the type of surface roughness.

In micro-meteorological applications very often a sub-layer Stanton number B is used which is defined as

$$B^{-1} = St_k^{-1} - Cd^{-1/2} \tag{3}$$

where Cd is a drag coefficient. The term kB^{-1} is also used frequently where k is the von Kármán constant. Experiment has shown that the roughness Stanton number St_k can be parameterized as

$$St_k^{-1} = cRe_*^m Pr^n \tag{4}$$

with surface roughness Reynolds number defined as $Re_* = u_* z_0/\nu$ and Pr as the Prandtl number (0.71 for air). The parameter z_0 is the roughness length for momentum transport while ν is the kinematic viscosity of air ($\approx 1.5\ 10^{-5}\ m^2 s^{-1}$).

Figure 1 below shows the analysis given by [3] where $kB^1 = \ln(z_0/z_{0c})$ is plotted as a function of the Reynolds surface-roughness number. The general scalar roughness length z_{0c} becomes z_{0h} for heat transport and z_{0v} for water vapor transport. The equations for the bluff roughness elements (bare soil) are given by

$$kB_v^{-1} = 2.25(Re_*)^{1/4} - 2 \quad ; \quad kB_h^{-1} = 2.46(Re_*)^{1/4} - 2 \tag{5}$$

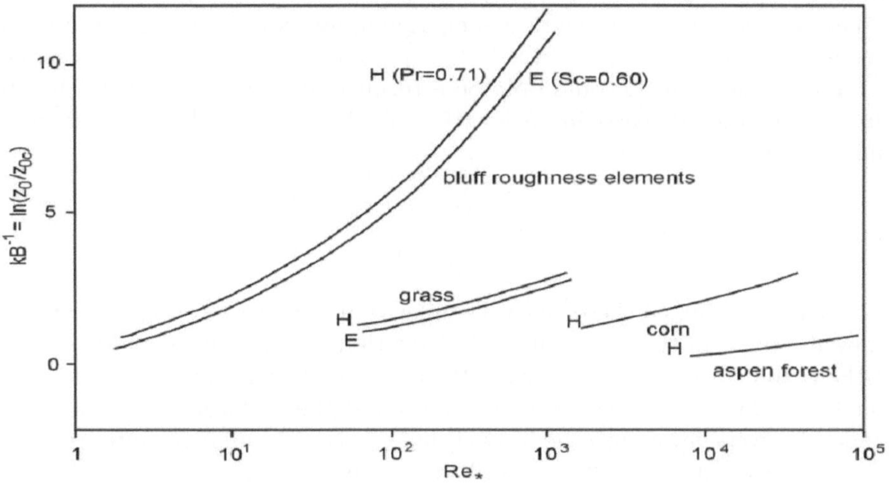

Fig. 1. Summary of analysis given by [3] where $kB^{-1} = \ln(z_0/z_{0c})$ is plotted as a function of the Reynolds surface roughness number ($Re_* = u_* z_0/\nu$). H refers to sensible heat with Prandtl number $Pr = 0.71$, while E refers to water vapor transport with Schmidt number $Sc = 0.6$. Also shown are some results for vegetated areas (grass, corn and forest).

The results obtained with these relations have become embedded in current practice to determine latent and sensible heat fluxes by remote sensing techniques [11, 12]. The equations (5) show the close correspondence between the results obtained by [3] for transport of water vapor and sensible heat. Figure 1 also shows some results for vegetated areas. However, the analysis in this paper is limited to the non-vegetated bluff-roughness case. It should also be noted that only rough flow ($Re_* > 1$) is considered here. The set of equations (5) were developed by [1] and [4] through a stochastic analytical model, whereas [5] developed a more complex, partly numerical and partly analytical model. It is the objective of this paper to show how quick results may also be obtained with a completely numerical finite difference approach. First a short review of the theory leading to the basic analytical solution is given in section 2. The structure of the model is discussed in section 3 while the results of the simulations are given in section 4. Finally, the results are compared with previous work, ending with a few conclusions and recommendations.

2 Analytical solution of basic stochastic model

Atmospheric transport by turbulent flow takes place through a wide range of eddy sizes varying from the internal to the external scale. However, the closer to the surface the smaller these eddies become. Ultimately a limit is reached where the Reynolds number becomes too low and no more eddies can be generated. The energy is then absorbed by diffusion at molecular level. Kolmogorov [6] put forward the hypothesis that the properties of the smallest eddies are determined by the rate of energy dissipation (ϵ) in the flow and the kinematic viscosity of the fluid (ν). The assumption was made that a change in internal length scale could only be a result of change in ϵ and ν. Thus by purely dimensional reasoning a length scale λ was defined by Brutsaert [4] and Obukhov [7] as

$$\lambda = (\nu^3/\epsilon)^{1/4} \tag{6}$$

and a time scale θ as

$$\theta = (\nu/\epsilon)^{1/2} \tag{7}$$

Brutsaert [1, 4] linked these general similarity relations to a stochastic surface renewal model. In this model it is assumed that water vapor molecules arrive at the turbulent eddies by diffusion through a stationary air layer (the interfacial boundary layer) between the eddies and the surface. These eddies are randomly swept away into the fully turbulent stream and constantly replaced by new ones. The procedure is well known in chemical engineering [10, 19] and appears to provide a reasonable physical and statistical picture of the transport processes across the interfacial sub-layer.

The renewal probability p is given by

$$p = se^{-st} \tag{8}$$

where s is the eddy renewal rate (Hz).

Transport through the stationary boundary layer was modeled by the one-dimensional heat conduction equation [9]

$$\frac{\partial T}{\partial t} = D\frac{\partial^2 T}{\partial z^2} \tag{9}$$

where T is the specific humidity and D is the thermal diffusivity which is equal to $\kappa/(\rho C_p)$. The conductivity κ for air is approximately 0.025 $Wm^{-1}K^{-1}$. For the case of water vapor transport T is replaced by q the specific humidity and D is the molecular diffusivity of water vapor.

Solution of (9) under the boundary conditions

$$\begin{array}{llll} T = T_a & ; \quad 0 < z < \infty & ; \quad t = 0 & \\ T = T_0 & ; \quad z = 0 & ; \quad t \geq 0 & \quad (10) \\ T = T_a & ; \quad z \rightarrow \infty & ; \quad t \geq 0 & \end{array}$$

gives

$$\left(\frac{\partial T}{\partial z}\right)_{z=0} = (T_a - T_0)/\sqrt{D\pi t} \tag{11}$$

and the average value H_{avg} can be found by integration

$$H_{avg} = -\kappa s \int_0^\infty e^{-st}\left(\frac{\partial T}{\partial z}\right)_{z=0} dt \tag{12}$$

yielding the solution

$$H_{avg} = -(T_a - T_0)\sqrt{s\kappa\rho C_p} \tag{13}$$

The renewal rate s is parameterized by assuming that it is inversely proportional to the time scale $\theta = (\nu/\epsilon)^{1/2}$ (equation 7). Moreover, close to the surface the rate of energy dissipation is equal to

$$\epsilon = \frac{u_*^3}{kz_0} \tag{14}$$

Combining equations (7) and (14) gives

$$s = C_2\left(\frac{u_*^3}{z_0}\right)^{1/2} \tag{15}$$

where C_2 is a constant, equal to about 4.84 when the Brutsaert empirical values are used. Thus (13) and (15) give the solution of the stochastic model as a function of the air-surface temperature difference $(T_0 - T_a)$, the friction velocity u_* and the roughness length for momentum transport z_0. The model is assumed to be valid between $z = 0$ and $z = e^2 z_0 (\approx 7.4 z_0)$. Note that the roughness length for heat transport z_{oh} does not appear as a parameter in the solution. This parameter is defined as the height where the temperature attains its surface value, when the logarithmic profile is extrapolated downward. However, in the interfacial sub-layer the Reynolds analogy is no longer valid as mentioned already in the introduction, so the physical interpretation of the roughness length for heat transport must be treated with caution. The same holds for all other scalar roughness lengths.

The Brutsaert model outlined above is simple because it assumes that the replacement of eddies takes place right at the surface. A more general version of the surface renewal model was already proposed by Harriott [5, 13] and improved later by Thomas and Fan [14] who considered eddies not only with random arrival times but also with variable approach distances, i.e., variable thickness of the interfacial boundary layer. The statistics of this more general process is governed by gamma distribution functions such as

$$p(t) = \frac{1}{\Gamma(\alpha)\beta^\alpha} t^{\alpha-1} \exp\left(-\frac{t}{\beta}\right) \tag{16}$$

The mean t of the distribution is equal to $\alpha\beta$, while the variance is equal to $\alpha\beta^2$. Relation (16) reduces to the Brutsaert inter-arrival time distribution

(Eqn. 8) for $\alpha = 1$ and $\beta = 1/s$. If the time t is replaced by the distance z, the same gamma distribution can used to model the approach distances h_p. The eddies are assumed to arrive to a distance h_p (see Figure 2) with a uniform temperature (T_a) causing an instantaneous temperature drop from $T(t, z)$ to T_a at distances larger than h_p. In practice, an outer model boundary is assumed with constant temperature T_a at a distance of about $7.4z_0$. The wall itself is assumed to have a constant temperature T_0. Conduction of heat takes place from the wall according to the general heat conduction equation (6) after this event until the next event when a fresh eddy arrives at another distance h_p. Two random values need to be determined after each event: an inter-arrival time t_i and an approach distance h_p.

Harriott [5] did not present a fully analytical solution but rather a mixture of analytical expressions with numerical simulations. The time intervals $t_1, t_2, \ldots, t_n$ and a corresponding sequence of approach distances $h_{p1}, h_{p2}, \ldots, h_{pn}$ were generated by a computer. After each arrival the existing temperature curve $T(t, z)$ is truncated at the new approach distance h_p, i.e., $T(t, z) = T_a$ for $z > h_p$. The solution between the eddy arrivals is given by the analytical solution of the heat conduction equation (9).

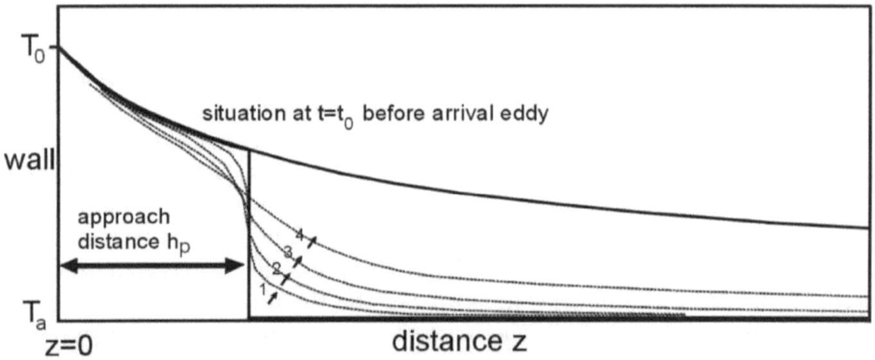

Fig. 2. The figure shows the steps in the modeling according to Harriott [5]. Up to t_0 the time evolution of the temperature is governed by the conduction of heat equation. At $t = t_0$ an eddy arrives up to distance h_p, leading to a truncated temperature distribution. After that the temperature evolves again according to the heat conduction equation and steps 1, 2, 3 and 4 (dashed lines) mark the successive changes in temperature. This continues until the arrival of a fresh eddy at another approach distance. Note that the average slope of the temperature curve at $z = 0$ determines the heat transport.

3 Numerical model implementation

The model proposed here makes use of the transient heat conduction equation (9) and the general gamma distribution (16). The one-dimensional flow equation can be solved by the implicit finite difference method as discussed below in section 3.1, whereas the generation of random deviates for arrival times and approach distances is described in section 3.2.

3.1 Finite difference model setup

The one-dimensional transient system of equation (9) with boundary conditions (10) can easily be written in finite differences [15]. In this case the implicit system of equations can be written as a tridiagonal matrix equation that is easily and quickly solved with the Thomas algorithm.

In order to make the model as realistic as possible, a grid was used consisting of 1000 elements with properties as summarized in Table 1. A temperature solution array of 1000 elements is produced at each time step and an average H was determined for each set of parameters after each simulation run.

3.2 Gamma distribution

A basic procedure to generate random deviates with a gamma distribution is given by [16]. However, since this procedure assumes that $\beta = 1$, a more general procedure $gamdev(\alpha, \beta)$ was developed which returns random deviates as a function of both α and β. The average approach distance $h_{p,avg}$ was used instead of β as a parameter during the simulation runs. Parameter β is then calculated as $h_{p,avg}/\alpha$.

The Brutsaert model [1] can be implemented by drawing the arrival rates with $gamdev(1,1)$ and resetting the entire temperature T array to the lower temperature upon arrival of the eddies. The more general procedure consists of drawing arrival rates as in the Brutsaert model with $gamdev(1,1)$ and then, after selecting $h_{p,avg}$ and α, drawing an approach distance h_p with $gamdev(\alpha, \beta)$.

4 Results

Table 1 below shows the basic parameter set with their chosen values. The parameters such as κ, ρ, C_p and ν depend to a minor extent on temperature. However, this has been ignored in the simulations. The following parameters were varied during the simulations: the friction velocity u_*, the average approach distance $h_{p,avg}$ and the gamma distribution parameter α. In section 4.1 the model validation against the analytical Brutsaert solution is briefly described, after which the simulations with variable approach distance are summarized in section 4.2.

symbol	parameter	value	unit	remark
L	Model length	0.01	m	
Δz	Grid step length	0.00001	m	
Δt	Time step	0.00001	s	(ranging from 0.00001 to 0.001 s depending on u*)
T_0	Surface temperature	300	K	
T_a	Air temperature	310	K	
κ	Air conductivity	0.025	$Wm^{-1}K^{-1}$	
ρ	Air density	1	kgm-3	
C_p	Air specific heat	1100	$Jkg^{-1}K^{-1}$	
ν	Kinematic viscosity	1.50E-05	m^2s^{-1}	
Pr	Prandtl number	0.71		
k	von Karman number	0.41		
z_0	roughness length	0.001	m	taken as constant

Table 1: Model parameters with their selected values.

4.1 Model validation with the Brutsaert analytical model

As mentioned before, the Brutsaert model [1] can be implemented by drawing the arrival rates with $gamdev(1,1)$ and resetting all temperature values to the constant air temperature at $z = L$ immediately after arrival of the eddies. This offers the opportunity to validate the stochastic numerical model against the analytical solution of the simple case with approach distance zero. The analytical solution is given by (13) with the renewal rate s given by (15). This renewal rate s depends mainly on the friction velocity u_* because z_0 is taken as a constant equal to 0.001 m. Figure 3 below shows the roughness Stanton number St_k as a function of the surface roughness number Re_* with a range from 6 to 200, corresponding to a range in friction velocity from 0.1 to 2 ms^{-1}. It is clear that the stochastic numerical model results compare well with the analytical approach by Brutsaert [1, 3, 4]. It should be noted that both the analytical and numerical model make use of relation (15) with constant C_2 having a value of 4.84 based on reported experimental values [1].

4.2 Model simulations with variable approach distances

In addition to varying Re_* as in Figure 3, α and $h_{p,avg}$ were also changed systematically. Parameter α was given the values 1, 2, 4, 9, 16. Increase in α means a decrease in the variance of the gamma distribution. The average approach distance was assigned the values 0.0001 m, 0.0002 m, 0.0005 m, 0.0010 m, 0.0015 m, 0.0020 m and 0.0040 m and finally, the gamma distribution parameter β was calculated as $h_{p,avg}/\alpha$.

Some results are illustrated in Figures 4 and 5 below. Figure 4 shows the simulation results for the inverse roughness Stanton number St_k^{-1} as a function of the approach distance at $Re_* = 13.34$ ($u_* = 0.2$ ms^{-1}). The curves show a marked increase in the St_k^{-1} value when the approach distances become larger. The curves also indicate that the heat transfer coefficient does not depend strongly on α, especially at low values of the approach distance h_p.

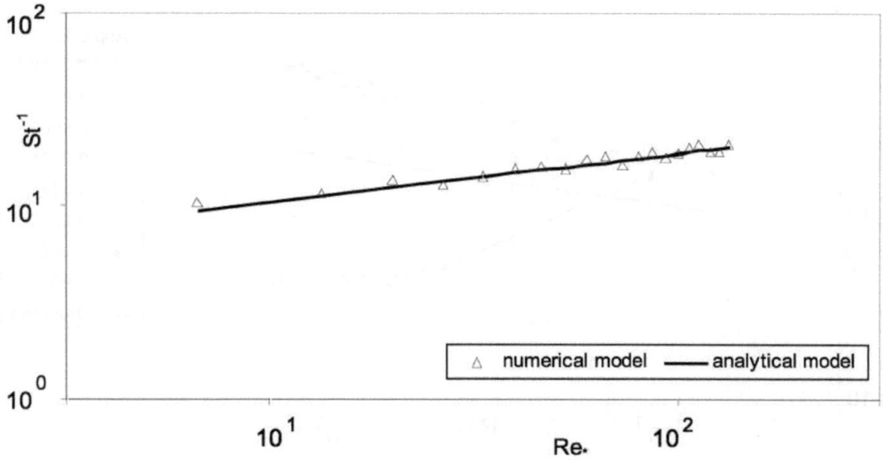

Fig. 3. Inverse Stanton number (St^{-1}) as a function of surface roughness (Re_*) for both the numerical model simulation and the analytical solution by Brutsaert [1].

This seems to be the case for all values of u_*. Because it appears that changes in α only have a minor influence on the heat transfer coefficients, a value of $\alpha = 1$ is chosen to show the general response of St_k to Re_* and h_p.

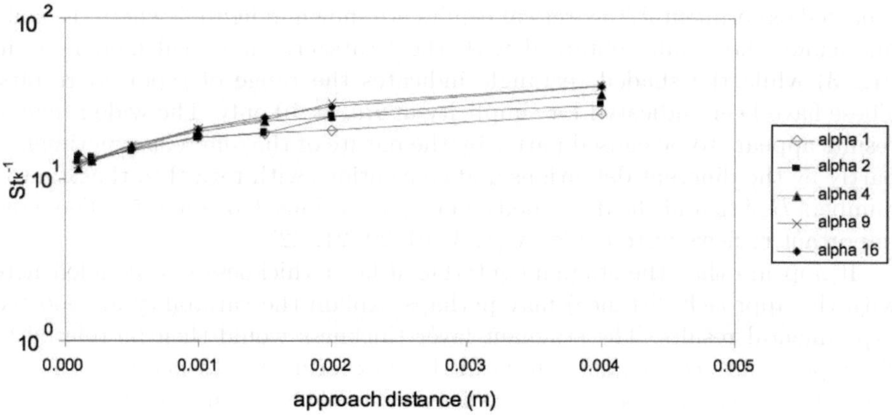

Fig. 4. Simulation results for the inverse roughness Stanton number St_k^{-1} as a function of the approach distance (thickness interfacial boundary layer) at $u_* = 0.2\ ms^{-1}$ ($Re_* = 13.3$).

Figure 5 shows the simulation results for St_k^{-1} versus Re_* for $\alpha = 1$. The curves show that a strong decrease of the heat transfer coefficient St (increase in St^{-1}) occurs with larger approach distance. All variations show

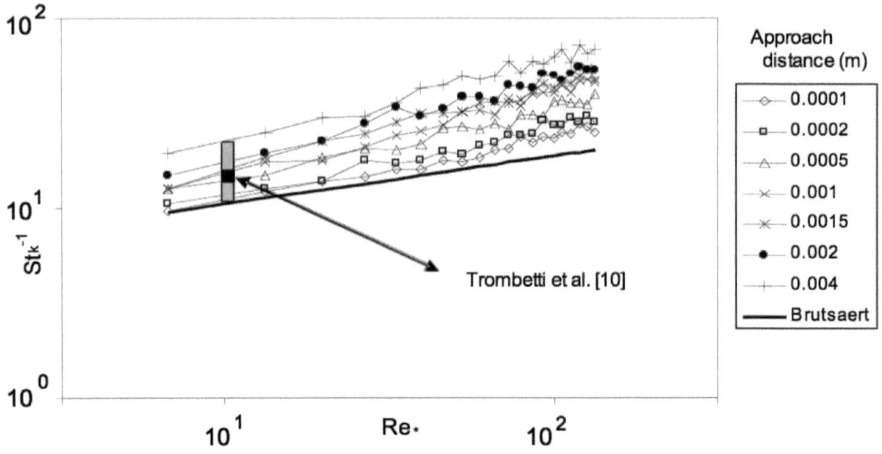

Fig. 5. The figure shows the inverse roughness Stanton number as a function of Re_* for several model approach distances. The shaded bar indicates the range of reported experimental results, for simplicity only shown at $Re_* = 10$ [1, 3, 10, 20, 21, 22]. The solid line shows the results obtained with the Brutsaert analytical model (Equation 20). The black square indicates the offset from the Brutsaert line resulting from the analysis by Trombetti et al. [10].

a decrease from the simple Brutsaert model with $h_p = 0$ (section 4.1). The reported experimental/theoretical results are shown in figure 5 where the solid line shows the results obtained with the Brutsaert analytical model (as in Fig. 3) while the shaded rectangle indicates the range of reported results. These have been indicated for simplicity at $Re_* = 10$ only. The wide range of results appears to be caused partly by the nature of the different experiments, partly by the different definitions and conventions with regard to the Stanton numbers B, St_k and the drag coefficient C_d (relations 2, 3, 4 and 5). The most important reviews were made by [1, 3, 10, 20, 21, 22].

It appears that the stagnant interfacial layer thickness (as modeled here with the approach distance) may perhaps explain the variability in reported experimental results. The stagnant layer thickness would then be related to the type of surface roughness used in these experiments. Inspection of Fig. 5 suggests that the approach distance lies on average between 0.0002 and 0.0005 m based on the experimental evidence. The Brutsaert model [1] is

$$St^{-1} = 7.3\ Re_*^{1/4} Pr^{1/2} \quad (Brutsaert) \tag{17}$$

where the constant 7.3 is mainly based on the experiments reported by [1]. However, the value of the constant is probably as high as 9.3 based on the review by [10] and therefore it is suggested to adapt relation (20) to the following relation which is also more in accordance with [22]

$$St^{-1} = 9.3\ Re_*^{1/4} Pr^{1/2} \tag{18}$$

This leads to a slightly different relation for kB^{-1} from the one shown in (5)

$$kB^{-1} = 3.21 Re_*^{1/4} - 2 \tag{19}$$

In summary, the simulations show that the heat transfer from the surface is strongly dependent on the approach distance. To a lesser extent it depends on the variance in the distribution. In all cases the simulated inverse Stanton number is higher than in the simple analytical stochastic model [1] and this model with $h_p = 0$ should therefore be seen as a special case of the more general case with $h_p \geq 0$. Although there is not enough recent experimental evidence to draw definite conclusions, most of the historical data seems to corroborate this.

5 Discussion

The stochastic model proposed here makes use of the transient heat conduction equation (6) and the general gamma distribution (16). The one-dimensional flow equation can be solved by implicit finite difference methods. This leads to a tridiagonal matrix equation that is inverted with the Thomas algorithm [15]. The gamma distribution then determines when and to what depth the boundary conditions need to be updated. The procedure to implement the gamma distribution in the model is a generalization of the procedure described in [16]. The algorithm is simple to implement and makes it possible to generate large ensembles for statistical analysis in a short period of time.

Good correspondence was achieved between the analytical solution of Brutsaert's model with $h_p = 0$ and the stochastic numerical solution. The simulations with the variable approach distance showed the large influence of the approach distance on the energy transfer. The heat transfer coefficient depends to a lesser extent on the variance in the distribution as modeled with parameter α. In all cases the numerically simulated heat transfer is lower than in the simple analytical stochastic model as developed by [1]. This model with $h_p = 0$ should be seen as a special case of the more general case with $h_p \geq 0$. Although there is not enough recent experimental evidence to draw definite conclusions, most of the historical data appears to confirm this.

The solutions for both the analytical and numerical models depend on the parameters z_0 (surface roughness), u_* (friction velocity) and the surface-air temperature difference $(T_0 - T_a)$. They do not depend on z_{0h}, the scalar roughness length for heat transport. Indeed, as already noted by [3] (and many other authors for that matter) this auxiliary parameter is used merely to facilitate parameterization of the boundary layer; in effect it is redundant.

The uncertainty still surrounding the parameterization of heat and water vapor transfer near the Earth's surface suggests to verify the Stanton number values for natural environments by experiment.

References

[1] Brutsaert W (1975) A theory for local evaporation from rough and smooth surfaces at ground level, Water Resour Res 11(4): 543-550
[2] Brutsaert W (1979 Heat and mass transfer to and from surfaces with dense vegetation or similar permeable roughness, Bnd-Layer Met. 16:365-388
[3] Brutsaert W (1982) Evaporation into the atmosphere. Reidel Pub Co, Dordrecht, The Netherlands
[4] Brutsaert W (1965) A model for evaporation as a molecular diffusion process into a turbulent atmosphere. J Geophys Res 70(20): 5017-5024
[5] Harriott P (1962a) A random eddy modification of the penetration theory. Chemical Engineering Science 17:149-154.
[6] Kolmogorov AN (1962) A refinement of previous hypotheses concerning the local structure of turbulence in a viscous incompressible fluid at high Reynolds number. J Fluid Mech 13:82-85
[7] Obukhov AM (1971) Turbulence in an atmosphere with a non-uniform temperature. Bnd-Layer Met. 2:7-29
[8] Kays WM, Crawford ME (1993) Convective heat and mass transfer. McGraw-Hill, USA
[9] Carslaw HS, Jaeger JC (1986) Conduction of Heat in Solids. Oxford University Press, UK
[10] Trombetti F, Caporaloni M, Tampieri F (1978) Bulk transfer velocity to and from natural and artificial surfaces. Bnd-Layer Met. 14: 585-595
[11] Kustas WP, Humes KS, Norman JM, Moran MS (1996) Single- and Dual-Source Modeling of Energy Fluxes with Radiometric Surface Temperature. J Appl Meteor 35: 110-121
[12] Su Z (2005) Estimation of the surface energy balance. In: Encyclopedia of hydrological sciences : 5 Volumes. / ed. by M.G. Anderson and J.J. McDonnell. Chichester Wiley & Sons 2:731-752
[13] Harriott P (1962b) A review of Mass Transfer to Interfaces. Can J Chem Eng 4:60-69
[14] Thomas LC, Fan LT (1971) Adaptation of the surface rejuvenation model to turbulent heat and mass transfer at a solid-fluid interface. Ind Eng Chem Fundam 10 (1): 135-139
[15] Wang HF, Anderson MP (1982) Introduction to Groundwater Modeling, Finite Difference and Finite Element Methods. W.H. Freeman and Company. San Francisco, USA

[16] Press WH, Flannery BP, Teukolsky SA, Vetterling WT (1986) Numerical Recipes, The Art of Scientific Computing. Cambridge University Press
[17] Crago R, Hervol N, Crowley R (2005) A complementary evaporation approach to the scalar roughness length. Water Res. Res. 41: W06117
[18] Verhoef A, De Bruin HAR, Van den Hurk BJJM (1997) Some practical notes on the parameter kB-1 for sparse vegetation. J Appl Met 36: 560
[19] Bird RB, Stewart WE, Lightfoot EN (1960) Transport Phenomena. Wiley and Sons, USA
[20] Owen PR, Thomson WR (1962) Heat transfer across rough surfaces. Journal Fluid Mech 15: 321-334
[21] Chamberlain (1968) Transport of gases to and from surfaces with bluff and wave-like roughness elements. Quart J. Royal Met Soc 94: 318-332
[22] Dipprey DF, Sabersky RH (1963) Heat and momentum transfer in smooth and rough tubes at various Prandtl numbers. Int Journal Heat Mass Transfer 6:329-353

Stromatactic patterns formation in geological sediments: field observations versus experiments

Jindrich Hladil[1] and Marek Ruzicka[2]

[1] Institute of Geology, Czech Academy of Sciences, Rozvojova 269, 16500 Prague, Czech Republic hladil@gli.cas.cz
[2] Institute of Chemical Process Fundamentals, Czech Academy of Sciences, Rozvojova 135, 16502 Prague, Czech Republic ruzicka@icpf.cas.cz

Summary. We demonstrate a novel purely hydrodynamic concept of formation of stromatactic cavities in geological sediments, originated by Hladil (2005a,b). First, the characteristic features of these cavities are described, as for their geometry and occurrence in the sedimentary rocks, and the several existing contemporary concepts of their formation are briefly reviewed. Then the new concept is introduced, and laboratory experiments described that were designed to validate it. Finally, the result obtained are presented and discussed, and the prospect for the future research is outlined. Note that the stromatactic patterns are three-dimensional cavities which are formed inside the rapidly thickening suspension/sediment. These are not the surface-related patterns like ripples or dunes.

1 Introduction

Here, the problem of the stromatacta origin is formulated in the perspective of the currently existing theories and their weaknesses. The name *Stromatactis* was originally used as a biological name (Dupont 1881), because these objects were then believed to be remnants of organisms buried in the sediments. Despite the later counter-evidence, this name stromatactis was continuously used for this specific type of filled cavities. Singular and plural forms are not settled yet. One consistent choice seems to be stromatactis and stromatactites. The other, we prefer, is stromatactum and stromatacta (adj. stromatactic). A simple new concept is presented and discussed in this paper.

1.1 What are stromatacta?

Stromatacta (abbreviated as ST) are, plainly said, petrified "holes" in sedimentary rocks. They are very specific cavities (voids, structures, patterns) occurring usually in carbonate sedimentary materials. Since the origin of these

Bernard J. Geurts et al. (eds), Particle Laden Flow: From Geophysical to Kolmogorov Scales, 85–94.

particular voids is unclear, sedimentologists spent a great effort in studying this phenomenon during the last ca. 120 years, but with a little success to date.

One can define ST as cavities occurring usually in sedimentary carbonate material, filled with fine-grained infiltrated internal sediment or with isopachous calcite cements. ST display the following basic features. The chamber is domical in shape. It has a smooth, flat, or wavy, well-defined sharp base. In contrast, its arched roof is highly irregular, ornamented with many cuspate or digitate protrusions, from large to very small, spanning a range of length scales. The vertical intersection with a plane resembles a fractal curve. Their (width):(height) aspect ratio typically varies from 3:1 to 6:1, say. Typically, they occur in swarms, either interconnected, forming a reticulate network, or isolated, but also individual occurrences can be found. The horizontally interconnected structures (usually a series of dish-shaped openings with chimneys) are usually flatter in comparison with other forms. The width of stromatacta can range from millimeters or centimeters, to decimetres, or even metres. The individual ages of their origin with sedimented beds span (with some lacunae in the documentation) a giant part of the geological time-scale, with the first possible occurrences in Paleoproterozoic (billions of years). The best fossil examples of ST are from Middle Paleozoic formations, and we have good access to localities in the Barrandian area, for instance. However, their occurrence is broad range, in many parts of the world, where the relevant carbonate deposition facies are present, see Figure 1.

By the above properties, summarized in Table 1, ST strongly differ from other types of cavities occurring in natural sedimentary materials. The other types can be, for instance, the following: shelter cavities, large inter- and intragranular pores (e.g., with shells), voids related to gas bubbles (accumulated under impermeable "umbrellas"), openings with sheet cracks, and variety of secondary hollowed structures).

1.2 How stromatacta formed?

It is a more than 120-year puzzle, not fully resolved until now. Various suggestions, speculations, hypothesis and theories have been offered by many authors, to explain the way ST were created. The most common version, which found its place also in textbooks, is that ST are cavities that remained after decay of certain organic precursors (soft-bodied organisms like sponges, microbial mats, extracellular polymers, etc.). Other concept says that ST are cavities after erodible mineral aggregates. Another concept stems from a selective dissolution of the base material or specifically in conditions of hydrothermal vents, accompanied with leaching, precipitation, corrosion, etc. Yet another concepts consider ST to be structures formed by non-uniform compaction of the sediment (maturation, de-watering), opening of shear fissures by gravitational sliding, over-pressured cracks, gas-hydrate decomposition, etc.

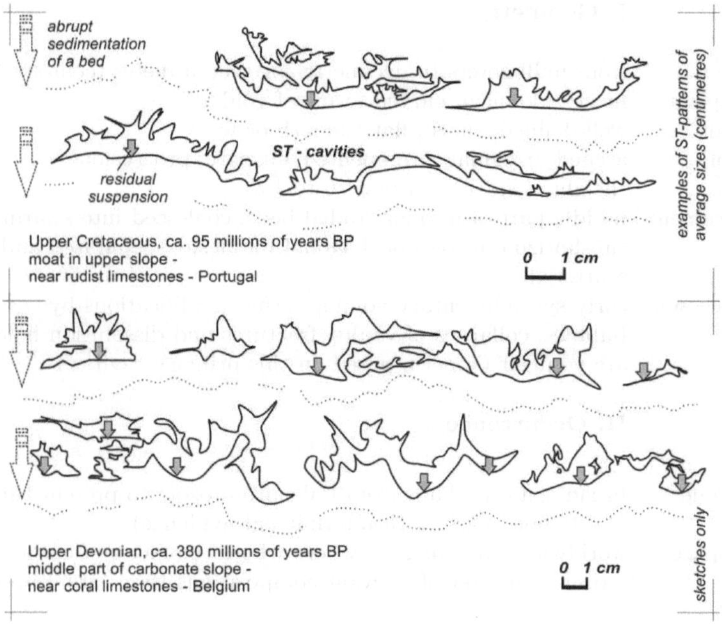

Fig. 1. Stromatacta shapes.

The advantages and disadvantages of the many solutions suggested to the ST problem are discussed in the special geological literature; this discussion is so intriguing and so voluminous that cannot be presented here (e.g., Bathurst 1982; Boulvain 1993; Neuweiler et al. 2001; Aubrecht et al. 2002; Hladil 2005b; Hladil et al. 2006). Striking is the severe contrast between the great diversity of the possible solutions, linked to particular conditions and specific presumptions, and the universality of the ST shape geometry and occurrence in space and time. All the concepts mentioned above share the two following features:

- They need many specific assumptions (diversity of dead bodies and/or material-related heterogeneities, up to heterogenetic/polygenetic nature of these cavities)
- None can be proved experimentally, here and now

1.3 Hydrodynamic concept

The marked discrepancy between the universality of the problem and the particularity of the suggested solutions lead us, more and more, to a new concept of ST formation. To overcome the deficiencies of the existing approaches, the new concept should comply with the following two demands:

- It should be based on a universal physical mechanism;

	I. Geometry
Size	from millimeters to decimetres or even a meter (coalesced)
Shape	mostly domical, curved forms of voids
Base	well-defined, sharp, flat, smooth or wavy
Roof	arched, irregular, ornamented, cuspate, protrusions
Aspect	(height):(width) $\sim$ from 1:6 to 1:3
Placement	middle part of normal graded beds; coalesced into swarms, sub-horizontal rows or diagonal meshes, or separated and scattered
Successions	early synsedimentary voiding, other modifications by bubbles, collapses of vaults, fractures and dissolution holes are younger (superimposed on this primary cavities)
	II. Occurrence
In time	broad span, perhaps from Paleoproterozoic to present times $\sim$ 4 billion years, with intermittent evidence)
In space	worldwide, spanning elsewhere where the facies of carbonate sediment matter of relevant compositions were (are?) present

Table 1. Basic features of stromatactic cavities

- It should be experimentally provable.

The universality means that the mechanism must be robust, operating both in the past and in the present, under various circumstances, being less sensitive to evolution of hydrosphere and forms of life. The experimentability means that the concept must have manageable length and time scales, and can therefore be subjected to laboratory tests. This is a general requirement on any active and controllable "geological experiment" (not the case when we passively observed a natural process).

The simple and universal way of stromatactic patterns formation could thus consist of a purely physical process. We have suggested that the process is the hydrodynamic process of rapid sedimentation of complex polydisperse mixtures of nonspherical anisometric rough grains of common geological materials, under suitable hydrodynamic conditions, where the stromatacta (voids, cavities) are formed inside the body of the deposit material, growing below the freely sedimenting dispersion (Hladil 2005a,b; Hladil et al. 2006). The first advantage of this hypothesis is the universality: the sedimentation is a physical process driven by gravity (presented since the Earth originated) and requires only the presence of very complex mixtures of fine granular material dispersed in fluids (readily available on many places). The second advantage is the fact that we can easily perform the necessary sedimentation experiments in laboratory, where the typical length and time scales are quite manageable (seconds and minutes, hours). Also the consolidation, durability and secular early diagenetic changes can be imitated (in ranges of days to several months). As for the required "suitable conditions", it is presently known that when a dispersed mixture of hydrodynamically interacting material particles is exposed to an external force field, the typical (= highly probable) behavior is the formation of certain structures and patterns, possibly proceeding via a sequence of generic instabilities. In general, these conditions in the case of ST are presently little known or unknown, as well as the hydrodynamic mechanism leading to the formation of void structures in the sediment (deposit). These two key issues present the object of our current research. In particular, some suitable conditions have already been found, and ST-like structures have been obtained in laboratory experiments. There also are some early hints towards discerning the physics behind the structures formation.

Thus, within this framework, an old geological problem has been reduced to the fluid mechanics of multiphase systems. Indeed, the new concept is rather simple: the ST are formed by pure sedimentation. For somebody working in fluid mechanics, this would perhaps be the very first choice. On the contrary, for somebody being involved in sedimentary geology and diagenesis for a long time, it must have taken certain time to find enough courage and to pass through painful catharsis to dismiss a significant part of the geological details about ST and ST-like patterns. Within the framework of the hydromechanic concept, these details are not essential for the ST formation and are not relevant to the basic formation mechanism. They can produce only "second-order effects" on the resulting patterns, and affect them mainly in the quantitative way.

2 Experiments

Here, several relatively simple experiments are described that were made with the purpose to prove or disprove the hypothesis about the purely hydrodynamic way of ST formation. The goal was to find some suitable conditions for ST production in laboratory containers.

2.1 Experiment E1: Complex system

Our first goal is to find, whether it is possible to produce ST artificially. We ask if the unknown hydrodynamic mechanism leading to ST formation on geological temporal and spatial scales in nature can also operate in a laboratory cell. To this end, we prepared our sedimenting mixture very similar to what is believed to be the genuine suspension of the past.

The first measurements were done with the most complex mixture of particles. The attempt was at preparing an artificial mixture whose composition would be nearly identical with the composition of the original mixture in which the ST were formed in the far past. The grain composition of the

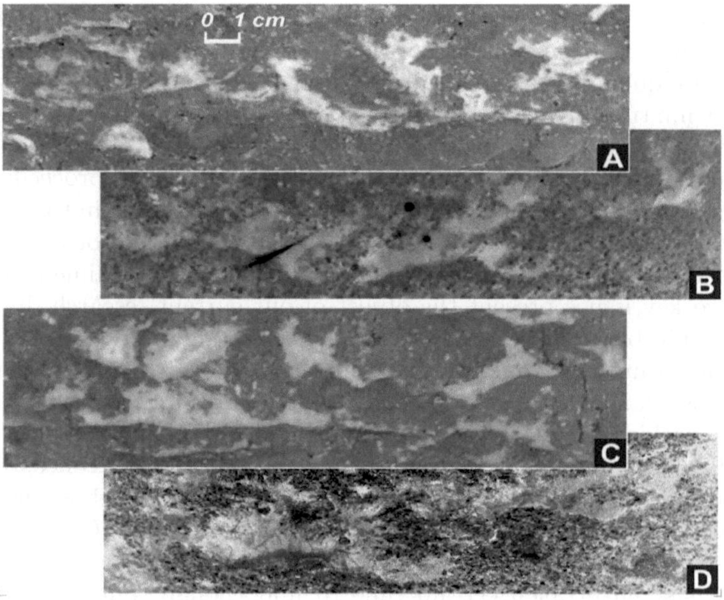

Fig. 2. Stromatacta of natural origin (A, C) and produced in laboratory in experiment E1 (B, D).

original material (carbonate rocks, limestone) was resolved by the image analysis of pictures obtained by the optical and electron microscopy of fine cuts of the ST bearing rocks. With this knowledge, the rocks were ground into a polydisperse substrate and an artificial but 'nature-identical' mixture was prepared by sieving off the undesired fractions. The product had the following qualities: density: 2700-2900 kg/m^3, size: 1 μm − 1 mm, polydispersity, polymodality (2-4 peaks), shape: anisometric, angular, irregular, surface: not smooth, rough, textured, abrasive.

The inorganic artificial mixture was mixed with a small amount of a organic component (sludge and slurry) (organic:inorganic ≈ 1:5) and sea-water (solid:liquid ≈ 1:5). This material was then homogenized by stirring and shaking and let settle in a transparent rectangular plexiglas column of volume ca. 1.5 L. The sedimentation process was recorded with a high-speed video, and the resulting 2D images of the near-wall sediment evolution and the patterns formation were analyzed, mainly visually. In addition, the 3D structures of the interior of the sedimentation deposit were investigated, after cutting it when frozen in liquid nitrogen. The 3D observations were in accord with the 2D observations.

The result was positive. The stromatacta-like structures were formed in the middle part of the layer of the polydisperse sediment bed (seen vertically), in size of millimeters to centimeters, see Figure 2. They were voids filled with the water only. Thus, a similar mechanism like that far before was likely operating. These results were reproduced both in "a glass of water" or in narrow flat cells, and also using other kinds of vessels and troughs of various shapes and size (dimensions from $\sim$ *cm* to less than 1 *m*), to exclude the possible effect of the container geometry.

Not only the ST patterns were reproduced but they also possessed certain stability features on different time scales (2 hours - end of visible compaction; 1 day - the bed resists rotation by 90¡; few days - decomposition of organic material starts, bubbles; few months - slight contractions, solidification and fracturing, etc.). The most significant feature was the formation of remarkable internal sediment on the floor of the cavities. This sediment deposited during the first minutes and has a continuous fining-upward structure, being settled from the relict suspensions of the finest particles. These markers of stromatacta strongly differ from all other internal fills, if they subsequently infiltered into these voids in sediment.

2.2 Experiment E2: Reduced system

These measurements were performed to provide an answer to the question how much we can reduce the system complexity, i.e., what are the minimum requirements to obtain the wanted stromatactic cavities.

First, the measurements described under E1 were repeated with a simpler system, only the artificial mixture + tap water. The result was positive, and the ST structures were obtained, with slightly reduced size. Second, a modified artificial mixture was treated to remove its ability to produce ST. By reducing the polydispersity and polymodality of the grain size distribution by sieving off certain fractions (small and large grains) we obtained "limiting mixture" or "matrix" with zero-capacity for ST formation. This matrix was used in the next experiments, especially for testing the effects of coarse-grained accessories.

2.3 Experiment E3: Simple system

These measurements were focused on increasing the system complexity, to initiate the production of the stromatactic patterns.

We prepared simple mixtures by combining kinds of larger particles ($\sim$ 1 *mm*) of regular shapes (3 cubes, 2 spheres, 2 cylinders). Also, we combined these larger particles with the matrix (grains $\sim 100\mu m$, or smaller and larger, alternatively), scoria (almost monodisperse), limestone (slightly polydisperse), see Table 2.3. The containers were flat quasi-2D cells (25 x 20 x 0.75 or 1.5 *cm*), and cylinders (6 *cm* dia). Certain combinations produced the structures, and certain did not. The results are in Figure 3. It follows that certain proportions among polydispersity, non-sphericity, anisometricity, roughness and abrasiveness of the material is needed to obtain the stromatactic structures. In this way, the important components of the unknown physical mechanism underlying the ST formation can be disclosed.

Materials	large grains					fine matrix			
name	C1	C2	C3	Y	B	S	L	[-]	
shape	cube	cube	cube	sphere	sphere	matrix	matrix	[-]	
size	1.8	3	5	2	1	0.1-0.3	0.01-0.8	[mm]	
density	1990	1990	1990	1720	2500	2020	2680	[kg/m^3]	
surface	rough	rough	rough	rough	smooth	rough	rough	[-]	
dispersity	mono	mono	mono	mono	mono	mono	poly	[-]	
Results									**ST ?**
too simple			x						**NO**
base state		x	x						**NO**
activation	x	x	x						**YES**
activation		x	x	x					**YES**
inhibition		x	x		x				**NO**
inhibition		x	x			x			**NO**
activation		x	x				x		**YES ***

* also with cylinders (DxL = 3x3 and 4x6 mm)

Table 2: Experiments on stromatacta formation.

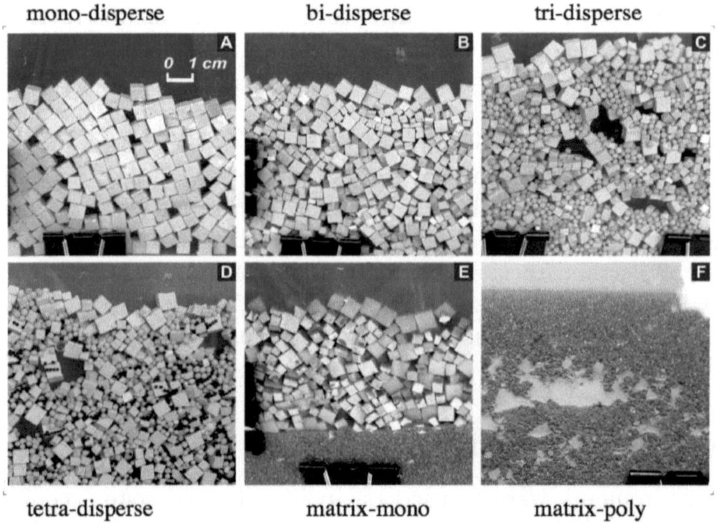

Fig. 3. Experiment E3. Effect of mixture complexity on its ability to produce void structures, see Table 2. A - cube C3. B - cube C3+C2. C - cube C3+C2 and sphere Y. D - cube C3+C2 and sphere Y+B. E - cube C3+C2 and matrix S. E - cube C3+C2 and matrix L.

The experiments proved that it is possible to generate ST-like cavities in laboratory experiments, even with particle mixtures much simpler than the common geological materials. Thus, the patterns formation in the sedimentary deposits deserve attention not only as a geological phenomenon, but also as a kind of a hydrodynamic instability in dispersed systems, where the process of the particle sedimentation is strongly coupled with the process of building of the void structures in the sediment below. This kind of instability seems to be typical for systems with polydisperse, nonspherical rough particles.

We believe that the hydrodynamic concept of the ST formation is thus proved in general, despite the fact that the experiments we have accomplished so far, suffer from the following. The comparison between the field observations and the laboratory results (shapes of ST) is only visual (no image processing and quantifications). The experiments are not exhaustive and represent a preliminary random mapping of the parameter space. No dimensional analysis or scaling arguments were employed in the design of the experiments, namely the sizes, shapes, densities of the particles and the physico-chemical properties of the fluids.

3 Theory

Currently, we do not have a theory for the phenomenon observed - the formation of ST structures during the sedimentation process, although certain hints in this direction follow from the observations. Starting with a simple phenomenology, we can consider the process as a sequence of several overlapping steps: mixing suspension, sedimentation, deposit formation, structure formation and structure duration. It can naturally be parameterized by time, or for convenience, also by the particle concentration. Certain physical aspects relevant for each step are indicated as:

I. Mixing suspension: mixing complex materials, uniformity/nonuniformity, length/time scales.
II. Sedimentation: polydisperse, nonspherical, textured particles, instabilities: planar waves, coarse graining, convective, lateral ..., finite-size container, liquid counter-current, partial fluidization.
III. Deposit formation: product of/strongly coupled with sedimentation rheology of dense suspensions and granular media.
IV. Structure formation in sediment: product of/strongly coupled with sedimentation and deposition contact forces, dry friction, yield stress, arching, doming, bridging.
V. Structure duration: compaction, aging, soil mechanics

One way of solution leads via increasing complexity of a simple base particle system, following the spirit of experiment E3. Here, the relevant control parameters related to the particles seem to be these: non-sphericity, polydispersity and surface roughness. Within this three-parameter space, certain niches should exist where the ST-like structures will typically be formed. Quantitative relations can then be obtained between the ST properties (dimensions, shape features) and the parameters, in form of correlations, with help of dimensional analysis. The underlying physics can be elucidated employing the huge potential of the theory of granular media (aspects III-V above), offering a sound basis for exploring the rheology of dense particulate matter.

The other way goes through systematic reducing complexity of the genuine ST-forming sedimenting systems (experiments E1 and E2). At a certain point, the formation ability will be lost since the key ingredients or their suitable proportion disappear. Taking this point as the base state, quantitative relations can be found between the ST properties and the control parameters, expanding the latter beyond the base state. The physical mechanisms should be understood by investigation into the hydrodynamic stability of sedimenting polydisperse mixtures. Namely, it concerns the phenomenon of the lateral instability of continuing the work started by Whitmore, Weiland, Batchelor and van Rensburg, and accounting also for the role of inertial effects (aspects I-III above). The ultimate goal could be reached by coupling the granular rheology aspects with the hydrodynamic stability aspects.

Acknowledgments

The financial support by the Grant Agency of Academy of Sciences CR (Project: IAAX00130702) and the Grant Agency of the Czech Republic (GACR Grants: 104/05/2566, 104/06/1418, 104/07/0111) is acknowledged.

References

[1] Aubrecht R, Szulc J, Michalik J, Schloegl J, Wagreich M (2002) Middle Jurassic stromatactis mud-mound in the Pieniny Klippen Belt (Western Carpathians). Facies 47:113-126

[2] Bathurst RGC (1982) Genesis of stromatactis cavities between submarine crusts in Palaeozoic carbonate mud buildups. J Geol Soc London 139:165-181

[3] Boulvain F (1993) Sedimentologie et diagenese des monticules micritiques "F2j" du Frasnien de l'Ardenne. Service Geologique de Belgique, Papier Professionel 2(260):1-427

[4] Dupont E (1881) Sur l'origine des calcaires devoniens de la Belgique. Bulletin de l'Academie royale des Sciences, des Lettres et des Beaux-Arts de Belgique, 3-ieme serie, 2(9-10):264-280

[5] Hladil J (2005a) Stromatactis in glass of water: An experiment simulating formation of particular cavities in limestone sediments (in Czech). Vesmir (Prague) 84 (7):388-394

[6] Hladil J (2005b) The formation of stromatactis-type fenestral structures during the sedimentation of experimental slurries - a possible clue to a 120-year-old puzzle about stromatactis. Bulletin of Geosciences 80(3):193-211

[7] Hladil J, Ruzicka M, Koptikova L (2006) Stromatactis cavities in sediments and the role of coarse-grained accessories. Bulletin of Geosciences 81(2):123-146

[8] Neuweiler F, Bourque PA, Boulvain F (2001) Why is stromatactis so rare in Mesozoic carbonate mud mounds? Terra Nova 13:338-346

Part II

Lagrangian statistics, simulation and experiments of turbulent dispersion

Anomalous diffusion in rotating stratified turbulence

Yoshi Kimura[1] and Jackson R. Herring[2]

[1] Graduate School of Mathematics, Nagoya University,
Furo-cho, Chikusa-ku, Nagoya 464-8602 JAPAN
`kimura@math.nagoya-u.ac.jp`

[2] National Center for Atmospheric Research,
P.O.Box 3000, Boulder, Colorado 80307-3000, USA
`herring@ucar.edu`

Diffusion in rotating and stratified fluids is one of the central subjects in geophysical and astrophysical dynamics. As a fundamental property of diffusion, particle dispersion has been studied extensively in various fields of engineering, physics and mathematics. In this paper, we report features of the dispersion of Lagrangian fluid particles in rotating stratified flows using the Direct Numerical Simulations (DNS) of the Navier-Stokes equations. And for calculation of particle dispersion, we use the cubic spline interpolation method by Yeung and Pope [1].

Taylor's picture of particle dispersion is summarized as follows. Suppose the equation of the motion of a particle and its formal solution are given as

$$\frac{d\mathbf{x}(t)}{dt} = \mathbf{u}(\mathbf{x}, t) \quad \rightarrow \quad \mathbf{x}(t) = \mathbf{x}(0) + \int_0^t \mathbf{u}(\mathbf{x}(t'), t')dt', \tag{1}$$

where $\mathbf{x}(t)$ and $\mathbf{u}(\mathbf{x}, t)$ are position and velocity of the particle, respectively. From the formal solution, the variance of a displacement vector, $\mathbf{r}(t) \equiv \mathbf{x}(t) - \mathbf{x}(0)$ is calculated as

$$\langle \mathbf{r}(t)^2 \rangle = \int_0^t dt' \int_0^t \langle \mathbf{u}(t')\mathbf{u}(s) \rangle = 2 \int_0^t dt' \int_0^t ds \, \langle \mathbf{u}(t')\mathbf{u}(t'-s) \rangle \tag{2}$$

The Lagrangian velocity auto-correlation function $R_L(s)$ is

$$\langle \mathbf{u}(t')\mathbf{u}(t'-s) \rangle \equiv \langle \mathbf{u}(t')^2 \rangle R_L(s) = \langle \mathbf{u}^2 \rangle R_L(s). \tag{3}$$

In (3), we assume that the velocity field is stationary. Using (3), (2) may be written as

$$\langle \mathbf{r}(t)^2 \rangle = 2 \langle \mathbf{u}^2 \rangle \int_0^t dt' \int_0^t R_L(s)ds = 2 \langle \mathbf{u}^2 \rangle \int_0^t (t-s)R_L(s)ds. \tag{4}$$

Bernard J. Geurts et al. (eds), Particle Laden Flow: From Geophysical to Kolmogorov Scales, 97–101.

There are two asymptotic time regimes for the above relation: one is for $t \ll t_B$ in which we can set $R_L(0) \sim R_L(s) \sim 1$, and

$$\langle \mathbf{r}(t)^2 \rangle = 2\langle \mathbf{u}^2 \rangle \int_0^t (t-s)ds = \langle \mathbf{u}^2 \rangle t^2 \qquad (t \ll t_B), \tag{5}$$

where t_B is the time after which R_L drops rapidly. The time dependence in this regime is called a ballistic mode. The other one is for $t \gg t_B$ in which we first write (4) as

$$\langle \mathbf{r}(t)^2 \rangle = 2\langle \mathbf{u}^2 \rangle \left[t \int_0^\infty R_L(s)ds - \int_0^\infty sR_L(s)ds \right], \tag{6}$$

and then assume that the integrals $\int_0^\infty R_L(s)ds$ and $\int_0^\infty sR_L(s)ds$ converge to constants T_L and C, respectively. The former corresponds to a time scale which is often called as Lagrangian time scale. Then the time dependence in this regime becomes

$$\langle \mathbf{r}(t)^2 \rangle = 2\langle \mathbf{u}^2 \rangle (T_L t - C) \qquad (t \gg 1). \tag{7}$$

This regime the time dependence of which describes the diffusion time scale is the so-called Brownian mode. A particle dispersion law different from the above canonical Taylor's picture is often called anomalous diffusion, and the objective of this paper is to show that stably stratified flows with rotation, which provide large anisotropy in the particle motion, exhibit departures from the canonical picture.

Our methodology is as follows[2]. We simulate the Navier-Stokes equation in the Boussinesq approximation,

$$(\partial_t - \nu\nabla^2)\mathbf{u} = -(\mathbf{u}\cdot\nabla)\mathbf{u} - \nabla p + \theta\hat{\mathbf{z}} + 2\Omega\hat{\mathbf{z}}\times\mathbf{u} \tag{8}$$

$$(\partial_t - \kappa\nabla^2)\theta = -N^2 w - (\mathbf{u}\cdot\nabla)\theta \tag{9}$$

$$\nabla\cdot\mathbf{u} = 0 \tag{10}$$

where $\mathbf{u}$ is the velocity whose (x, y, z) components are (u, v, w), and θ is the temperature fluctuations about the linear (stable) mean temperature profile $d\overline{T}/dz \equiv -N^2$. N is the Brunt–Väisälä frequency, $\sqrt{g\alpha(\partial\overline{T}/\partial z)/T_0}$ and Ω is the angular velocity of rotation. Particle trajectories are computed by solving (1) using the same time–marching scheme as with the velocity. The DNS consists of an initial Gaussian random isotropic velocity field which has a 3D energy spectrum given by

$$E(k) = 16\sqrt{\frac{2}{\pi}} u_0^2 k_0^{-5} k^4 \exp(-2(k/k_0)^2). \tag{11}$$

We report particle dispersion in the vertical and horizontal directions separately for various values of stratification and rotation parameters.[3]. Similar

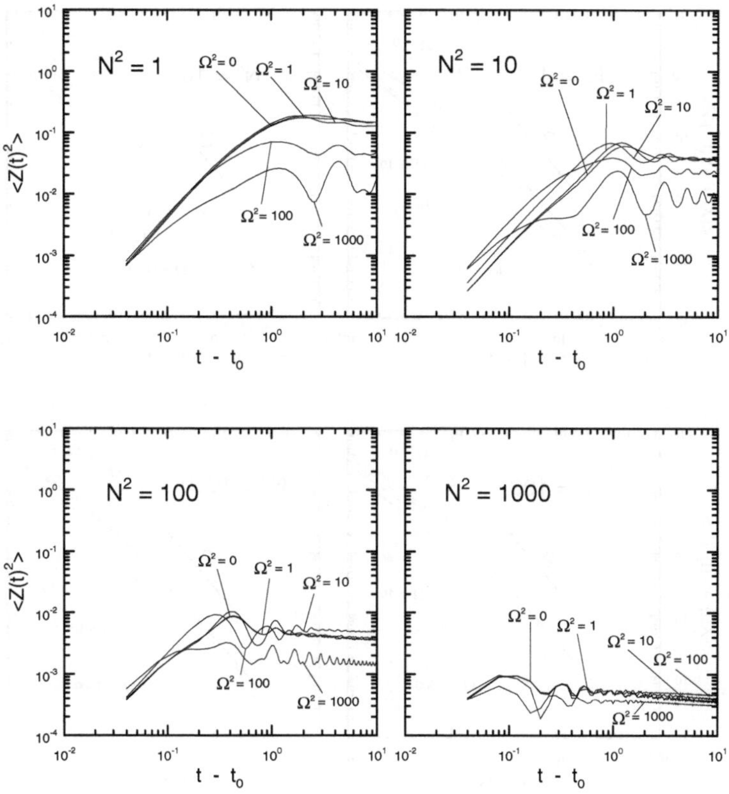

Fig. 1. Single Particle Dispersion in vertical direction at various parameters.

results are recently obtained by Liechtenstein, *et. al*[4]. When only stratification is active, particles stop migration in the vertical direction after they pass the time interval of the ballistic mode. Such suppression of diffusion was theoretically predicted by Csanady[5] with an heuristic stochastic modeling of the pressure gradient term in the Navier-Stokes equation. In [2], we extended the Langevin analysis incorporating the numerical data of energy decay to reproduce the overshoot of the dispersion plot at early time and slightly decreasing afterwards in particular for the strongly stratified cases. If rotation is also imposed, we can observe that [1] the dispersion plots show more oscillation with rotation, and [2] the suppression of the vertical dispersion is enhanced. This may be true only asymptotically when $\Omega \to \infty$. For intermediate values of Ω, rotation may enhance dispersion. For example, compare the plot of $\Omega^2 = 1$ to that of $\Omega^2 = 10$ for $N^2 = 100$.

In the horizontal direction, contrary to the vertical, the dispersion shows less dramatic but potentially interesting changes from the Taylor's picture for isotropic turbulence. Observations are summarized as follows: [1] With both

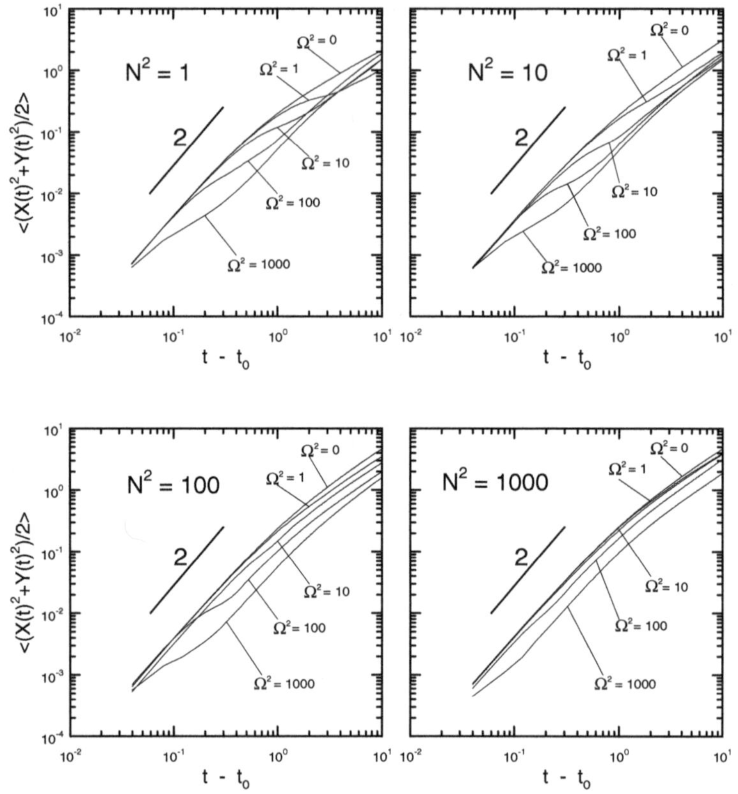

Fig. 2. Single Particle Dispersion in horizontal direction at various parameters.

stratification and rotation t^2 behavior is more and more evident not only for the initial ballistic mode but also at later times. [2] In particular for weaker stratification, there is a transition period of time between the ballistic mode and the later t^2 regime. [3] For a fixed value of stratification, the transition period shifts to earlier time with rotation. [4] With strong stratification, the transition period disappears.

The long lasting asymptotic t^2 dispersion range may be universal for flows which contains both regular and chaotic regions[6]. In this sense we expect persisting coherent structures (waves/vortices) in rotating stratified turbulence. In the transition period, the dependence of the horizontal dispersion on time is sub-diffusive ($i.e.\ t^\alpha, \alpha < 1$). The physical mechanism is an open problem, but we may predict that this sub-diffusive mode corresponds to the situation that particles are somehow trapped in the elliptic regions[7],[8]. Trapped particles are usually characteristic for 2D steady or slowly varying flows, and for 3D flows, such particles are seldom observed. The reason for the untrappedness in 3D flows may attribute to the fact that it is often quite difficult to find closed streamlines in such flows. If particles tend to be trapped in rotating stratified flows, this would imply possibility of abundant closed streamlines in rotating stratified flows. For dispersion in quasigeostrophic turbulence, reader is referred to [9] and the references therein.

References

[1] Yeung, P.K., & S. B. Pope (1989) J. Fluid Mech. 207: 531–586.
[2] Kimura, Y. & Herring, J.R. (1996) J. Fluid Mech. 328: 253–269.
[3] Kimura, Y. & Herring, J.R. (1999) Particle dispersion in rotating stratified turbulence in *Proceedings of FEDSM99 Fluid Engineering Division Summer Meeting 1999 July 18-23, 1999, San Francisco, California, USA*
[4] Liechtenstein, L., Godeferd, F. S. & Cambon, C. (2005) Journal of Turbulence, 6 (N24): 1–18
[5] Csanady, G.T (1964) J. Atmos. Sci., 21, 439–447.
[6] Mezić, I. & Wiggins, S. (1994) Phys. Fluids 6: 2227–2229.
[7] Kraichnan, R.H. (1970) Phys. Fluids 13: 22–31.
[8] Young, W.R. (1988) J. Fluid Mech., 193: 129–149.
[9] Bracco, A., von Hardenberg, J., Provenzale, A., Weiss, J.B. & McWilliams, J.C. (2004) Phys. Rev. Lett., **92** 084501.

Geometry and statistics in homogeneous isotropic turbulence

Aurore Naso[1] and Alain Pumir[2]

[1] Department of Applied Physics, University of Twente, P.O. Box 217, 7500 AE Enschede, The Netherlands
[2] Institut Non Linéaire de Nice (U.M.R. C.N.R.S 6618), Université de Nice Sophia Antipolis, 1361 Route des Lucioles, F-06560 Valbonne, France

Summary. In this paper, we consider a phenomenological model, incorporating the main features of hydrodynamic fluid turbulence, aimed at predicting the structure of the velocity gradient tensor, M, coarse-grained at a spatial scale r. This model (M. Chertkov, A. Pumir and B.I. Shraiman, Phys. Fluids **11**, 2394 (1999)) is formulated as a set of stochastic ordinary differential equations depending on three dimensionless parameters. The joint probability distribution functions of the second and third invariants of M, as well as the scaling laws of the average enstrophy, strain and energy transfer are computed by using a semi-classical method of resolution of the model. These results are compared with direct numerical simulations (DNS) data. The semi-classical solutions correctly reproduce the DNS data behavior provided the parameter that controls nonlinearity reduction induced by pressure is finely tuned.

1 Introduction

A fascinating aspect of turbulent flows comes from the existence of a broad range of excited length scales. The successful prediction of a $k^{-5/3}$ spectrum of velocity fluctuations [1] and the discovery that simple scaling laws do not adequately describe turbulence has been a major source of inspiration for theoretical work on turbulence. In this context, the study of turbulence has relied for many years on measurements and analysis of a signal, such as a velocity component, at one or two points of the flow. Much effort has been devoted to the investigation of the structure functions and of their scale dependence. But, although these quantities are very suitable for the study of scaling laws, they do not provide much information on dynamical processes. However, the existence of coherent vortical structures has a profound impact on turbulent motion [2, 3, 4], and the strain determines the local stretching of material lines, and of vorticity itself. The velocity gradient tensor $m_{ab} = \partial_a v_b$, as function of which the vorticity and the strain can be expressed, is therefore of crucial importance to understand the dynamics of turbulence.

Bernard J. Geurts et al. (eds), Particle Laden Flow: From Geophysical to Kolmogorov Scales, 103–113.

Coarse-graining of the velocity gradient tensor over a region of scale r, by defining $M_{ab} = \frac{1}{V} \int_{\Gamma} m_{ab} d^3x$, where Γ is a region of characteristic size r, provides an averaged description of the velocity gradient tensor at scale r, very appropriate in a number of contexts, for example in describing the energy transfer in turbulence [5], or more generally in Large Eddy Simulations [6].

To understand the dynamics of the matrix M, one must address the question of the Lagrangian evolution of the volume Γ, advected by the flow. The model proposed by M. Chertkov, A. Pumir and B.I. Shraiman [7] couples the evolution of the coarse-grained velocity gradient, M, and the geometry of the volume, represented by the moment of inertia tensor g of Γ. The tensor g, defined by $g_{ab} \equiv \sum (r - \bar{r})_a (r - \bar{r})_b$, also called the Cauchy-Green tensor [8], is characterized by its eigenvalues g_i, with the convention that $g_1 \geq g_2 \geq g_3$, and the associated eigenvectors, $\hat{\mathbf{e}}_i$. This information in turn provides a simplified approximation of the geometry of the set of points, by an ellipsoid of semi-axes $\sqrt{g_i}$ in the direction $\hat{\mathbf{e}}_i$. The problem can be posed as a set of stochastic differential equations which depends on three dimensionless parameters, characterizing the reduction of the nonlinearity induced by the pressure term, the re-isotropization effect of the small scale velocity field, and the influence of the small scales on the coarse-grained velocity derivative tensor.

The solutions of this system can be formally written in a path integral representation. Because of the large number of degrees of freedom, a straightforward Monte-Carlo approach is unreliable to evaluate these solutions. As a first step, we will rather use the semi-classical approximation. Effectively, this amounts to reducing the contribution of all possible paths to the contribution of the path that has the largest statistical weight. The formal procedure is equivalent to the saddle-point approximation for computing simple integrals. Semi-classical solutions of this model are presented and compared with direct numerical simulations (DNS) data [7, 9]. Our results are presented in the plane of Q and R, the second and third invariants of the matrix M ($Q = -Tr(M^2)/2$ and $R = -Tr(M^3)/3$), which leads to a synthetic presentation of the topology of the flow [10]. The DNS data show that the joint probability distribution function of Q and R, the invariants of M, becomes increasingly skewed as the scale r decreases in the inertial range. The model results correctly reproduce this behavior provided the parameter that controls nonlinearity reduction is finely tuned [9]. The influence of the other parameters in the model is much weaker.

The model is defined in Section 2. The method of resolution we used is presented in Section 3. In Section 4 are presented the model solutions in the semi-classical approximation. Concluding remarks are given in Section 5.

2 Definition of the model

In this section, we give the definition of the model, and introduce the parameters it depends on. The derivation as well as a more detailed description of the model can be found in [7, 9].

The model is formulated in terms of the matrix M, defining the coarse-grained velocity gradient tensor, and of the moment of inertia tensor g, which characterizes the geometry of the Lagrangian volume. In this model, the Lagrangian evolutions of M and g are described in terms of the following set of phenomenological equations:

$$\frac{dM}{dt} + (1-\alpha)\left(M^2 - \Pi\, TrM^2\right) = \eta \tag{1}$$

$$\frac{dg}{dt} - gM - M^t g - \beta\sqrt{Tr(MM^t)}\left(g - \frac{1}{3}Tr(g)\ Id\right) = 0 \tag{2}$$

$$\Pi \equiv g^{-1}/Tr\left(g^{-1}\right) \tag{3}$$

$$\langle \eta_{ab}\left(\rho;t\right)\eta_{cd}\left(0;0\right)\rangle = \gamma\left(\delta_{ac}\delta_{bd} - \frac{1}{3}\delta_{ab}\delta_{cd}\right)\frac{\varepsilon}{\rho^2}\delta\left(t\right) \tag{4}$$

where $\rho^2 = Tr(g)$.

To justify these phenomenological equations, it is convenient to consider a tetrahedron whose vertices are Lagrangian particles, whose positions are $\mathbf{r}_i$ $(i = 1..4)$. Because of the assumed homogeneity of the flow, the motion of the center of mass $\boldsymbol{\rho}_0 = \sum_i \mathbf{r}_i$ is immaterial, so the geometry of the tetrahedron is described by a set of 3 reduced coordinates, $\boldsymbol{\rho}_i$. It is more convenient to define the tensor ρ_i^a where a is the spatial index, and the moment of inertia tensor $g = \rho^t \rho$.

The evolution equation for the coarse-grained velocity derivative tensor, M, is reminiscent of the evolution equation for the velocity gradient tensor $m_{ab} \equiv \partial_a u_b : \frac{dm}{dt} + m^2 = H$, where H is the pressure Hessian, plus viscous corrections. Numerical observations [11, 7] indicate that the pressure term tends to diminish the nonlinear effect : $H \sim \alpha m^2$, where the parameter α in equation (1) parameterizes the reduction of the nonlinearity. The incompressibility condition, $Tr(M) = 0$ is satisfied, thanks to the $\Pi\, Tr(M^2)$ term ($Tr(\Pi)=1$, by construction). This choice of Π as a 'projection operator' to impose incompressibility is dictated by the fact that (i) the pressure term does not do any work, as it should be the case in an incompressible flow, and (ii) in the deterministic case ($\eta = 0$), the solutions of the system do not blow-up in a finite time. The stochastic term represents the effect of the rapidly fluctuating small scales on the pressure term. It is assumed to be Gaussian, white in time, with a scaling form as a function of spatial scales compatible with Kolmogorov scaling. The dimensionless factor γ simply measures the intensity of this noise term.

The equation describing the evolution of the moment of inertia tensor (the 'geometry') of the set of points, equation (2), can be understood by decomposing the velocity field as an overall, shape preserving displacement, a straining flow, coherent over the scale ρ ($Tr(g) = \rho^2$) of the object, and an incoherent, fluctuating component of the velocity field. Each of these three terms can be formally written by a filtering of the velocity field. The global displacement is ignored here, as we are considering a homogeneous flow. The straining term is responsible for the coupling between M and g in equation (2). One way to model the incoherent part of the velocity field over the set of points is to add a fluctuating, white in time noise term, as proposed in [7, 12]. This term can be written as the sum of a longitudinal component contributing to the overall growth of the volume, and of a transversal part which counteracts the tendency of the stretching term $g \cdot M$ to generate very anisotropic shapes. The former can be neglected with respect to the $g \cdot M$ term. In order to simplify the search of solutions, we simply replace the latter by its mean field representation, the β-term in equation (2).

To summarize, the model reduces to a set of stochastic differential equations depending on one dimensional parameter, ε (the energy dissipation rate), and on 3 dimensionless parameters: α, the reduction of nonlinearity due to the pressure effect, β, the re-isotropization effect of the small scales and γ, the intensity of the noise appearing in the M-equation. We present in the following the dependence of the model solutions on α. The influence of the other parameters on these solutions has been shown to be much weaker [9].

3 Method of resolution of the system

3.1 Path integral formulation

The system (1-4) defines a stochastic problem. A Fokker-Planck equation for the Eulerian probability distribution function (PDF) of the velocity fluctuations can be derived from it. Formally, this equation reads:

$$\partial_t P(M, g, t) = LP(M, g, t) \tag{5}$$

The expression of the operator L can be found in [7]. We look for the PDF solutions of equation (5) satisfying 3 conditions: (i) stationarity, (ii) a normalization condition, (iii) a boundary condition. This last is a large-scale condition: the PDF of the velocity fluctuations must be Gaussian at the integral scale L, this constraint being consistent with many experimental observations [13].

The PDF solutions of (5) satisfying these 3 conditions can be expressed in terms of the Green's function of the system. In this formulation, the PDF of M and g at a time $t = 0$ can be expressed as a function of the corresponding

PDF of M' and g', at the integral scale at $t = -T$. The Green's function can be written with a path integral formulation, in which all the trajectories starting from (M', g') at $t = -T$ and ending at (M, g) at $t = 0$ must be considered. $P(M, g)$ is then defined as:

$$P(M,g) = \int dM' \int dT \int_{M''(-T)=M'}^{M''(0)=M} [\mathcal{D}M''] \int_{g''(-T)=g'}^{g''(0)=g} [\mathcal{D}g'']$$
$$\exp\left[-\frac{Tr\,[M'[M']^t]}{\varepsilon^{2/3}L^{-4/3}} - \mathcal{S}\,(M'';g'')\right]$$
$$\times\delta\left(\frac{dg''}{dt} - g''M'' - [M'']^t g'' - \beta\sqrt{Tr(M''[M'']^t)}\left(g'' - \frac{Tr(g'')}{3}\,Id\right)\right) \tag{6}$$

where $\mathcal{S}$ is the "classical" action on each trajectory, whose expression can be derived from the system (1-4):

$$\mathcal{S} = \int_0^T dt \frac{\left[\dot{M} + (1-\alpha)\left(M^2 - \Pi\, TrM^2\right)\right]\left[\dot{M} + (1-\alpha)\left(M^2 - \Pi\, TrM^2\right)\right]^t}{2\gamma\varepsilon/\rho^2} \tag{7}$$

Phenomenologically, one starts from a point (M', g'), at the integral scale, and integrates the system in time until a given scale in the inertial range. In principle, all these trajectories should be taken into account, which could be done numerically by using a Monte Carlo algorithm. However, because of the large number of degrees of freedom of the system (large dimension of the phase space) it is impossible to obtain reliable results with a straightforward Monte Carlo approach, especially at small scales. This is why we will use as a first step a simplifying assumption to solve equation (6).

3.2 Semi-classical approximation

Equation (6) can be solved in the semi-classical approximation, in which one considers only the trajectory on which the action is minimal. This approximation is in principle valid in the limit of small noise amplitude. At fixed initial and final conditions, this trajectory can be calculated by integrating the Euler-Lagrange equation. In our calculations, aimed at determining the probability distributions in the (R, Q) plane as a function of scale, the Euler-Lagrange equation requires a number of a-priori unknown boundary conditions. Specifically, imposing (R, Q) at scale r leaves 11 unknown free parameters to completely determine the initial conditions M, $\dot{M}$, g and $\dot{g}$. For each value of (R, Q), we will use the saddle approximation, by computing the logarithm of the probability distribution in equation (6) as a function of the 11 other parameters, and looking for its maximum with the help of the algorithm "*amebsa*" [14]. Technical details about the method we used can be found in [9].

4 Results of the semi-classical calculations

The results presented in this section have been calculated by solving the model in the semi-classical approximation with the method described in 3.2. The model depends on 3 independent parameters: α, the reduction of nonlinearity, β, which prevents the growth of anisotropy of the Lagrangian volume, and γ, the amplitude of the noise acting on M. Only the α dependence of the model solutions is presented here, the influence of β and γ on them being much weaker [9].

4.1 Scaling laws

We discuss first the scaling properties of quantities such as the square of the strain, $\langle Tr(S^2)\rangle$ and of the vorticity, $\langle\omega^2\rangle$, as well as of the energy transfer term, $\langle -r^2 Tr(M^2 M^t)\rangle$ [7, 9]. According to the Kolmogorov scaling, $\langle \Delta v\rangle \propto r^{1/3}$, therefore $M(r)$ should scale as $r^{-2/3}$, and as a consequence the second order moments of strain S (symmetrical part of M) and of vorticity ω (anti-symmetrical part of M) should behave as $r^{-4/3}$, whereas third order quantities such as the energy transfer divided by r^2, $\langle -Tr(M^2 M^t)\rangle$, should scale as r^{-2} (equivalently, the mean energy transfer should be independent of the scale). We have checked that these scaling laws are satisfied by DNS data [9].

The semi-classical solutions of the model have been found to exhibit the following behaviors:

- the enstrophy $\langle\omega^2\rangle$ scales as $r^{-4/3}$ for any value of α;
- the square of strain $\langle Tr(S^2)\rangle$ scales as $r^{-4/3}$ only if α is greater than or similar to 0.35 (see Fig. 1(a));
- the energy transfer, $\langle -r^2 Tr(M^2 M^t)\rangle$, is positive only if α is lower than or similar to 0.5 (see Fig. 1(b)). In that case this third order moment of M respects the Kolmogorov scaling.

Finally, the value of α needed to obtain the right scaling of energy transfer, strain and vorticity from the solutions of the model has to be rather precisely tuned. This can be also seen by studying the computed probability distribution functions in the (R, Q) plane.

4.2 Joint probability distribution function of the Q, R invariants

Before presenting the joint PDF of the Q, R invariants solutions of the model, we show here for comparison purposes $P(R, Q)$ calculated by DNS at two different scales. Briefly, we have used a standard pseudo spectral code, described in [15]. The run discussed here has a 256^3 resolution. We have made sure that the highest wavenumber in the simulation, k_{max}, is large enough to describe the smallest length scales in the flow : $k_{max}\eta \geq 1.4$. Our Reynolds number is $R_\lambda = 130$, and the ratio between the integral scale and η is $L/\eta \sim 100$. Because the end of the inertial range is at a scale $\sim 10\eta$, the inertial range in our simulation corresponds roughly to a factor 10 in scale. In Fig. 2 are represented these distributions, at the integral scale and at a scale lying in the inertial range: $r = L/8$. At the integral scale (Fig. 2(a)), the distribution is symmetric with respect to the $R-$axis. As the scale is decreased, the $P(R,Q)$ distribution becomes more and more skewed (Fig. 2(b)).

In Fig. 3 (resp. Fig. 4) are represented joint PDF of the Q, R invariants calculated in the semi-classical approximation of the model at two different scales, for $\alpha = 0.4$ (resp. $\alpha = 0.2$). The qualitative agreement of these distributions with DNS data (Fig. 2) is good for the largest value of α (Fig. 3). The main difference between the PDF evolution as scale decreases

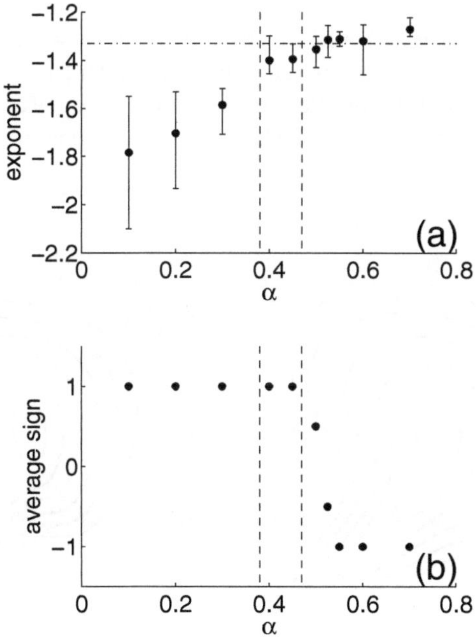

Fig. 1. Dependence with respect to α of (a) the scaling law exponent of $\langle Tr(S^2) \rangle$ and (b) the sign of the energy transfer $\langle -r^2 Tr(M^2 M^t) \rangle$ (β=0.4, γ=0.25). In (a) the dot-dashed line indicates the Kolmogorov prediction $-4/3$. In (b) is plotted for each value of α the average of the sign of the energy transfer at the different scales considered. The range of values of α leading to a qualitatively acceptable behavior of the model solutions is delimited by the vertical dashed lines.

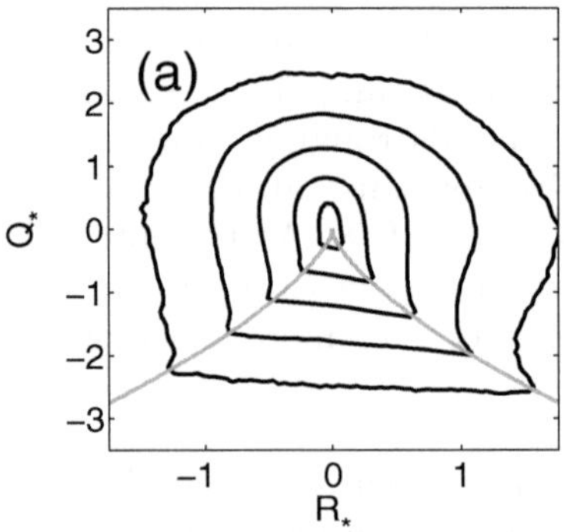

Fig. 2. PDF, calculated by DNS, of Q,R invariants normalized to the variance of strain: $Q_* = Q/\langle Tr(S^2)\rangle$, $R_* = R/\langle Tr(S^2)\rangle^{3/2}$. (a) $r = L$ and (b) $r = L/8$. The isoprobability contours are logarithmically spaced, and separated by factors of 10. The grey line is the zero discriminant line [10].

at the value $\alpha = 0.4$ (Fig. 3) and at the low value $\alpha = 0.2$ (Fig. 4) concerns the growth of the probability distribution along the positive R-side of the zero discriminant (PRZD) line. At small values of α (Fig. 4) the tail along this line grows considerably when the scale r decreases, significantly more than what is observed in DNS (Fig. 2).

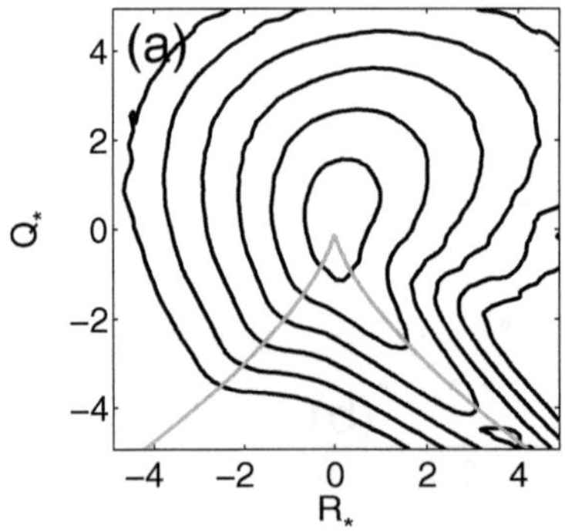

Fig. 3. PDF of Q,R invariants normalized to the variance of strain: $Q_* = Q/\langle Tr(S^2)\rangle$, $R_* = R/\langle Tr(S^2)\rangle^{3/2}$. These distributions are solutions of the model in the semi-classical approximation, with α=0.4, β=0.4 and γ=0.25: (a) $r = L/4$ and (b) $r = L/16$. The isoprobability contours are logarithmically spaced, and separated by factors of 10. The grey line is the zero discriminant line [10].

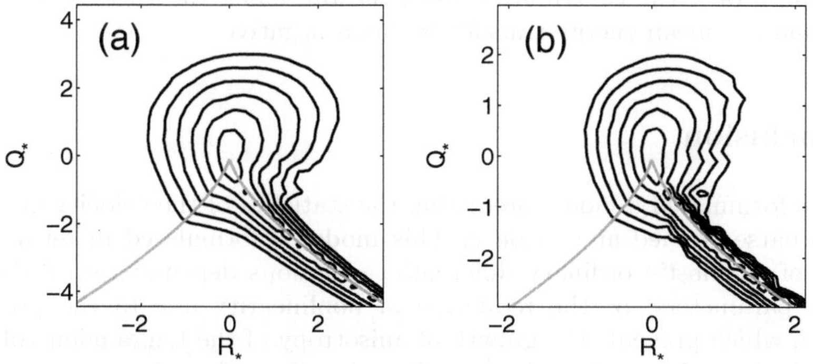

Fig. 4. PDF of normalized Q_*,R_* invariants, for α=0.2, β=0.4 and γ=0.25: (a) $r = L/4$ and (b) $r = L/16$. Same conventions as in Fig. 3.

Although the semi-classical solutions of the model reproduce in several essential ways the probability distribution function in the (R, Q) plane obtained by DNS, they lead to a number of quantitative incorrect features. The main example is the enhanced probability distribution in the $R > 0, Q > 0$ quadrant. The method of resolution (semi-classical) discussed here should ultimately be improved. But the knowledge gained in obtaining semi-classical solutions has helped us in designing a better approximations scheme. Preliminary results in this direction show a better agreement of the model solutions with DNS data [16].

4.3 Discussion

The growth of the PDF tail along the PRZD-line predicted by the model for small values of α is generally consistent with the results of [6]. In this work, it was noticed that the small scales of the flow tend to slow down the effect of the nonlinearity, estimated at the given scale. This is precisely the origin of the α-term in our model [7]. A large enough value of the parameter α is needed to sufficiently reduce the growth of the probability along the PRZD-line as r decreases, consistent with the model results.

The fact that the strain $\langle Tr(S^2)\rangle$ grows faster when r decreases at small values of α and that the energy transfer is positive only for small values of α is consistent with the distribution of these quantities is the R, Q-plane. It is indeed shown in [7, 9] that the strain density is large in the neighborhood of the PRZD-line, and that the energy transfer is positive near this neighborhood, whereas it can be negative in other regions of the plane. Therefore, for small values of α the strong growth of the PDF tail along the PRZD-line as the scale is decreased leads to a fast growth of the strain intensity, whereas, for large values of α the contribution from the tail along the PRZD-line is too small, and the mean energy transfer becomes negative.

5 Conclusion

We have formulated a model predicting the statistics of the velocity gradient tensor coarse-grained at a scale r. This model is formulated in terms of a system of stochastic ordinary differential equations depending on 3 dimensionless parameters: α, the reduction of nonlinearity due to the pressure effect, β, which prevents the growth of anisotropy of the Lagrangian volume, and γ, the intensity of the noise modeling the effect of small scales.

Semi-classical solutions of this model have been presented. In particular, the model solutions reproduce the behavior of DNS data provided the α parameter is finely tuned ($\alpha \sim 0.45$).

The model can also be extended to study flows with more complicated large-scale structures by simply modifying the large-scale condition. We have already applied it to shear turbulence [17], and plan to study the effect of large-scale rotations or contractions on the M-dynamics.

References

[1] U. Frisch, *Turbulence: The legacy of AN Kolmogorov* (Cambridge University Press, 1995).
[2] E. Siggia, J. Fluid Mech. **107,** (1981) 375.
[3] S. Douady, Y. Couder and M.E. Brachet, Phys. Rev. Lett. **67,** (1991) 983.
[4] J. Jimenez, A. Wray, P. Saffman and R. Rogallo, J. Fluid Mech. **255,** (1993) 65.
[5] S. Chen, R.E. Ecke, G.L. Eyink, X. Wang and Z. Xiao, Phys. Rev. Lett. **91,** (2003) 214501.
[6] F. Van der Bos, B. Tao, C. Meneveau and J. Katz, Phys. Fluids **14,** (2002) 2456.
[7] M. Chertkov, A. Pumir and B.I. Shraiman, Phys. Fluids **11,** (1999) 2394.
[8] E. Jeong and S. Girimaji, Theoret. Comput. Fluid Dynamics **16,** (2003) 421.
[9] A. Naso and A. Pumir, Phys. Rev. E **72,** (2005) 056318.
[10] B.J. Cantwell, Phys. Fluids A **4,** (1992) 782.
[11] V. Borue and S.A. Orszag, J. Fluid Mech. **366,** (1998) 1.
[12] A. Pumir, B.I. Shraiman and M. Chertkov, Phys. Rev. Lett. **85,** (2000) 5324.

[13] B. Lüthi, J. Berg, S. Ott and J. Mann, Proceedings of the Euromech colloquium 477 (2006).
[14] W.H. Press, S.A. Teukolsky, W.T. Vetterling and B.P. Flannery, *Numerical Recipes in C* (Cambridge University Press, 1992) 451.
[15] A. Pumir, Phys. Fluids **6,** (1994) 2071.
[16] A. Naso, A. Pumir and M. Chertkov, in preparation (2006).
[17] A. Naso, M. Chertkov and A. Pumir, J. Turb. **7,** (2006) N41.

Refined vorticity statistics of decaying rotating three-dimensional turbulence

L.J.A. van Bokhoven[1], C. Cambon[2], L. Liechtenstein[2], F.S. Godeferd[2], and H.J.H. Clercx[1]

[1] Fluid Dynamics Laboratory, Department of Applied Physics, Eindhoven University of Technology, P.O. Box 513, 5600 MB Eindhoven, The Netherlands
`l.j.a.v.bokhoven@tue.nl`

[2] Laboratoire de Mécanique des Fluides et d'Acoustique, École Centrale de Lyon, 69134 Écully Cedex, France

Summary. The influence of background rotation on all nontrivial triple correlations of vorticity (VTCs) has been studied for an unbounded incompressible homogeneous turbulent flow, using pseudo-spectral direct numerical simulation. It is found that the time evolutions of the VTCs are in agreement with exact theoretical predictions for rotating turbulence presented herein. Furthermore, the effects of viscosity, initial value of the velocity derivative skewness $S_{\partial_i u_i}(t_{\mathrm{ini}})$, and background rotation rate on the vertical vorticity skewness S_{ω_3} have been thoroughly investigated. The initial growth rate of S_{ω_3} is found to be proportional to $t^{0.75\pm0.1}$ for all considered cases, in agreement with recent experimental results by Morize et al. [Phys. Fluids 17, 095105 (2005)]. Also, it is found that higher background rotation rates – implying more linearity – result in lower final values of S_{ω_3}, while lower viscosities and higher $S_{\partial_i u_i}(t_{\mathrm{ini}})$ both yield higher final values of S_{ω_3}.

1 Introduction

Rotating turbulence plays an important role in fields as diverse as geophysics, astrophysics and engineering (e.g. turbomachinery and reciprocating engines with swirl and tumble). In these applications, effects of rotation are often combined with those of mean shear, mean strain and solid boundaries. Thus, the problem of homogeneous rotating turbulent flow without physical boundaries must be considered as a canonical flow, even though it is very far from applications. Despite the mentioned idealizations a better knowledge of the dynamics of homogeneous rotating turbulent flow is a prerequisite to understand more complex flow cases.

Here, we restrict ourselves to the simplest case of homogeneous rotating turbulence, namely unbounded divergence-free flow with zero mean-flow in the co-rotating frame of reference. In this case it is most convenient to adopt a Cartesian coordinate system (x_1, x_2, x_3) rotating at constant angular velocity

Bernard J. Geurts et al. (eds), Particle Laden Flow: From Geophysical to Kolmogorov Scales, 115–127.

$\boldsymbol{\Omega}$. We choose $\Omega = \Omega_3 > 0$ without loss of generality. In this non-Galilean coordinate frame, rotation introduces the well-known centrifugal and Coriolis forces. Since the centrifugal force can be incorporated in the pressure term, only the Coriolis force appears explicitly in the Navier–Stokes equations

$$\partial_t u_i + u_j \partial_j u_i = -\partial_i p - 2\Omega \epsilon_{i3k} u_k + \nu \partial_j^2 u_i \; , \tag{1}$$

where u_i represents the fluctuating velocity in the co-rotating reference frame, p is the corrected pressure divided by a mean reference density, and ϵ is the antisymmetric Ricci tensor. The incompressibility of the flow is expressed by

$$\partial_i u_i = 0 \; . \tag{2}$$

The modification of turbulence dynamics by rotation is due to the presence of inertial waves in rotating flows [8]. These wave motions arise from the linearized equations of motion (1). In contrast with most turbulent shear flows, however, there is no 'production' of turbulent kinetic energy since the Coriolis force produces no energy. Instead, energy is redistributed by nonlinear mechanisms such as resonant wave interactions. Accordingly, single-point modeling is almost irrelevant to describe the dynamics of rotating turbulence whereas the spectral, or two-point, approach is well adapted (see e.g. [5]).

When a Coriolis force is suddenly imposed on initially isotropic turbulence, the following three main effects are shown. First, the energy cascade is partly inhibited which is linked to a strongly reduced dissipation rate (such an effect can be mimicked by an empirical modification of the dissipation equation [5]). Second, because the dispersion relations for inertial waves are anisotropic, the initial isotropy is broken. This breaking of isotropy is reflected by an incomplete transition from 3D to 2D structure. Third, an asymmetry appears between cyclonic and anti-cyclonic fluctuating vertical (along the rotation axis) vorticity.

The effects mentioned above are intimately connected and result from both linear and nonlinear effects which interplay in a subtle way to drive the dynamics of rotating turbulence. In this paper, we focus on the third effect, using Direct Numerical Simulation (DNS) as well as rational analysis of relevant spectral equations. As firstly pointed out by Bartello et al. [1], the dominance of cyclonic vorticity can be quantified by the vertical vorticity skewness $S_{\omega_3} \equiv \langle \omega_3^3 \rangle / \langle \omega_3^2 \rangle^{3/2}$ with the vorticity $\omega_i \equiv \epsilon_{ijk} \partial_j u_k$ and the brackets $\langle \cdot \rangle$ denoting ensemble averaging, since the third order vorticity correlation $\langle \omega_3^3 \rangle$ can distinguish by its sign cyclonic prevalence ($\omega_3 > 0$) from anticyclonic prevalence ($\omega_3 < 0$). Bartello et al. [1] found a clear growth of the vertical vorticity skewness using Large Eddy Simulation (LES) with hyper-viscosity. Although the results of Bartello et al. [1] are questionable, because vorticity represents small scales and is not correctly captured in a LES, they reflect an actual feature in rotating flows, as confirmed experimentally by Morize et al. [14].

Clearly, third order statistics play an important role when addressing the asymmetry between cyclonic and anticyclonic vorticity, and more generally it can be shown that in the *axisymmetric* case single-point triple vorticity correlations (VTCs hereinafter) involve only two key correlations, viz. $\langle \omega_1^2\omega_3 + \omega_2^2\omega_3\rangle$ and $\langle \omega_3^3\rangle$, as follows directly from

$$\langle \omega_i\omega_j\omega_n\rangle = +\frac{1}{2}\langle \omega_3^3\rangle\,(5n_in_jn_n - \delta_{ij}n_n - \delta_{in}n_j - \delta_{jn}n_i) + \\ +\frac{1}{2}\langle \omega_1^2\omega_3 + \omega_2^2\omega_3\rangle\,(\delta_{ij}n_n + \delta_{in}n_j + \delta_{jn}n_i - 3n_in_jn_n)\ , \quad (3)$$

where δ_{ij} is the Kronecker tensor, and the normal vector $n_i = \delta_{i3}$ without loss of generality – index 3 refers to the axial (vertical) direction.

Since the dynamics of rotating turbulence is driven by both linear and nonlinear effects, one may wonder what *linear* theory can say about the VTCs. The so-called Rapid Distortion Theory (RDT) is relevant here, provided that it is developed in Fourier space [4, 6] in order to render tractable the non-local relation between pressure and velocity fluctuations. Accordingly, a general solution in terms of Green's functions can be derived for the VTCs $\langle \omega_i\omega_j\omega_n\rangle$ at any time t. This general solution however, is only useful if a full description of all triads of the initial spectral vorticity distribution is at hand.

Furthermore, for very short times, Gence and Frick [9] showed that all odd statistical moments of the turbulent vorticity field (with exception of the first one) are instantaneously sensitive to the influence of the mean rotation while all even moments of that vorticity do not change at the first order in time. (Note that for isotropic turbulence all odd order tensors are zero while a priori non-zero for the tensors of even order.) In particular, it can be shown for the VTCs that the time derivative at $t = 0^+$ is proportional to both the rotation rate and the initial vorticity intensity production rate $e_0 \equiv \langle \omega_i\omega_j d_{ij}\rangle$ at $t = 0$ with d_{ij} the (symmetric) strain rate tensor, so that some VTCs grow instantaneously if the rotation rate is positive. Mathematically, this can be expressed as

$$\frac{\mathrm{d}}{\mathrm{d}t}\langle \omega_1^2\omega_3\rangle\Big|_{0+} = \frac{\mathrm{d}}{\mathrm{d}t}\langle \omega_2^2\omega_3\rangle\Big|_{0+} = \frac{2}{15}e_0\Omega\ ,$$

$$\frac{\mathrm{d}}{\mathrm{d}t}\langle \omega_3^3\rangle\Big|_{0+} = \frac{2}{5}e_0\Omega\ ,$$

where the background rotation is abruptly initialised at $t = 0$. This result also reflects the fact that the complex nonlocal effect of pressure fluctuation, apparent from the linearized Poisson equation, is not involved at the first order of a Taylor expansion in time, so that a simplified solution can be found in physical space.

This paper is organized as follows. A brief description of the numerical algorithm and procedures is given in Sect. 2. In Sect. 3 we present the time evolution of relevant statistical quantities, such as the velocity derivative skewness and vorticity skewness. In particular, we show that our numerical results are consistent with (3). We conclude in Sect. 4 with a short discussion.

2 Numerical procedures

Equations (1) and (2) are solved directly using a pseudo-spectral collocation method expressed in Fourier space following a classical scheme as in e.g. [16, 17, 12]. The computational domain uses periodic boundary conditions, and consists of N^3 points in physical space. Furthermore, the velocity field is completely de-aliased using a 2/3-truncation method in Fourier space. At $t = 0$, the velocity field is initialised with the narrow band energy spectrum $E(k) \propto k^4 \mathrm{e}^{-2(k/k_i)^2}$, with peak energy at wavenumber k_i, typically one tenth of the maximum wavenumber of the de-aliased field. The directions of the initial velocity vectors are distributed randomly in space and so do not correlate with each other after initialization. From this initially random isotropic field an isotropic precalculation is performed up to time t_{ini} in order to allow for higher order velocity correlations to develop and isotropic energy dynamics to build up. Various Eulerian temporal statistics are calculated during the simulation at predefined time steps.

At $t = t_{\mathrm{ini}}$, the background rotation $\mathbf{\Omega} = (0, 0, \Omega)$ is suddenly imposed upon the well-developed isotropic turbulent velocity field. Discontinuities in the statistical averages of the derivatives at time t_{ini}, that are due to the suddenly imposed anisotropic body force, are ignored. In order to check consistency with the theoretical result (3), the minimal set of components relevant to describe all VTCs in rotating turbulence, viz. $\langle\omega_1^3\rangle$, $\langle\omega_1^2\omega_3\rangle$, $\langle\omega_1\omega_3^2\rangle$, $\langle\omega_1\omega_2\omega_3\rangle$, $\langle\omega_2^2\omega_3\rangle$, $\langle\omega_3^3\rangle$, is calculated for $t > t_{\mathrm{ini}}$.

We have performed anisotropic computations with different (nondimensional) kinematic viscosities ν and for various background rotation rates (as expressed by the Coriolis parameter $f \equiv 2\Omega$) in order to quantify how these parameters affect the temporal evolution of the VTCs. An overview of the performed runs and their parameters is presented in Table 1. We remark that the Taylor-based Reynolds number $Re_\lambda \equiv \langle u_i^2\rangle^{1/2}\lambda/\nu$, with $\lambda \equiv \sqrt{15\langle u_i^2\rangle/\langle\omega_i^2\rangle}$ the Taylor micro-scale, is not conserved in decaying turbulence, so that Re_λ depends on t_{ini}.

The time t_{ini} at which anisotropy is introduced is a crucial parameter. In a conventional DNS (such as ours), any third order correlation is almost zero initially in connection with the (close to) Gaussian initialization. During the subsequent precalculation certain triple correlations develop, for instance creating a significant gradient skewness $S_{\partial_i u_i} \equiv -\sqrt{135/98}\langle\omega_i\omega_j\partial u_i/\partial x_j\rangle/\Omega^{3/2}$ [3], with $\Omega \equiv \langle\omega_i^2\rangle/2$ the global enstrophy. The gradient skewness can asymptote to a positive value, reflecting a nondimensional constant rate of enstrophy production by nonlinear vortex stretching. Other triple correlations such as the cubic vertical vorticity $\langle\omega_3^3\rangle$ remain zero because of isotropy. At $t = t_{\mathrm{ini}}$, the isotropy is broken by rotation and certain VTCs will develop whereas the gradient skewness $S_{\partial_i u_i}$ will be rapidly and severely damped. The exact mechanism that causes certain VTCs to grow in time is still unknown. Nevertheless, the work by Gence and Frick [9] strongly suggests that the growth of certain VTCs is caused by *linear* mechanisms. If a purely linear mechanism is

Table 1. Overview of the performed runs. For each run are given the number of computation points in physical space N, the initial Reynolds number $Re(0) \equiv \nu^{-1}$ (with ν the nondimensional kinematic viscosity), the duration of the precalculation t_{ini}, the Coriolis parameter f, the initial Taylor-based Reynolds number $Re_\lambda(t_{\text{ini}})$, and the initial Taylor-based Rossby number $Ro_\lambda(t_{\text{ini}})$. Note that our values for $Re_\lambda(t_{\text{ini}})$ are somewhat smaller than the typical values of 60 to 80 often used in DNS of forced homogeneous turbulence with 128^3 resolution.

Run	N	$Re(0)$	t_{ini}	f	$Re_\lambda(t_{\text{ini}})$	$Ro_\lambda(t_{\text{ini}})$
A	144	600	5.0	5.0π	26.7	0.073
B1	288	1200	2.0	5.0π	39.2	0.36
B2	288	1200	4.0	5.0π	33.2	0.14
B3	288	1200	6.0	5.0π	30.3	0.084
B4	288	1200	8.0	5.0π	28.6	0.057
C1	288	1200	4.0	0.5π	33.2	1.41
C2	288	1200	4.0	2.5π	33.2	0.28
C3	288	1200	4.0	5.0π	33.2	0.14
C4	288	1200	4.0	10π	33.2	0.071
D	540	3000	2.0	5.0π	46.6	0.37

assumed, then the maximum growth of the cubic vertical vorticity $\langle\omega_3^3\rangle$ will depend on the level of non-Gaussianity reached at the end of the precalculation, with two extremes: 1) a very short precalculation implying small departure from Gaussianity at $t = t_{\text{ini}}$ so that the final level of $\langle\omega_3^3\rangle$ will be small too, and 2) a long precalculation implying maximum level of non-Gaussianity at $t = t_{\text{ini}}$ so that the final level of $\langle\omega_3^3\rangle$ will be maximum.

Since the velocity derivative skewness clearly illustrates the departure of Gaussianity of the velocity fluctuations ($S_{\partial_i u_i} \approx 0.5$ in isotropic turbulence [13]), it is an obvious parameter from which to derive time t_{ini}. Accordingly, it seems obvious to choose time t_{ini} such that $S_{\partial_i u_i}$ is maximum. However, during a precalculation of decaying turbulence a significant part of the turbulent kinetic energy is lost due to dissipation. In order to have a sufficiently turbulent flow field at the end of the isotropic precalculation, time t_{ini} is therefore chosen such that the velocity derivative skewness has almost reached its final value. How the choice of time t_{ini} affects the relevant statistics is discussed in Sect. 3.

3 Numerical results

In what follows we present the time evolution of various Eulerian quantities during the anisotropic computation unless stated otherwise. Wherever convenient we use the scaled, shifted time $\tau \equiv (t - t_{\text{ini}})f/(2\pi)$ to simplify the comparison between runs for which anisotropy is introduced at distinct times t_{ini}, and also to simplify the comparison with recent experimental results [14].

We first investigate how the viscosity affects the time evolution of the kinetic energy and relevant nondimensional numbers in presence of background rotation. Figure 1(a) shows the decay of the kinetic energy E_k for various viscosities ν. This decay is monotonic in time, and proportional to power-law $t^{-3/2}$ for $t \ll t_{\rm ini}$ (power-law behavior not shown in Fig. 1(a) because τ is shifted by $-t_{\rm ini}$ with respect to t) while expected to be proportional to $t^{-1/2}$ in the asymptotic limit $t \gg t_{\rm ini}$ (this power-law behavior is not yet reached by $\tau = 5$). The change in the decay rate is caused by the background rotation which inhibits energy transfer to the viscous scales.

Next, consider the time dependence of the nondimensional numbers Re_λ and $Ro_\lambda \equiv \langle u_i^2 \rangle^{1/2}/2\Omega\lambda$, see Figs. 1(b) and (c), respectively. In contrast with the decrease of Re_λ during the isotropic precalculation (not shown), Fig. 1(b) shows that sufficiently strong background rotation (as in cases A, C3 and D) results in a *growth* of Re_λ. This implies that the Taylor micro-scale λ grows faster than that the kinetic energy decays. The Taylor-based Rossby number decays monotonically (Fig. 1(c)), which is characteristic for decaying rotating turbulence [11, 5, 10]. Furthermore, $Ro_\lambda < 1$ for all times and in all cases which means that background rotation is dominant over nonlinear effects for all times in all cases.

Similar time evolutions of E_k, Re_λ and Ro_λ have been obtained for the remaining runs (not shown) with one exception: for the lowest background rotation rate considered here (case C1) $Ro_\lambda(t_{\rm ini}) > 1$, illustrating that background rotation is not dominant over nonlinear effects (weakly rotating turbulence), and Re_λ initially still decreases.

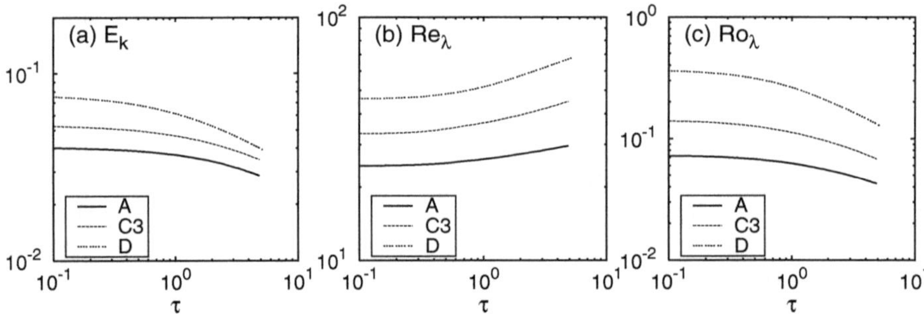

Fig. 1. (a,b,c) Time evolution of the kinetic energy E_k, the Taylor-based Reynolds number Re_λ, and the Taylor-based Rossby number Ro_λ, respectively, for various viscosities.

We now discuss the effect of rotation on the energy spectrum $E(k)$ for cases A, C3 and D. Figure 2 shows $E(k)$ at time $t_{\rm ini}$ and time t_f, i.e. the end of the anisotropic calculation, for each of the mentioned cases. The corresponding isotropic (reference) energy spectra at time t_f are also plotted. The effect of rotation is to inhibit the direct energy cascade in the inertial range so that dissipation at the smaller wave-numbers is reduced, and energy appears to be conserved for the larger wave-numbers.

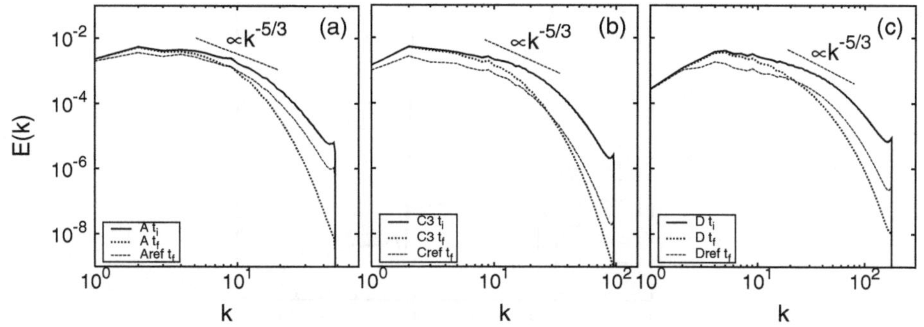

Fig. 2. (a,b,c) Energy spectrum of the decaying turbulence at $t = t_{\rm ini}$ and $t = t_f$ in case A, C3 and D, respectively. An isotropic reference spectrum is shown for $t = t_f$.

Velocity Derivative Skewness and Vorticity Skewness

The time evolution of the velocity derivative skewness $S_{\partial_i u_i}$ and the vorticity skewness S_{ω_3}, respectively, for cases A, C3 and D (different viscosities ν) is depicted in Fig. 3. Clearly, the velocity derivative skewness develops during the isotropic precalculation ($t \leq t_{\rm ini}$) and collapses when the anisotropic body force is applied ($t > t_{\rm ini}$). Furthermore, a higher viscosity (i.e. lower spatial resolution) requires longer precalculations in order to attain the same value of $S_{\partial_i u_i}$. In cases A and C3 the initial increase of S_{ω_3} [see Fig. 3(b)] is proportional to the power-law $t^{0.75\pm0.1}$. This power-law behavior is less apparent in case D which is most likely related to the short precalculation, as will be addressed further on. For later times [$\tau > \mathcal{O}(\Omega^{-1})$], S_{ω_3} wiggles about a positive value, reflecting the dominance of cyclonic vorticity. Although it seems likely that a lower viscosity ν results in a more developed vorticity skewness, the different final values of S_{ω_3} may also be ascribed to slight differences in $S_{\partial_i u_i}(t_{\rm ini})$, see below.

Consider the effects of the duration of the isotropic precalculation on the time evolution of S_{ω_3}. It follows directly from Fig. 4(a) that shorter precalculations yield smaller $S_{\partial_i u_i}(t_{\rm ini})$. The vorticity skewness S_{ω_3} [Fig. 4(b)] shows a power-law dependence of $t^{0.75\pm0.1}$ for $\tau < \mathcal{O}(\Omega^{-1})$, and wiggles about a positive value for $\tau > \mathcal{O}(\Omega)^{-1}$. Careful inspection of Fig. 4(b), however, could reveal that the initial power-law behavior is affected by the duration of the precalculation, in such a way that either a very short or a very long precalculation results in a smaller power-law exponent. Reverting to case D [Fig. 3(b)], it seems that the less apparent power-law behavior of S_{ω_3} is most likely related to the short precalculation. Returning to Fig. 4(b), the final value of

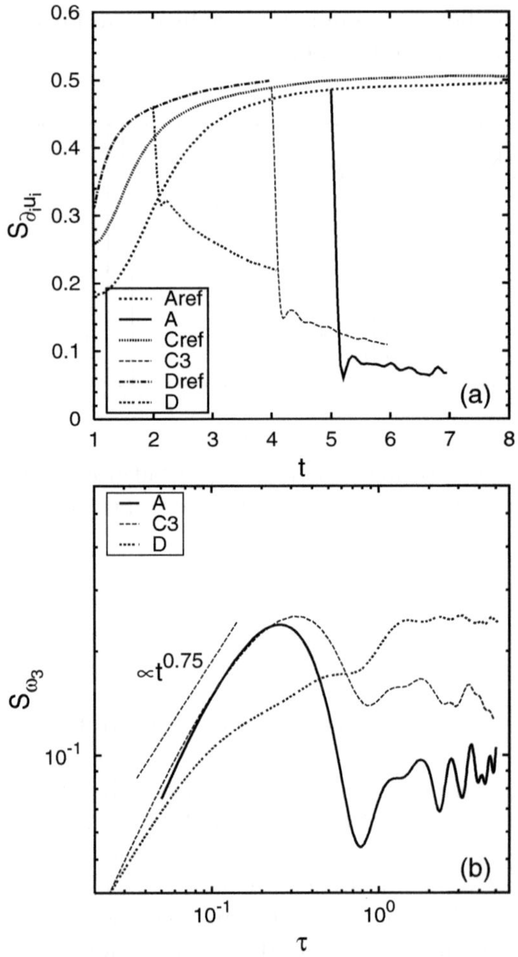

Fig. 3. (a) Time evolution of the velocity derivative skewness $S_{\partial_i u_i}$ during and after the isotropic precalculation for different viscosities ν. Background rotation in cases A, C3 and D is applied at $t_{\mathrm{ini}} = 5.0, 4.0$ and 2.0, respectively. For reference, the isotropic precalculations have been prolonged. (b) Log-log plot showing the vorticity skewness S_{ω_3} as a function of the scaled, shifted time τ for cases A, C3 and D.

S_{ω_3} appears to depend inversely on time t_{ini}, i.e. shorter precalculations yield higher final values of S_{ω_3}. The behavior observed in Fig. 3(b) may partly be ascribed to slight differences in $S_{\partial_i u_i}$ at time t_{ini}.

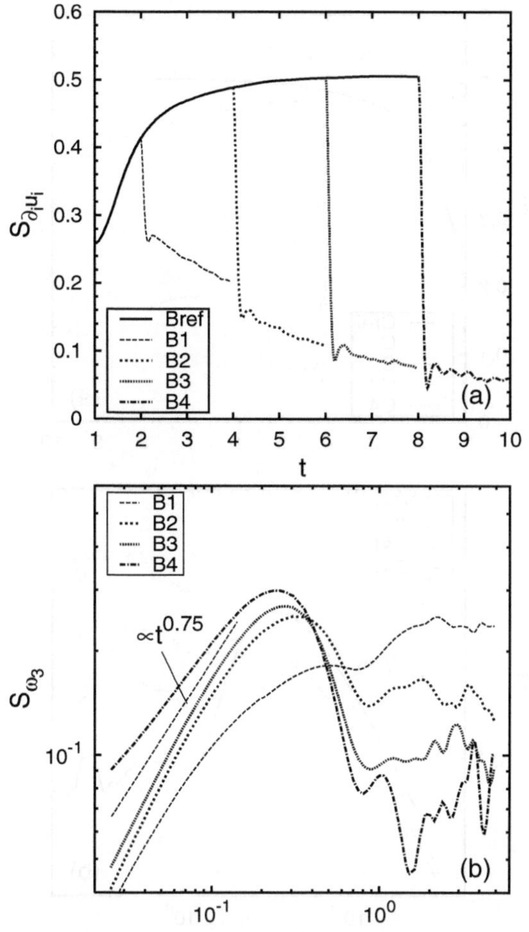

Fig. 4. As Fig. 3, but for different durations of the isotropic precalculation. Background rotation in cases B1-B4 is applied at $t_{\mathrm{ini}} = 2.0$, 4.0, 6.0 and 8.0, respectively.

Finally, Fig. 5 shows the time evolution of $S_{\partial_i u_i}$ and S_{ω_3} for various background rotation rates, viz. $f = 0.5$, 2.5, 5.0 and 10.0 (cases C1-C4). Clearly, a lower background rotation rate results in a larger final value of S_{ω_3}. This result expresses the fact that the asymmetry between cyclonic and anticyclonic structures is more pronounced at low rotation rates than at high rotation rates. It is remarked that similar results were extracted from lower resolution ($N = 144$) calculations.

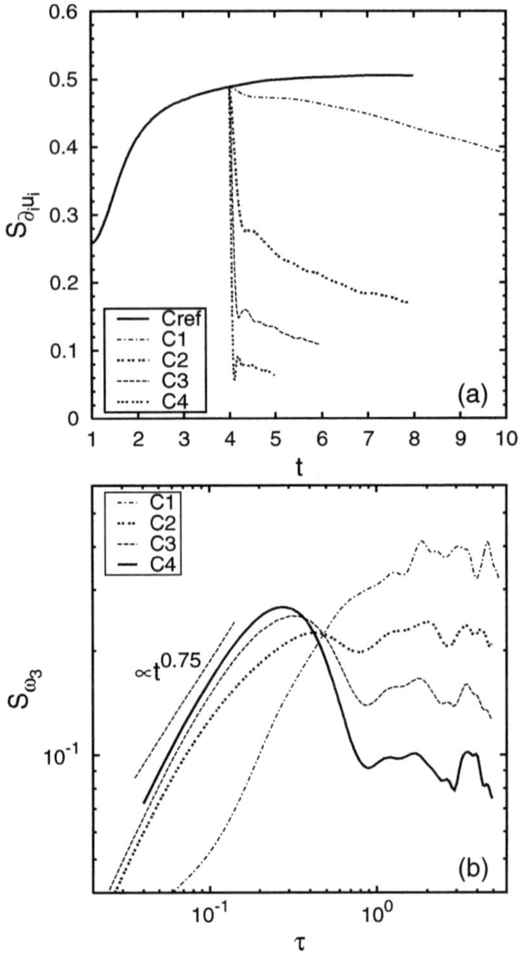

Fig. 5. As Fig. 3, but for different background rotation rates. Background rotation in cases C1-C4 is applied at $t_{\text{ini}} = 4.0$.

Third Order Vorticity Correlations

Figure 6 shows the time evolution of all nontrivial VTCs for various background rotation rates. The following three observations are made: 1) $\langle \omega_1^3 \rangle$, $\langle \omega_1 \omega_3^2 \rangle$ and $\langle \omega_1 \omega_2 \omega_3 \rangle$ are much smaller than unity and fluctuate around zero; 2) $\langle \omega_1^2 \omega_3 \rangle$, $\langle \omega_2^2 \omega_3 \rangle$ and $\langle \omega_3^3 \rangle$ are clearly nonzero; and 3) the ratio $\langle \omega_1^2 \omega_3 \rangle / \langle \omega_2^2 \omega_3 \rangle$ (not shown) is found to fluctuate around unity. These results are consistent with relationship (3).

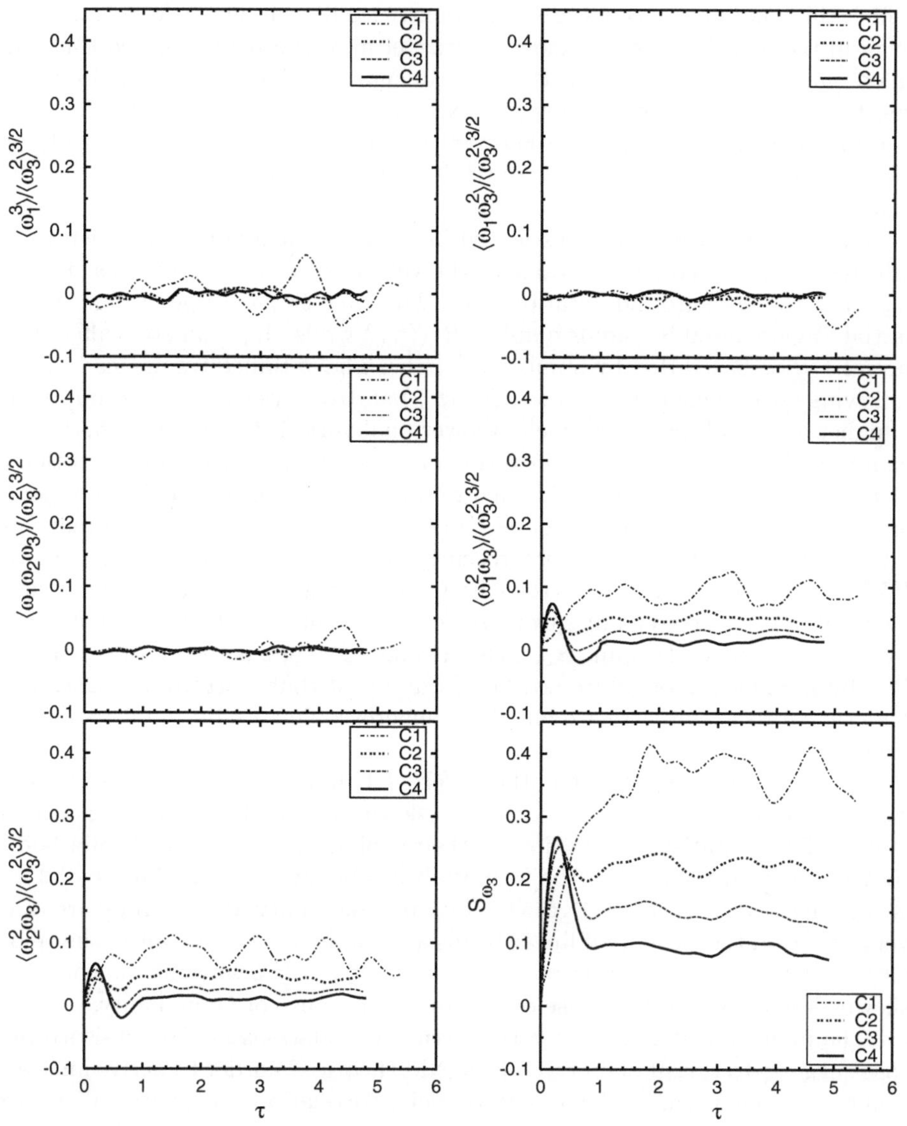

Fig. 6. Time evolution of the minimal set of VTCs in axisymmetric turbulence for various background rotation rates. All VTCs are normalized by $\langle\omega_3^2\rangle^{3/2}$.

4 Discussion

Our numerical results show that in most of the considered cases S_{ω_3} initially grows at a rate proportional to $t^{0.75\pm0.1}$. The latter power-law exponent is in good agreement with the 0.7 obtained from recent laboratory experiments [14, 15]. However, the amplitude of maximum S_{ω_3} and the (scaled) time at which this maximum occurs are significantly smaller in our DNS calculations than in the mentioned experiment. Morize et al. [15] already showed that the maximum of S_{ω_3} depends on the experimental configuration. Based on our results, a better agreement between DNS and experiment is expected for very low rotation rates (e.g. $f = 0.10\pi$), viz. a weak background rotation will not succeed to immediately destroy the triple correlation of the velocity derivative $\partial_3 u_3$.

Furthermore, we have investigated how the vertical vorticity skewness is affected by the viscosity, the value of the velocity derivative skewness at time $t_{\rm ini}$, and the background rotation rate. The first two parameters affect the initial Taylor-based Reynolds number $Re_\lambda(t_{\rm ini})$ while all parameters affect the initial Taylor-based Rossby number $Ro_\lambda(t_{\rm ini})$. The obtained results lead to the following general conclusion: lower $Re_\lambda(t_{\rm ini})$ and/or lower $Ro_\lambda(t_{\rm ini})$ – implying a higher degree of linearity – yield a lower final vorticity skewness. This result confirms that the asymmetry in terms of cyclonic and anticyclonic vorticity is most prominently present in an intermediate range of Rossby numbers, as also discussed by Jacquin et al. [11] for the anisotropic development of integral length-scales, and more recently by Bartello [2] in the context of VTCs. If the Rossby number is too small, nonlinearity is not important enough – even if pure linear dynamics can induce a transient growth of S_{ω_3}, that same dynamics results in damping S_{ω_3} at later times. The opposite case of very large Rossby number is not addressed here, but recall that isotropy is conserved, and therefore asymmetry excluded, at macroscopic Rossby numbers larger than one [11].

Another surprising result is the different behavior of different triple correlations. Even if the velocity derivative skewness and the vorticity skewness look similar as statistical descriptors, their evolution in presence of solid-body rotation is far from similar, the former always being damped while the latter is showing transient (linear) growth. Initial Gaussianity and isotropy are also very important, especially if linear terms are dominant. For instance, the velocity derivative skewness is zero only if Gaussianity holds, whereas the vorticity skewness is zero either because of isotropy or because of Gaussianity.

The multi-fold behavior of various triple correlations in the non-isotropic case suggests to revisit elaborated EDQNM theories in order to derive any relevant three-point triple velocity and vorticity correlation, which are difficult to extract experimentally and numerically. In previous studies, the EDQNM2-3 formalisms were used to derive a nonlinear energy transfer, but much more information, including VTCs, can be obtained. At least isotropic basic EDQNM can be used for initializing vorticity correlations in the general linear solution applied to VTCs, but more can be done. In this sense, anisotropic multi-point statistical theory remains a relevant alternative to DNS, allowing much higher Reynolds numbers and elapsed times (with in counterpart, less flexibility and need for statistical assumptions).

In addition, the subtle interplay between linear and nonlinear processes is altered in the presence of boundaries, Ekman pumping, or initially coherent structures: interesting insights to these effects can be found in recent studies by Zavala Sansón and Van Heijst [18], Morize et al. [14], and Davidson et al. [7].

References

[1] Bartello P, Métais O, Lesieur M (1994), J. Fluid Mech. 273:1–29.
[2] Bartello P (2006), private communication.
[3] Batchelor GK, Townsend AA (1947), Proc. Roy. Soc. A 191:534–550.
[4] Cambon C, Jacquin L (1989), J. Fluid Mech. 202:295–317.
[5] Cambon C, Mansour NN, Godeferd FS (1997), J. Fluid Mech. 337:303–332.
[6] Cambon C, Scott JF (1999), Annu. Rev. Fluid Mech. 31:1–53.
[7] Davidson PA, Staplehurst PJ, Dalziel SB (2006), J. Fluid Mech. 557:135–144.
[8] Greenspan HP (1968) The theory of rotating fluids. Cambridge University Press.
[9] Gence J-N, Frick C (2001), C. R. Acad. Sci. Paris Série IIB 329:351–356.
[10] Godeferd FS, Lollini L (1999), J. Fluid Mech. 393:257–308.
[11] Jacquin L, Leuchter O, Cambon C, Mathieu J (1990), J. Fluid Mech. 220:1–52.
[12] Liechtenstein L, Godeferd FS, Cambon C (2005), JoT 6:1–21.
[13] McComb WD (1990) The Physics of Fluid Turbulence, Oxford Engineering Science Series 25. Clarendon Press, Oxford.
[14] Morize C, Moisy F, Rabaud M (2005), Phys. Fluids 17:095105.
[15] Morize C, Moisy F, Rabaud M, Sommeria J (2006), Conf. Proc. TI2006, Porquerolles, France.
[16] Rogallo RS (1981), NASA Tech. Mem. 81315.
[17] Vincent A, Meneguzzi M (1991), J. Fluid Mech. 225:1–20.
[18] Zavala Sansón L, Van Heijst GJF (2000), J. Fluid Mech. 412:75–91.

Lagrangian passive scalar intermittency in marine waters: theory and data analysis

François G. Schmitt[1] and Laurent Seuront[1,2]

[1] CNRS, FRE 2816 ELICO, Wimereux Marine Station, Universite des Sciences et Technologies de Lille - Lille 1, 28 av Foch, 62930 Wimereux, France
francois.schmitt@univ-lille1.fr

[2] School of Biological Sciences, Flinders University, GPO Box 2100, Adelaide 5001, South Australia Laurent.Seuront@flinders.edu.au

Intermittency is a basic feature of fully developed turbulence, for both velocity and passive scalars. We consider here intermittency in a Lagrangian framework, which is also a natural representation for marine organisms. We characterize intermittency using multi-fractal power-law scaling exponents. In this paper we recall four theoretical relations previously obtained to link Lagrangian and Eulerian passive scalar multi-fractal functions. We then experimentally estimate these exponents and compare the result to the theoretical relations. Section 1 describes the non intermittent Lagrangian passive scalar scaling laws; section 2 introduces the multi-fractal generalization, and gives the four theoretical relations ; section 3 presents experimental results.

1 Non-intermittent Lagrangian passive scalar scaling laws

Marine particle dynamics is an important area in turbulence studies. Particles sampling is most easily achieved in the Eulerian sense, that is, in a reference frame fixed with respect to the moving fluid, such as moored buoy or a pier. However, plankton organisms such as viruses, bacteria, phytoplankton and copepods, perceive their surrounding environment in a Lagrangian way. Those are mostly advected by the flows. The related Lagrangian turbulent fluctuations in the flow velocity and passive scalars perceived by individual plankton organisms have critical implications for foraging, growth and populations dynamics, and ultimately for a better understanding of the structure and functioning of the pelagic realm. An absolute pre-requisite to the analysis of e.g. behavioral response to the fluctuations of purely passive scalars (e.g. temperature and salinity) and potentially biologically active scalars in their Lagrangian environment is the characterization of Lagrangian passive scalar intermittency. The main objective of the present work is thus to

Bernard J. Geurts et al. (eds), Particle Laden Flow: From Geophysical to Kolmogorov Scales, 129–138.

provide baseline information on passive scalar Lagrangian intermittency that could be compared to biological scalars (e.g. prey/mate abundance) in future studies. In the following we will consider scales belonging to the inertial range; i.e. larger than the Kolmogorov scale. For phytoplankters of smaller size, the influence of turbulence is still important, but limited to inertial range scales.

In this section, we will recall the basic scaling properties for Lagrangian passive scalar turbulence. We consider the inertial convective subrange, associated to large Peclet and Reynolds numbers, and hypothesize an homogeneous and isotropic turbulence, which is generally the case at small scales for 3D oceanic turbulence.

In the Eulerian framework, velocity and passive scalar fluctuations in homogeneous turbulence are classically characterized using Kolmogorov-Obukhov-Corrsin (KOC) [1, 2, 3] scaling laws (see [4] for details). For passive scalar scaling exponents, let us mention the important result indicating that even in case of uncorrelated velocity field, the passive scalar field is multiscaling (see [5] and [6] for a review). However such scaling exponents are quite far from experimental estimates, indicating that intermittency in velocity fluctuations has influence on temperature scaling exponents.

This framework has been extended to the Lagrangian framework for velocity fluctuations by Landau [7] and for passive scalar fluctuations by Inoue [8]. Let us note $V(x_0, t)$ and $\Theta(x_0, t)$ the velocity and passive scalar concentration of an element of fluid at time t, initially at a position $x(0) = x_0$. Hereafter these will be simply referred to as $V(t)$ and $\Theta(t)$ since we assume statistical homogeneity. We note also for the Lagrangian velocity and passive scalar time increments $\Delta V_\tau = |V(t+\tau) - V(t)|$ and $\Delta\Theta_\tau = |\Theta(t+\tau) - \Theta(t)|$. This gives Landau's relation for the velocity [7]:

$$\Delta V_\tau \sim \epsilon^{1/2}\tau^{1/2} \tag{1}$$

and Inoue's law for passive scalars [8]

$$\Delta\Theta_\tau \sim \chi^{1/2}\tau^{1/2} \tag{2}$$

where ϵ is the dissipation, $\chi = \Gamma_\theta\langle|\nabla\theta|^2\rangle$ is the scalar variance dissipation rate and Γ_θ is the scalar diffusivity of the fluid.

The Eulerian power spectra are of the form $E(k) \sim k^{-5/3}$ for velocity and passive scalars (k is the wave number). In contrast, for Lagrangian fields, the power spectra are also scaling, with a different exponent: $E(f) \sim f^{-2}$ for both velocity and passive scalars (f is the frequency). These laws provide velocity and passive scalar fluctuations in time, assuming constant and homogeneous values for the fields ϵ and χ. In reality, one of the characteristic features of fully developed turbulence is the intermittent nature of the fluctuations of associated fields, providing intermittent corrections for Eulerian and Lagrangian fields (see reviews in [4]). This is discussed in the next section.

2 Intermittent Lagrangian passive scalar multi-fractal relations: four predictions

2.1 Intermittent multi-fractal generalization

Previous scaling relations describe only the mean behavior of passive scalar fluctuations. Fully developed turbulence is now known to be associated to intermittency: fluctuations are random variables, whose scale dependence is usually characterized using statistical moments of various order $q > 0$. Following developments obtained in an Eulerian framework, Lagrangian intermittency has been characterized using scaling moments functions as:

$$\langle \Delta\Theta_\tau^q \rangle \sim \tau^{\xi_\theta(q)} \tag{3}$$

where Θ is the passive scalar concentration, $\Delta\Theta_\tau = \Theta(t+\tau) - \Theta(t)$ is the passive scalar increment, and $\xi_\theta(q)$ is the Lagrangian passive scalar scaling moment function [9]. Without intermittency the latter is linear: $\xi_\theta(q) = q/2$. In case of intermittency $\xi_\theta(q)$ is nonlinear and concave, and the non-intermittent value is valid only for $q = 2$: $\xi_\theta(2) = 1$, indicating also that there is no intermittency correction for the power spectrum exponent.

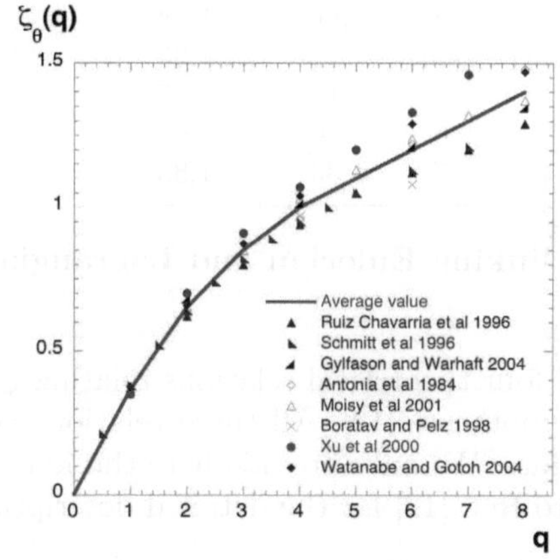

Fig. 1. The passive scalar Eulerian scaling exponent function $\zeta_\theta(q)$ estimated by various authors, and with an average fit (see Table 1).

It is interesting here to compare this Lagrangian scaling exponent $\xi_\theta(q)$ to the more classical Eulerian $\zeta_\theta(q)$ defined by:

$$\langle (\Delta\theta_\ell)^q \rangle \sim \ell^{\zeta_\theta(q)} \tag{4}$$

In the following we will also need another Eulerian quantity, which depends only on the passive scalar flux χ_ℓ, and which is called "mixed moment function" and is denoted here $\zeta_m(q)$. This may be written in the following way (see [10, 11]):

$$\langle (\Delta\theta_\ell^2 \Delta U_\ell)^{q/3} \rangle \sim \ell^{\zeta_m(q)} \tag{5}$$

We will also need experimental or numerical estimates of the functions $\zeta_\theta(q)$ and $\zeta_m(q)$. Several values have been provided in the literature, and an average estimate has been provided in [12]. These values are provided in Table 1, and the corresponding curves are displayed in Figure 1 and Figure 2.

Table 1. Average values of $\zeta_\theta(q)$, estimated from several published estimates [10, 13, 11, 14, 15, 16, 17, 18] (Column 1). Average values of $\zeta_m(q)$, estimated from several published estimates [11, 14, 19, 20, 18, 15, 21] (Column 2).

q	$\zeta_\theta(q)$	$\zeta_m(q)$
0.5		0.21
1	.365	0.39
1.5		0.56
2	.65	0.72
2.5		0.87
3	.85	1
3.5		1.12
4	.99	1.24
4.5		1.35
5	1.10	1.45
6	1.20	1.65
7	1.30	1.83

2.2 Four relations linking Eulerian and Lagrangian passive scalar scaling exponents

We recently obtained four theoretical relations relating $\zeta_\theta(q)$ and $\xi_\theta(q)$ based on different sets of hypotheses [12]. All these relations verify $\xi_\theta(2) = 1$, and differ for other moments. We only provide here the four theoretical relations and refer the reader to Ref. [12] for the detailed description of how they have been derived.

The first and simplest relation was obtained assuming a "characteristic time" relation for the de correlation of eddies, and a non-intermittent space-time relation:

$$\xi_\Theta(q) = \frac{3}{2}\zeta_\theta(q) \qquad \textit{Case I} \tag{6}$$

The second choice was to assume an "ergodic" hypothesis corresponding to an equality of the statistics of the passive scalar flux in Eulerian and Lagrangian frame, and a non-intermittent space-time relation:

$$\xi_\Theta(q) = \frac{3}{2}\zeta_m\left(\frac{3q}{2}\right) - \frac{q}{4} \qquad \textit{Case II} \tag{7}$$

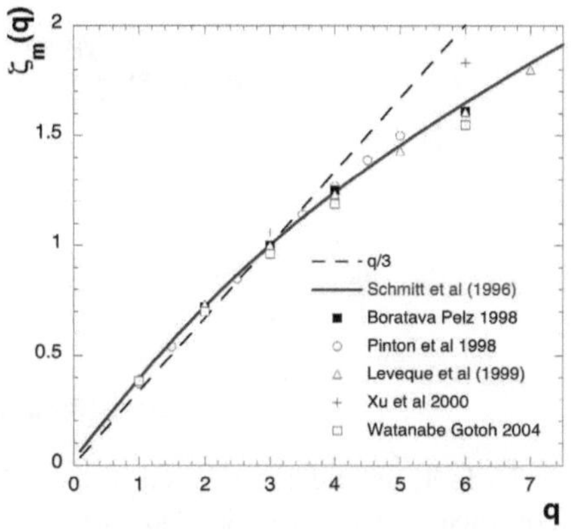

Fig. 2. The mixed velocity-temperature Eulerian scaling exponent function $\zeta_m(q)$ estimated by various authors (See Table 1).

The two last relations were both obtained assuming an intermittent space-time relation, and the characteristic time relation (case III), and the ergodic relation (case IV):

$$\xi_\Theta(q) = \zeta_m(q_0) \qquad \textit{Case III} \tag{8}$$

$$\frac{q}{2} = \frac{2q_0}{3} - \zeta_m(q_0) \tag{9}$$

$$\xi_\Theta(q) = \zeta_\theta(q_0) \qquad \textit{Case IV} \tag{10}$$

$$q = q_0 - 2\zeta_\theta(q_0) \tag{11}$$

Relations corresponding to Case I and Case II are linear, and the ones corresponding to Case III and Case IV are fully nonlinear: a given q value is associated to a q_0 value given by solving the second line, and the value of $\xi_\Theta(q)$ is given by the first line. The four curves will be shown below (Fig. 7) and compared to experimental data.

3 Analysis of Lagrangian marine temperature data

We have previously shown that temperature, salinity and phytoplankton fields recorded adrift in the Eastern English Channel during from February to December 1996 exhibit Eulerian and Lagrangian components separated by a length scale intrinsically linked to the size of the ship used to collect the field data [22]. For scales smaller and larger than the eddy turnover time associated to the size of the ship, we identified Eulerian and Lagrangian statistics, respectively. These results show that Eulerian and Lagrangian scaling and

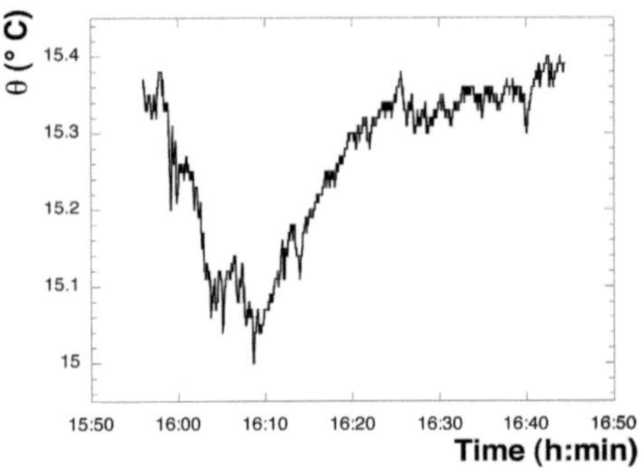

Fig. 3. The time series N.2 which has been analyzed in this study (1 Hz resolution, 50 minutes recording). The measuring device is also shown.

multi-scaling properties of temperature and salinity are very similar and fully compatible with the behavior of purely passive scalars, and with preliminary results obtained from ocean temperature sampled in the same area [23]. In contrast, phytoplankton biomass exhibited a specific behavior for both Eulerian and Lagrangian regimes. Briefly put, phytoplankton exhibited a non-passive behavior, a density-dependent control of phytoplankton distribution in relation with the biological seasonal cycle, and the scaling and multi-scaling laws of passive scalars and phytoplankton are closer in the Eulerian than in the Lagrangian framework. However, the size of the ship used during this preliminary experiment (i.e. 12 m) intrinsically limits the extent of the Lagrangian scaling range and is hardly compatible with the fluctuations occurring at the minute scales characteristic of plankton organisms. To investigate more thoroughly the Lagrangian fluctuations of purely passive scalars, we thus used a small (0.5 m) buoy equipped with a miniature temperature sensor (Alec Electronics, model MDS MkV/T). The temperature sensor is 8 cm long, 18 mm wide, a weight of 50 g; it has a sampling frequency of 1 Hz and autonomously record data through a lithium ion battery.

We have recorded 2 time series of 80 and 50 minutes duration, on 6 June, 2006, in the Eastern English Channel. The power spectrum of the series 2 is shown in Figure 3 (for series 1, the result is similar). It displays a very clear -2 power-law scaling for a large range of scales, as expected theoretically. Only higher frequencies display a departure from this scaling law, corresponding to the limit of sensor's precision.

As a next step, structure functions have been estimated for both time series (Eq. 3). The resulting scaling relation is shown in Figure 5. This shows that the scaling property displayed by the power spectrum (Fig. 4) is also respected for other order of moments in real space. For better precision, we have in the following estimated the scaling moment function $\xi_\theta(q)$ using an Extended Self-Similarity relation. This has been proposed originally in the Eulerian framework for the velocity field, using as a reference the third order

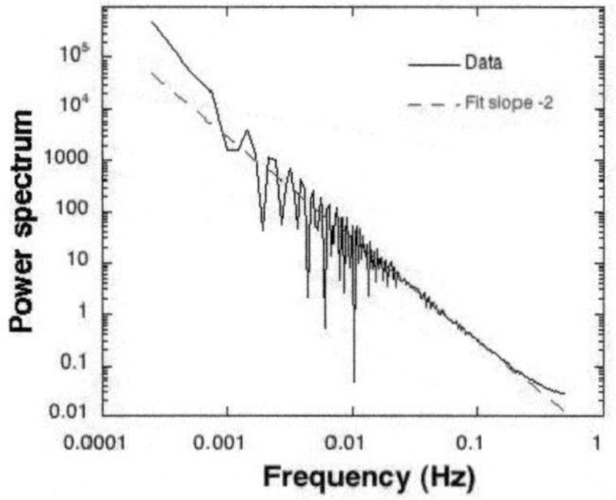

Fig. 4. The Fourier power spectrum of the data set N. 2, in log-log plot, together with a power-law fit of slope −2. An extremely nice power-law spectrum is visible.

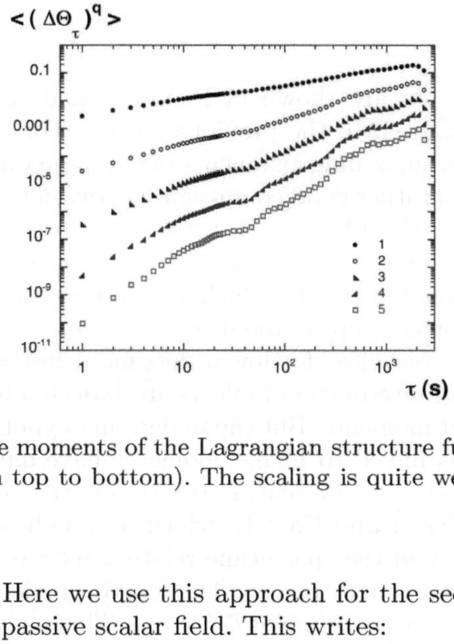

Fig. 5. Scaling of the moments of the Lagrangian structure functions, for moments of order 1 to 5 (from top to bottom). The scaling is quite well respected, even for larger moments.

moment (see [24]). Here we use this approach for the second order moment, for the Lagrangian passive scalar field. This writes:

$$\langle \Delta\Theta_\tau^q \rangle \sim \left(\langle \Delta\Theta_\tau^2 \rangle \right)^{\xi_\theta(q)} \tag{12}$$

The scaling exponent $\xi_\theta(q)$ estimated this way is more precise than the one obtained through a best-fit of the lines in Figure 5: see Figure 6, displaying a really nice scaling for scales larger than 4 s. We could obtain this way the following experimental estimates of $\xi_\theta(q)$, which were obtained from a fit of Figure 6 as shown by the straight lines in this Figure: $\xi_\theta(1) = 0.54$, $\xi_\theta(3) = 1.39$, $\xi_\theta(4) = 1.73$, and $\xi_\theta(5) = 2.02$. Due to the relatively small amount of data used here, we have not estimated higher moments.

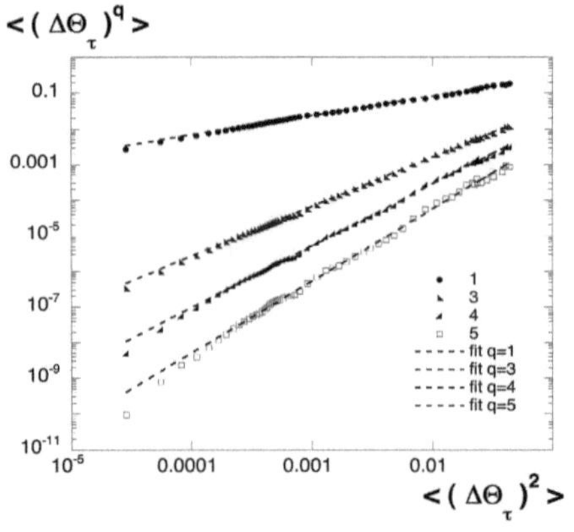

Fig. 6. Relative scaling of Lagrangian structure functions using the ESS approach: moments of order q versus moment of order 2, with $q = 1, 3, 4, 5$. There is an extremely nice scaling.

The resulting values are shown in Figure 7, and compared to the four theoretical cases discussed in the previous section. For Case I and Case IV, we take here for Eulerian scaling exponents the average values estimated above (Table 1), which are a rather good compromise between many published values (see Fig. 1). For Case II and Case III, providing a prediction for Lagrangian passive scalars as a relation to the mixed Eulerian exponents, we take for $\zeta_m(q)$ the values estimated in [11], which are close to other values reported in the literature for moments up to about 6.

Case I to III are very close for low orders moments, which can be understood by the fact that intermittency effects are expected to become important mainly for high order moments. But the underlying hypotheses are clearly different, and questions linked to higher moments have quite different outputs since scaling exponents are different. Furthermore, the deviation from linearity is stronger for Case I and Case II, which may indicate that to take into account intermittency in the space-time relation reduces the apparent intermittency of the Lagrangian estimates. The fourth prediction is quite far from the others, except the common point $\zeta_\theta(2) = 1$. This may be the consequence of the additional hypothesis which was needed to obtain Case IV prediction (see Ref. [12]).

We may see in Figure 7 that the experimental estimates do not fit any theoretical cases. First, experimental estimates are clearly nonlinear and concave, and since they were obtained for a quite large range of scales, with a very nice power-law scaling, this can be seen as a direct evidence of multi-fractal Lagrangian intermittency property. Case III can be considered as the closest to data; however, experimental estimates are far from this theoretical prediction for moments larger than 3. We may consider that much more data points may be needed to adopt a clear rejection of case III. Indeed, more and more

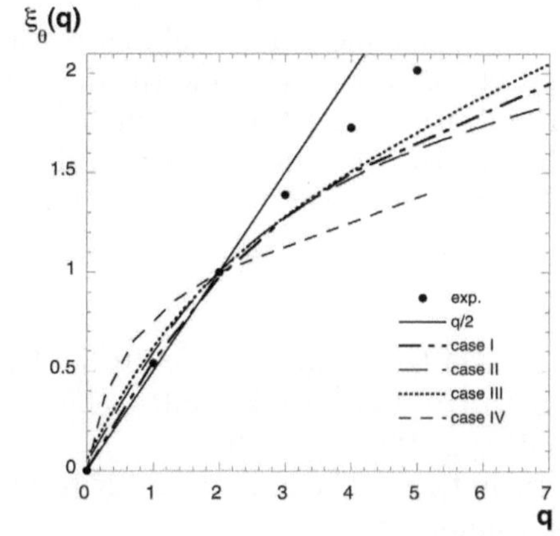

Fig. 7. Curves of $\xi_\theta(q)$ obtained from experimental values for $\zeta_\theta(q)$ and $\zeta_m(q)$ associated to four different theoretical relations between these scaling exponents (dotted lines). The straight line is the non-intermittent case of equation $q/2$. The experimental estimates are also shown (black dots).

extreme events which are encountered when increasing the sampling size, lead to more and more concave curves. A finite sampling is associated to a linear scaling exponent for moments larger than a critical order of moment. To have more confidence in large order of moments estimates, larger data sets will be needed in the future.

4 Conclusion

Using experimental data recorded in the marine environment of the Eastern English Channel, we have shown that temperature data behave as expected, as an intermittent passive scalar. We obtained a very good scaling behavior in the inertial range, with concave power-law exponents. For larger scales, we find a failure of isotropy and homogeneity, due to side effects, the influence of topography, or other reason: this large scale is visible in fig. 5, at scales of about 2000 s, about 30 minutes.

We have also compared these experimental values to four theoretical scaling moment functions that have previously been obtained using several sets of hypotheses. The theoretical curve that appears the closest to data has been obtained using an intermittent space-time relation and a characteristic time approach. The agreement with experimental data is, however, only good for small order of moments. For larger moments, there is a discrepancy. More data points might be needed to sample more intermittent events and achieve a more concave experimental curve.

While further data are needed to generalized the present observations, our results have salient potential consequences on our understanding of the physical nature of turbulent flows, and the matter fluxes in the ocean through biophysical interactions. For instance, the combination of the identified Lagrangian properties of purely passive scalars and the density-dependent control of phytoplankton distribution demonstrated elsewhere [22, 25] might open new perspectives in investigating the links between the scaling laws of biologically active scalars, phytoplankton concentrations and turbulence.

References

[1] Kolmogorov AN (1941) Izv Akad Nauk SSSR 30: 301
[2] Obukhov AM (1949) Izv Akad Nauk SSSR Geogr. Geofiz. 13: 58
[3] Corrsin S (1951) J Appl Phys 22: 469
[4] Frisch U (1995) Turbulence; The Legacy of AN Kolmogorov. Cambridge University Press, Cambridge
[5] Kraichnan RH (1994) Phys Rev Lett 72: 1016
[6] Falkovich G, Gawedzki K, Vergassola M (1994) Rev Mod Phys 73: 913
[7] Landau L, Lifshitz EM (1944) Fluid Mechanics. MIR, Moscow
[8] Inoue E (1952) J Meteorol Soc Japan 29: 246
[9] Novikov EA (1989) Phys Fluids A 1:326
[10] Antonia RA, Hopfinger E, Gagne Y, Anselmet F (1984) Phys Rev A 30: 2704
[11] Schmitt FG, Schertzer D, Lovejoy S, Brunet Y (1996) Europhys Lett 34: 195
[12] Schmitt FG (2005) Eur Phys J B 48:129
[13] Ruiz-Chavarria G, Baudet C, Ciliberto S (1996) Physica D 99: 369
[14] Boratav ON, Pelz RB (1998) Phys Fluids 10: 2122
[15] Xu G, Antonia RA, Rajagopalan S (2000) Europhys Lett 49: 452
[16] Moisy F, Willaime H, Andersen JS, Tabeling P (2001) Phys Rev Lett 86: 4827
[17] Gylfason A, Warhaft Z (2004) Phys Fluids 16: 4012
[18] Watanabe T, Gotoh T (2004) New J Phys 6: 40
[19] Pinton JF, Plaza F, Danaila L, Le Gal P, Anselmet F (1998) Physica D 122: 187
[20] Leveque E, Ruiz-Chavarria G, Baudet C, Ciliberto S (1999) Phys Fluids 11: 1869
[21] Mydlarski L (2003) J Fluid Mech 475: 173
[22] Seuront L, Schmitt FG (2004) Geophys Res Lett 31: L03306
[23] Seuront L, Schmitt F, Schertzer D, Lagadeuc Y, Lovejoy S (1996) Nonlin Proc Geophys 3: 236
[24] Benzi R, et al. (1993) Europhysics Letters 24 : 275
[25] Seuront L (2005) Mar Ecol Prog Ser 302: 93

Compositional and particulate gravity currents: a computational investigation

V. K. Birman and E. Meiburg

Department of Mechanical Engineering, University of California Santa Barbara, CA 93106, USA.

Summary. We present some high-resolution 2D numerical simulation results for gravity and turbidity currents in the lock exchange configuration. Results are provided for both Boussinesq and non-Boussinesq flows, and for horizontal as well as sloping bottom geometries. Furthermore, recent results for reversing buoyancy currents are discussed.

1 Introduction

Gravity currents, which form when a heavier fluid propagates into a lighter one in a predominantly horizontal direction, have been the subject of numerous investigations over the past half century. They are frequently encountered both in the environment and in engineering applications ([7], [14]). Gravity currents can be driven by density differences of the fluids involved, or by differential particle loading. In many situations (a freshwater river flowing into a saltwater ocean, atmospheric flows involving warm and cold air, and many others), the density differences are no more than a few percent, so that the Boussinesq approximation can be employed. However, there are circumstances when the density differences can be much more substantial (industrial gas leaks, tunnel fires, powder snow avalanches, turbidity currents, pyroclastic flows), and the full variable density equations have to be solved.

It is desirable to develop simplified models for the prediction of such flows. However, such models are based on a variety of assumptions regarding the nature of the flow whose validity needs to be established first. In this context, high-resolution numerical simulations can be of great value, as they offer access to several quantities that are hard to measure experimentally. The spatially and temporally resolved dissipation field represents one example in this regard. In the following, we will present a brief overview of our numerical simulation results for a variety of gravity and turbidity currents. For this purpose, we have focused on the lock-exchange configuration, which is the most commonly used geometry for studying gravity currents (see fig. 1).

Bernard J. Geurts et al. (eds), Particle Laden Flow: From Geophysical to Kolmogorov Scales, 139–150.

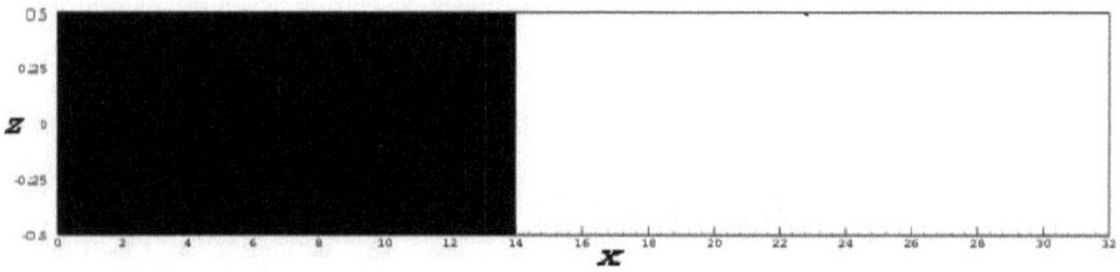

Fig. 1. Lock exchange configuration. A membrane initially divides the rectangular container into two compartments. The left chamber is filled with fluid (or suspension) of density $\rho = \rho_1$, while the right one contains a lighter fluid of density $\rho = \rho_2$. Upon release of membrane, a dense front moves rightwards along the lower boundary, while the light front propagates leftward along the upper boundary.

2 Basic equations

The 2D simulations employ a rectangular channel of height H and length L, cf. figure 1. The channel is filled with two miscible fluids initially separated by a membrane. While the left compartment holds a fluid (or suspension) of density ρ_1, the right reservoir is filled with a fluid of smaller density ρ_2. This initial configuration causes a discontinuity of the hydrostatic pressure across the membrane, which sets up a predominantly horizontal flow once the membrane is removed. The denser fluid moves rightward along the bottom of the channel, while the lighter fluid moves leftward along the top.

The full incompressible Navier-Stokes equations for variable density flows without use of Boussinesq approximation, read

$$\nabla \cdot \mathbf{u} = 0 \,, \tag{1}$$

$$\rho \frac{D\mathbf{u}}{Dt} = \rho \mathbf{g} - \nabla p + \nabla \cdot (2\mu \mathbf{S}) \,. \tag{2}$$

Here $\frac{D}{Dt}$ denotes the material derivative of a quantity, $\mathbf{u} = (u, v)^T$ indicates the velocity vector, p the pressure, ρ the density, and $\mathbf{S}$ the rate of strain tensor, while $\mathbf{g} = g\mathbf{e}_g$ represents the vector of gravitational acceleration. In the following, we will keep the kinematic viscosity ν constant for both fluids. In deriving the above continuity equation, it is assumed that the material derivative of the density vanishes, i.e., $\frac{D\rho}{Dt} = 0$. This common assumption requires small diffusivities of the species concentration. The conservation of species is expressed by the convection-diffusion equation for the concentration c of the heavier fluid. By assuming a density-concentration relationship of the form $\rho = \rho_2 + c\,(\rho_1 - \rho_2)$, we arrive at the following equation for the density field

$$\frac{D\rho}{Dt} = K \nabla^2 \rho \,, \tag{3}$$

where the molecular diffusivity K is taken to be constant. Note that the diffusive term needs to be kept in the above equation in order to avoid the development of discontinuities in the computation of the density field. This holds true even if diffusive effects are very small, as in the case of liquids. In order to nondimensionalize the above set of equations, the channel height H is taken as the length scale, while the density ρ_1 of the heavier fluid serves as the characteristic density. Velocities are scaled by the buoyancy velocity $u_b = \sqrt{g'H}$, in which g' denotes the reduced gravity ([14]), which is related to the dimensional gravitational acceleration g by $g' = g\frac{\rho_1-\rho_2}{\rho_1} = g\,(1-\gamma)$, where the density ratio is given by $\gamma = \frac{\rho_2}{\rho_1} < 1$. A characteristic pressure p is given by $u_b^2\rho_1$. We thus arrive at the following set of governing dimensionless equations

$$\nabla \cdot \mathbf{u} = 0 \,, \tag{4}$$

$$\rho\frac{D\mathbf{u}}{Dt} = \frac{1}{1-\gamma}\rho\mathbf{e_g} - \nabla p + \frac{1}{Re}\nabla \cdot (2\rho\mathbf{S}) \,, \tag{5}$$

$$\frac{D\rho}{Dt} = \frac{1}{Pe}\nabla^2\rho \,. \tag{6}$$

For compositional gravity currents, the convective velocity employed to form the substantial derivative $\frac{D\rho}{Dt}$ is equal to the fluid velocity. For particle-driven currents, we employ the fluid velocity plus a superimposed, constant settling velocity in the direction of gravity ([11, 12]). If one were to make use of the Boussinesq assumption, equation (5) would instead simplify to

$$\frac{D\mathbf{u}}{Dt} = \rho\mathbf{e_g} - \nabla p + \frac{1}{Re}\nabla^2\mathbf{u} \,. \tag{7}$$

Note that we cannot obtain eq (7) from eq (5) just by substituting $\gamma = 1$, as the hydrostatic pressure field absorbed into the variable p varies between the two cases. e_g is given by the unit vector ($\sin\theta$,0,- $\cos\theta$). The three governing dimensionless parameters in equations (4) - (6) are the density ratio γ, the Reynolds number Re, and the Peclet number Pe, respectively, which are defined as $Re = \frac{u_b H}{\nu}$ and $Pe = \frac{u_b H}{K}$. They are related by the Schmidt number $Sc = \frac{\nu}{K}$, so that $Pe = Re \cdot Sc$. It represents the ratio of kinematic viscosity to molecular diffusivity. For most pairs of gases, the Schmidt number lies within the narrow range between 0.2 and 5. By means of test calculations we established that the influence of Sc variations in this range is quite small, so that in the simulations to be discussed below we employ $Sc = 1$ throughout. It is to be kept in mind, however, that for liquids such as salt water, $Sc \approx 700$.

For the purpose of numerical simulations, we recast equations (4) - (6) into the vorticity-streamfunction formulation. In this way, the incompressibility condition (4) is automatically satisfied throughout the flow field. Let ψ be the streamfunction and ω the vorticity in the spanwise direction. Then the relations $\omega = \frac{\partial v}{\partial x} - \frac{\partial u}{\partial z}$, $u = \frac{\partial \psi}{\partial y}$, and $v = -\frac{\partial \psi}{\partial z}$ hold, and we obtain

$$\nabla^2 \psi = -\omega \ , \tag{8}$$

$$\begin{aligned} \frac{D\omega}{Dt} &= \frac{1}{Re}\nabla^2\omega - \frac{\rho_x}{(1-\gamma)\rho} + \frac{\rho_z}{\rho}\frac{Du}{Dt} - \frac{\rho_x}{\rho}\frac{Dv}{Dt} \\ &+ \frac{1}{\rho Re}\left\{2\rho_x\nabla^2 v - 2\rho_z\nabla^2 u + 4\rho_{xz}v_z + (u_z+v_x)(\rho_{xx}-\rho_{zz})\right\} \ . \end{aligned} \tag{9}$$

If the dynamic viscosity μ is held constant instead of the kinematic viscosity ν, (9) takes the form

$$\frac{D\omega}{Dt} = \frac{1}{\rho Re}\nabla^2\omega - \frac{\rho_x}{(1-\gamma)\rho} + \frac{\rho_z}{\rho}\frac{Du}{Dt} - \frac{\rho_x}{\rho}\frac{Dv}{Dt} \ . \tag{10}$$

3 Computational approach

The simulations employ equidistant grids in the rectangular computational domain. The grid size used for our simulations varies from 1032X100 to 2064X200 depending on Re. Spectral Galerkin methods are used in representing the streamwise dependence of the streamfunction and the vorticity fields

$$\psi(x,z,t) = \sum_l \hat{\psi}_l(z,t) sin(l\alpha x) \ , \quad \omega(x,z,t) = \sum_l \hat{\omega}_l(z,t) sin(l\alpha x) \ , \tag{11}$$

where $|l| < N_1/2$ and $\alpha = 2\pi/L$. N_1 denotes the number of grid points in the streamwise direction. Vertical derivatives are approximated on the basis of the compact finite difference stencils described by [8]. As in the Boussinesq investigation of [6], derivatives of the density field are computed from compact finite differences in both directions. At interior points, sixth order spatially accurate stencils are used, with third and fourth order accurate ones employed at the boundaries. The flow field is advanced in time by means of the third order Runge-Kutta scheme described by [6]. The material derivatives of the velocity components appearing in the vorticity equation (9) are computed by first rewriting them in terms of the local time derivative plus the convective terms. The spatial derivatives appearing in the convective terms are then evaluated in the usual, high order way. The local time derivative is computed by backward extrapolation as follows

$$\left[\frac{\partial \mathbf{u}}{\partial t}\right]^n = \left(\mathbf{u}^n - \mathbf{u}^{n-1}\right)/\Delta t \ . \tag{12}$$

This approximation is consistently utilized during the successive Runge-Kutta sub-steps. Test calculations demonstrated that the low order approximation of this term did not influence the results in a measurable way.

The Poisson equation for the streamfunction (8) is solved once per time step in Fourier space according to

$$\left(\hat{\psi}_l^{m+1}\right)'' - (l\alpha)^2\hat{\psi}_l^{m+1} = -\hat{\omega}_l^{m+1} , \tag{13}$$

with the prime denoting differentiation with respect to z.

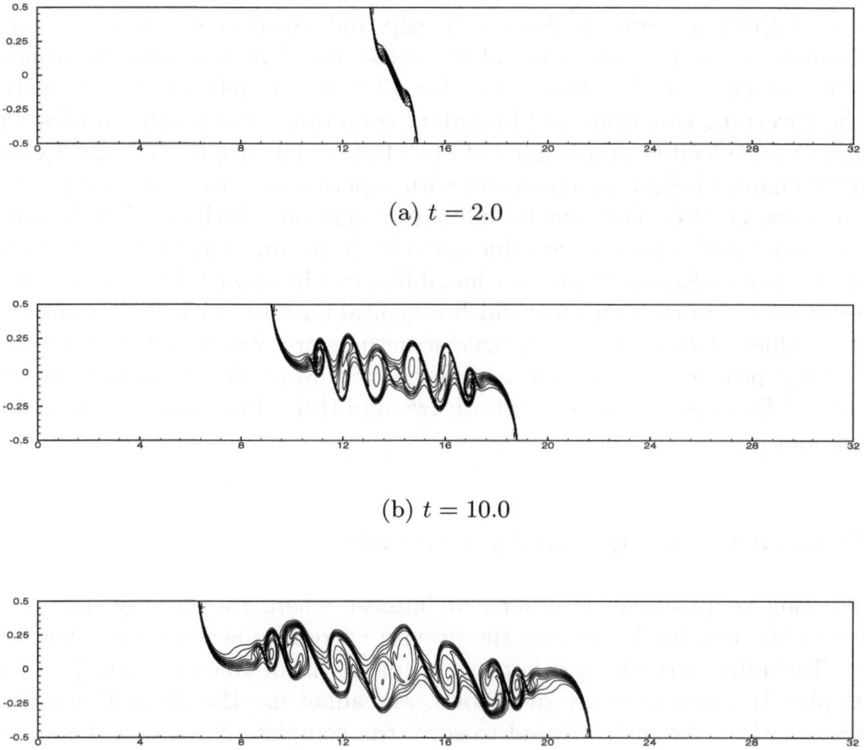

(a) $t = 2.0$

(b) $t = 10.0$

(c) $t = 16.0$

Fig. 2. Concentration contours from a Boussinesq simulation for $Re = 4,000$, at different times, showing the evolution of a typical Boussinesq gravity current.

4 Boussinesq gravity currents

The density difference between two fluids can range from very small to very large. In many geophysical situations such as sea water and fresh water the density difference is very small (within 5%). In cases of small density difference, density variations can be neglected in the inertia term, but retained in buoyancy term where they are multiplied with g. This approximation of the momentum equations is referred to as the Boussinesq approximation. It is accurate for fluids with densities within a few per cent of each other. The formation of a Boussinesq gravity current is shown in fig. 2 for $Re = 4,000$. Here, slip boundary conditions are employed both at the bottom and top walls. For a comparison between flows with slip and no-slip conditions, we refer the reader to [3]. The symmetry of the dense and light fronts for Boussinesq gravity currents can be clearly seen for all times. It reflects the symmetry of the governing equations and boundary conditions, and results in identical propagation velocities and heights of both fronts. The heights remain close to half the channel height, in agreement with experimental observations by other researchers, cf. [14]. The dimensionless propagation velocities of both fronts are near one half, which is the value given by [1] for an energy conserving current. A vigorous Kelvin-Helmholtz instability can be observed along the entire interface of the gravity current, which is typical for flows with $Re > 1,000$. At higher values of Re, more small-scale structures are observed, but the essential global properties remain nearly unchanged. More details on numerically simulated Boussinesq currents in both two and three dimensions are provided by [5] and [6].

5 Non-Boussinesq gravity currents

There can be practical situations of interest where the density difference between the two fluids forming the gravity current is larger than a few per cent. Turbidity currents and hot gas eruptions from volcanoes are just few examples. In order to study such flows, we cannot use the above Boussinesq approximation. We instead need to solve the complete Navier-Strokes equations involving variable density. For density ratios of $\gamma = 0.92$, 0.7, and 0.2, fig. 3 shows contour plots of simulations for $Re = 4,000$ at time $t = 10$. Fig. 3(a) still resembles the Boussinesq gravity current, although the symmetry is maintained only approximately. This loss of symmetry can be observed most clearly in the vortex pairing process. Fig. 3(b) clearly shows that already for $\gamma = 0.7$ the dense front moves significantly faster than the light front, and that it has traveled a longer distance than for the $\gamma = 0.92$ case. Also the height of the dense front is smaller than that of the light front. These observations can be seen even more clearly in fig. 3(c). We observe that the Kelvin-Helmholtz instability forms only along the dense front, while the light front is stable. All these observations are in agreement with experimental findings by [4] and [9].

Simplified models of non-Boussinesq gravity currents have been developed by patching an energy-conserving light front to a dissipative dense front via an expansion wave. The validity of this model is confirmed by the dissipation data provided by [3]. The results demonstrate that the dissipation in the light

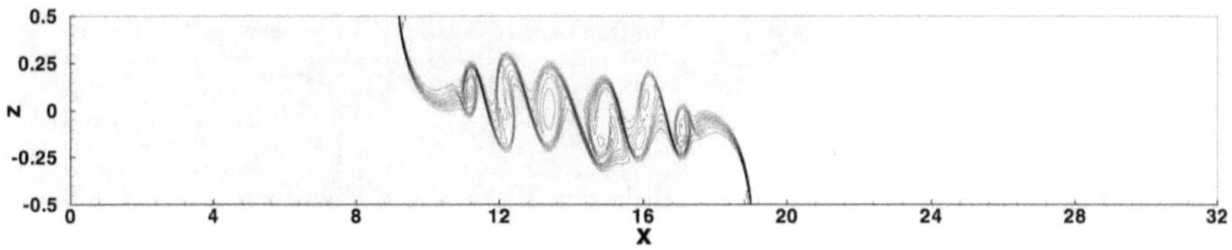

(a) $\gamma = 0.92$

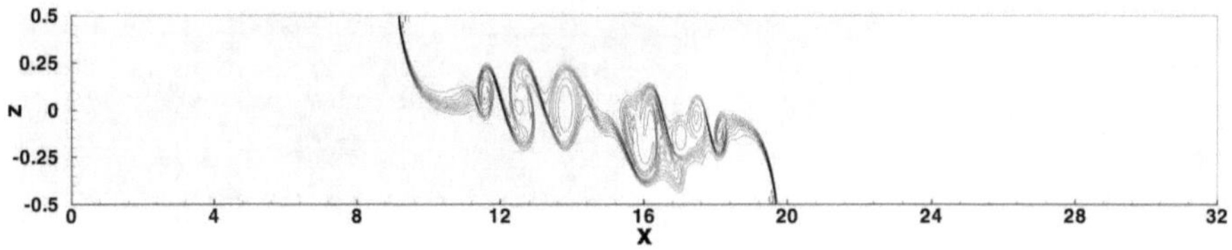

(b) $\gamma = 0.7$

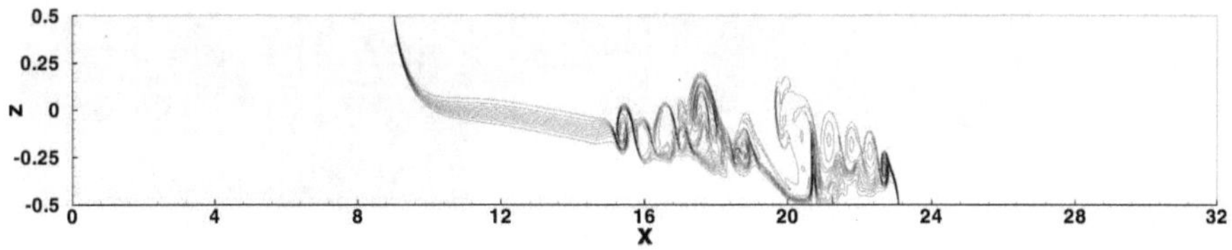

(c) $\gamma = 0.2$

Fig. 3. Concentration contours from a non-Boussinesq simulation for $Re = 4,000$ at time $t = 10$ for different density ratios γ, clearly showing the changes in the front speeds and the front heights.

front remains nearly constant as the density ratio changes from 1 to 0.2, while the dissipation level in the dense front increases. This is observed for all values of *Re* studied.

6 Gravity currents on slopes

Gravity currents in nature and industry frequently flow along slopes. Thus it is important to understand the effects of a sloping bottom on the global properties of gravity currents, cf. [2]. Gravity currents along inclines have been modeled in the past using wedge models, cf. [13]. The wedge shape is used

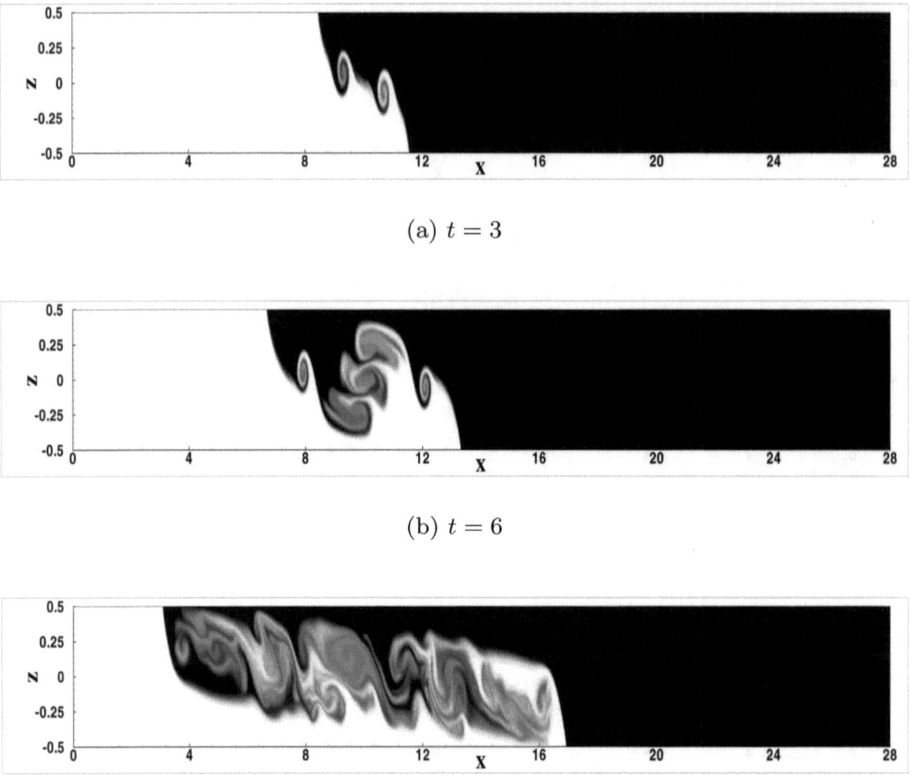

(a) $t = 3$

(b) $t = 6$

(c) $t = 12$

Fig. 4. Flow for $\gamma = 0.998$ at different times for $Re = 4,000$. The angle of the bottom is 30° to horizontal.

to model the long-term behavior of the gravity current, while it is proposed that in the short term the angle of the slope does not matter. Also it has been noted that friction along the bottom wall does not play an important role in this flow. Thus, in order to better understand the long and short-term behavior of gravity currents on a slope, we present highly resolved numerical simulation results in fig. 4. Full non-Boussinesq governing equations are used, although most results discussed have $\gamma \approx 1$.

Fig. 4(a) shows that at $t = 3$ the gravity currents are just beginning to form. There does not appear to be a strong effect of the slope angle during this early phase, as the shape of the currents is similar to the case of a horizontal flow. We observe slight differences in the front propagation velocity, however. Fig. 4(b) shows that soon thereafter much more vigorous mixing takes place along the interface, as compared to the horizontal case. Accelerating fluid layers form behind both fronts, similar to those studied experimentally by [15] in inclined channels. In fig. 4(c) we observe that these accelerated layers, whose dynamics are clearly affected by the angle of the slope, have reached the fronts and start to affect their propagation velocities.

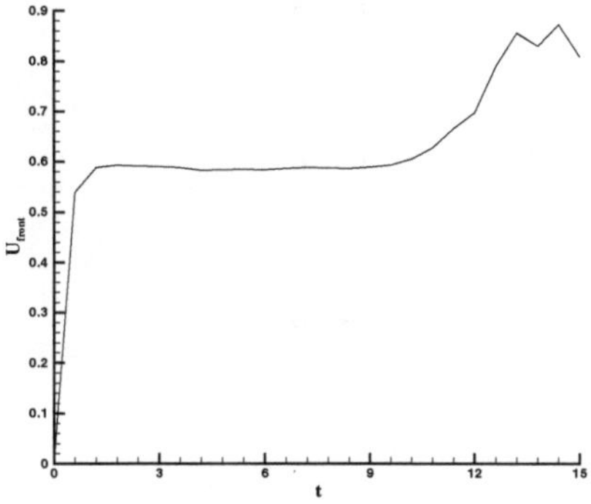

Fig. 5. The dense front velocity for the gravity current on a slope shown in fig. 4. It can be seen that close to $t = 12$ there is a jump in the velocity of the front. This suggests a transition between different phases.

Gravity currents on slopes thus exhibit two different phases: The first phase is characterized by early mixing and the formation of the accelerated fluid layers behind the fronts, while the second phase is dominated by more rapidly advancing gravity current fronts. The simulations give the time for the transition between the two phases as $t_\theta = \tan\theta^{-\frac{4}{9}}$. This is agreement with the results provided by [13] for the time after which the angle becomes important. Fig. 5 shows the jump in the front propagation velocity when the transition between the two phases occurs.

7 Reversing buoyancy gravity current

We consider dilute suspensions of particles with volume fractions well below one percent, so that particle-particle interactions can be neglected and coupling between particles and fluid motion is dominated by the transfer of momentum. Particles move with fluid velocity and settle out with Stokes settling velocity.

Reversing buoyancy gravity currents (Fig. 6) can be generated if the interstitial fluid density in particle-driven gravity currents is lower than the density of the ambient fluid, cf. [10]. If we set up such a current in the traditional lock-exchange configuration described above, its initial phase looks very similar to that of compositional gravity currents. However, as particles settle out of the

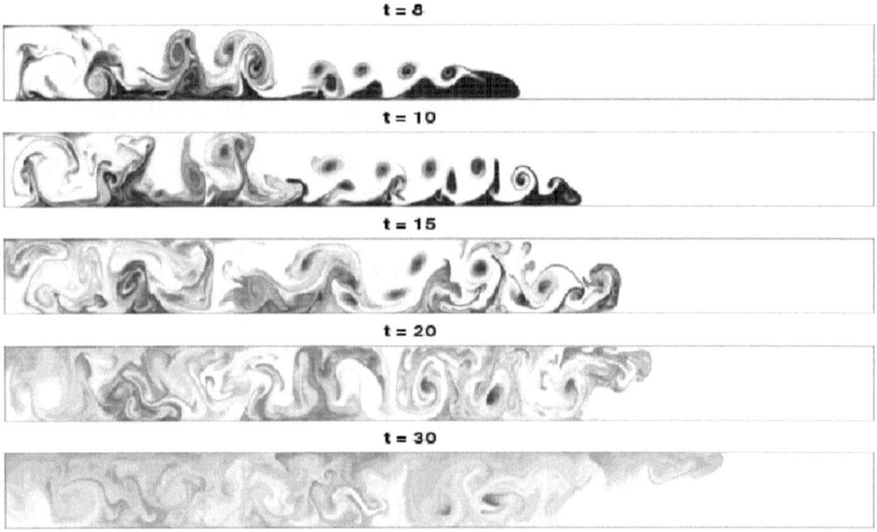

Fig. 6. The buoyancy reversal of a propagating gravity current is shown using density contour plots for times $t = 8, 10, 15, 20$ and 30. As the gravity current loses particles, its bulk density decays to a value lower than that of the ambient fluid, so that it lifts up from the bottom wall, and subsequently proceeds along the top wall.

lighter interstitial fluid, the bulk density of the gravity current decreases below the density of the ambient fluid, while its front velocity decays. Fig. 7 shows the tip location of such a reversing buoyancy gravity current. The slowdown in the tip of the front can be clearly seen at $t \approx 10$.

As particles settle, plumes of interstitial fluid rise from the gravity current behind the tip. The shape of these plumes is similar to the well-known mushroom associated with the classical Rayleigh-Taylor instability. As the forward velocity of the front propagating along the bottom wall decreases to zero, the rising interstitial fluid forms a gravity current front that propagates along the top wall of the channel. This gravity current eventually assumes a constant front speed. Significant mixing of both fluid and particles in the vertical direction takes place due to the rise of the interstitial fluid.

8 Summary

Highly resolved numerical simulations can provide substantial insight into the mechanisms governing gravity and turbidity currents. Here we have given a partial overview over the effects of non-Boussinesq density ratios, sloping channels and reversing buoyancy. Both flows driven by fluid density differences and flows driven by particle loading are considered. The simulation results are generally in good agreement with experimental observations regarding quantities

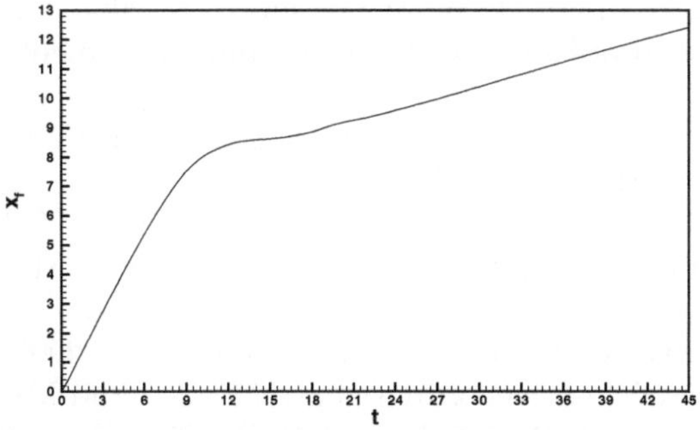

Fig. 7. Front location of the reversing buoyancy gravity current as function of time. The current initially propagates along the bottom as a dense front, with decreasing velocity until time $t \approx 10$. At this time the front comes to a halt. The current subsequently rises to the top wall of channel and propagates along it with constant velocity after $t \approx 10$.

such as the front heights and their propagation velocities. They provide access to quantities that sometimes are difficult to obtain from experiments, thereby allowing us to test simplified theoretical models proposed in the literature.

Acknowledgments

This work has been supported by the National Science Foundation.

References

[1] Benjamin, T. B. 1968 Gravity currents and related phenomena. J. Fluid Mech. 31, 209.

[2] Birman, V. K., Battandier, B. A., Meiburg, E. & Linden, P. F. 2006 Lock exchange flows in sloping channels. Submitted.

[3] Birman, V. K., Martin, J. E. & Meiburg, E. 2005 The non-Boussinesq lock-exchange problem. part 2. high resolution simulations. J. Fluid Mech. 537, 125.

[4] Grobelbauer, H. P., Fannelop, T. K. & Britter, R. E. 1993 The propagation of intrusion fronts of high density ratio. J. Fluid Mech. 250, 669.

[5] Haertel, C., Meiburg, E. & Necker, F. 1999 Vorticity dynamics during the start-up phase of gravity currents. Il Nuovo Cimento C 22, 823.

[6] Haertel, C., Meiburg, E. & Necker, F. 2000 Analysis and direct numerical simulation of the flow at a gravity-current head. part 1. flow topology and front speed for slip and no-slip boundaries. J. Fluid Mech. 418, 189.

[7] Huppert, H. E. 1986 The intrusion of fluid-mechanics into geology. J. Fluid Mech. 173, 557.

[8] Lele, S. K. 1992 Compact finite difference schemes with spectral-like resolution. J. Comput. Phys. 103, 16-42.

[9] Lowe, R. J., Rottman, J. W. & Linden, P. F. 2005 The non-boussinesq lock exchange problem. part 1. theory and experiments. J. Fluid Mech. 537, 101-124.
[10] Maxworthy, T. 1999 The dynamics of sedimenting surface gravity currents. J. Fluid Mech. 392, 27.
[11] Necker, F., Haertel, C., Kleiser, L. & Meiburg, E. 2002 Highresolution simulations of particle-driven gravity currents. Int.J.Multiphase Flow 28, 279-300.
[12] Necker, F., Haertel, C., Kleiser, L. & Meiburg, E. 2005 Mixing and dissipation in particle-driven gravity currents. J. Fluid Mech. 545, 339-372.
[13] Ross, A. N., Linden, P. F. & Dalziel, S. B. 2002 Three-dimensional gravity currents on slopes. J. Fluid Mech. 453, 239.
[14] Simpson, J. E. 1997 Gravity Currents in the Environment and the Laboratory. Cambridge University Press.
[15] Thorpe, S. A. 1968 A method of producing a shear flow in a stratified fluid. J. Fluid Mech. 32, 693.

The effect of stable stratification on fluid particle dispersion

M. van Aartrijk and H.J.H. Clercx

Fluid Dynamics Laboratory, Eindhoven University of Technology, The Netherlands
m.v.aartrijk@tue.nl, h.j.h.clercx@tue.nl

Summary. The dispersion of fluid particles in statistically stationary stably stratified turbulence is studied by means of direct numerical simulations. Due to anisotropy of the flow, horizontal and vertical dispersion show different behavior. Single-particle dispersion in horizontal direction is similar to that in isotropic turbulence for short times, but shows a long-time growth rate larger than $\propto t$. In vertical direction three successive regimes can be identified: a classical t^2-regime, a plateau which scales as N^{-2} and a diffusion limit $\propto t$. By forcing the flow and performing long-time simulations we were able to observe this last regime, which was predicted but not observed before in purely stratified forced turbulence. A model based on the assumed shape of the velocity autocorrelation function correctly predicts these three regimes. The vertical mean-squared separation of particle pairs shows two plateaus that are not present in isotropic turbulence. They can be linked with characteristics of the flow. Also here the diffusion limit is found.

1 Introduction

Dispersion of particles in stratified flows plays an important role in geophysical environments. These particles can be active or passive, like aerosols in the atmospheric boundary layer or micro-organisms in coastal areas and estuaries. We focus on the latter application where competition can exist between tidally- or wind-driven turbulent mixing on one hand, and density stratification on the other hand which in general suppresses dispersion, at least in the vertical direction. Including all biological and physical parameters in modeling particle dispersion is complicated and as a starting point in this work the effect of stratification on dispersion of fluid particles will be studied. Not much work has been carried out on particle dispersion in stratified turbulence. Both Kimura & Herring [1] and Liechtenstein et al. [2] studied dispersion in decaying stratified turbulence by means of direct numerical

Bernard J. Geurts et al. (eds), Particle Laden Flow: From Geophysical to Kolmogorov Scales, 151–163.

simulations (DNS) and Nicolleau & Vassilicos [3] used kinematic simulations (KS) to study dispersion in non-decaying stratified turbulence. Some examples of experimental studies on particle dispersion in stratified turbulence can be found in Pearson et al. [4]. We use here DNS of the Boussinesq equations and study statistically stationary homogeneous stratified turbulence. By applying forcing we have been able to follow particles for sufficiently long times in order to obtain long time series for calculating Lagrangian statistics.

The present study deals with the effect of stable stratification on turbulent flows, so the density of the fluid is decreasing with height. Under the influence of strong stable density stratification two sorts of motion occur simultaneously: propagating internal gravity waves and a non-propagating nonlinear component connected with quasi-horizontal motions [5]. We want to keep the stratified turbulence statistically stationary, thus we have to add energy in our simulations to account for energy losses due to viscous dissipation. There is still some discussion how these types of anisotropic flows should be forced, whether forcing should be applied to all three directions or just in the horizontal plane, and at which length scales the turbulence should be forced [6][7][8]. We have forced only the largest scales and equally in all three directions.

This study specifically aims to look at fluid particle dispersion. For homogeneous isotropic turbulence a lot of work has been reported on both single-particle and particle-pair dispersion, experimentally but mainly using numerical simulations [9][10]. Theoretically, the spreading of particles goes like t^2 for short times and is proportional to t in the long-time diffusion limit. When stable background stratification is present, dispersion in the vertical direction is suppressed. As a result a plateau is found for one-particle diffusion in the vertical direction, which scales proportional to N^{-2} with N the buoyancy frequency [1][3]. This plateau is reached around $t = \frac{2\pi}{N}$. For particle-pair dispersion the theoretical scaling laws for isotropic turbulence depend on the initial separation between the particles. When stratified turbulence is considered, vertical dispersion is suppressed like for single-particle statistics. Nicolleau & Vassilicos [3] retrieved a plateau again at $t \simeq \frac{2\pi}{N}$ and observed the beginning of a second plateau for large times.

Attempts to model single-particle dispersion in stratified flows started with the work of Csanady [11]. More recently, next to the above mentioned plateau, models by both Pearson et al. [4] and Das & Durbin [12] predict a linear growth of the mean-squared displacement for large times.

2 Numerical method

2.1 DNS of the Boussinesq equations

The motion of an incompressible fluid in a stably stratified environment is fully described by the Navier-Stokes equations in combination with an equation that imposes the divergence-free constraint and one for the density. After applying the Boussinesq approximation and using the hydrostatic balance $\frac{\partial \overline{p}}{\partial z} = -(\rho_0 + \overline{\rho})g$ they can be written as

$$\nabla \cdot \mathbf{u} = 0, \tag{1}$$

$$\frac{\partial \mathbf{u}}{\partial t} = -\mathbf{u} \cdot \nabla \mathbf{u} - \frac{1}{\rho_0} \nabla p' - \frac{\rho'}{\rho_0} g \hat{\mathbf{z}} + \mathbf{F}_{u,ext} + \nu \nabla^2 \mathbf{u}, \tag{2}$$

$$\frac{\partial \rho'}{\partial t} = w N^2 \frac{\rho_0}{g} - \mathbf{u} \cdot \nabla \rho' + F_{\rho,ext} + \kappa \nabla^2 \rho'. \tag{3}$$

Herein is $\mathbf{u} = (u, v, w)$ the velocity in the x-, y- and z-direction respectively, with $\hat{\mathbf{z}}$ pointing upwards. Furthermore, ρ is the density, p the pressure, g the gravitational acceleration, ν the molecular viscosity and κ the scalar diffusivity. $\mathbf{F}$ denotes external forces for $\mathbf{u}$ or ρ. The buoyancy frequency, or Brunt-Väisälä frequency, is defined as $N^2 = -\frac{g}{\rho_0}\frac{\partial \overline{\rho}}{\partial z}$ and the ratio of $\frac{\nu}{\kappa} = Sc$ is the Schmidt number. In this work we only present results for cases where $Sc = 1$. Fluctuating components are indicated with a prime and in the following an overbar is used for an averaged quantity. The density $\rho = \rho_0 + \overline{\rho}(z) + \rho'(x, y, z, t)$ is split in three components: a typical value (ρ_0) plus a time-independent background profile ($\overline{\rho}$) plus a fluctuating part (ρ'). The equations of motion are solved using a three-dimensional parallel pseudo-spectral DNS code (see [13] for details). DNS enables to solve the Navier-Stokes equations exactly at all relevant scales in the flow without making use of any model. Main drawback is that only relatively low resolutions can be used, thus flows with relatively low Reynolds numbers can be solved due to its high computational costs. Here we used a resolution of 128^3, to be able to track particles for long times. However, as a check some cases are studied at a higher resolution (256^3) and they gave similar results. Periodic boundaries are implemented in all three directions, allowing the use of a Fourier representation of the velocity and scalar field. Time-stepping of the linear viscous term is performed using exact integration, whereas the other terms are treated by a third-order Adams-Bashforth method. In a pre-computation a divergence-free homogeneous isotropic turbulent velocity field is created using forcing of the flow by injecting energy at the largest scales, equally in all three directions. A general description of the forcing scheme is

$$F^{n+1} = (1 - \alpha)F^n + A\phi \tag{4}$$

where $(1-\alpha)F^n$ (with $\alpha = [0,1]$) denotes a memory effect of the forcing, A is the forcing amplitude and ϕ consists of both a random value for the forcing amplitude, taken from a Gaussian with zero mean, and a random phase. F^{n+1} and F^n are the forces at forcing-times $n+1$ and n, respectively. The value of the force is updated every five time-steps. Forcing is only applied to the largest scales of the velocity field with wavenumber modes $0 < k \leq 2\sqrt{2}k_0$ (k_0 the smallest wavenumber), and serves to keep the total kinetic energy statistically stationary.
The initial conditions of the simulations of forced stratified turbulence are a velocity field of homogeneous isotropic turbulence together with a zero-valued scalar fluctuation field. At $t = 0$ a linear stable background stratification is switched on, which is kept constant throughout the simulation. Five different cases are studied with different background stratification. The forcing method mentioned above is used of which the amplitude A is adapted to keep the kinetic energy statistically stationary. Some properties of the five cases are given in table 1. The relative importance of stratification can be expressed

Table 1. Some properties of the five cases, with increasing stratification level.

case	N^2 (s^{-2})	Fr	T_L (s)	u_{rms} (ms^{-1})	$\sqrt{\overline{w'^2}}$ (ms^{-1})
N0	0	-	9.5	$1.1 \cdot 10^{-2}$	$1.04 \cdot 10^{-2}$
N1	$9.55 \cdot 10^{-3}$	0.57	12	$1.0 \cdot 10^{-2}$	$8.06 \cdot 10^{-3}$
N10	$9.55 \cdot 10^{-2}$	0.14	11.9	$1.06 \cdot 10^{-2}$	$4.36 \cdot 10^{-3}$
N100	0.955	0.06	9	$1.17 \cdot 10^{-2}$	$4.24 \cdot 10^{-3}$
N1000	9.55	0.03	9	$1.4 \cdot 10^{-2}$	$6.48 \cdot 10^{-3}$

by the Froude number, which is defined as $Fr = u_{rms}/NL_z$, and it gives the ratio of inertial forces to buoyancy forces. The root-mean-square velocity is given by $u_{rms}^2 = \frac{2}{3}E_{kin}$ and L_z is the vertical integral length-scale given by $\frac{\pi}{2}\hat{E}_{zz}(0)/\overline{w'^2}$, with $\hat{E}_{zz}(0)$ the spectral energy in the modes with wavenumber $k_z = 0$. A measure for the turbulence intensity is $Re_\lambda \approx 85$ for the initial field.

2.2 Particle tracking

A natural way to describe turbulent dispersion is the Lagrangian frame of reference, in which the observer is moving with the particle. Here we study fluid particles, which are infinitely small fluid elements that exactly follow the flow. Particle trajectories are derived from

$$\frac{d\mathbf{x}_p}{dt} = \mathbf{u}_p \tag{5}$$

with $\mathbf{x}_p$ the particle position and $\mathbf{u}_p = \mathbf{u}(\mathbf{x}_p)$ its velocity. The velocity at the particle position can be derived from knowledge of the Eulerian velocity field by use of interpolation. In our code we implemented cubic spline interpolation of the velocity field at the particle position. Next the particle trajectories are obtained by numerical integration of equation 5. Time integration is performed using the same third-order Adams-Bashforth technique as for the Eulerian velocity field.

Velocity and position time series of 16384 particles are collected. These particles are grouped in triangular pyramid structures, to be able to study both single-particle and particle-pair statistics. The initial position of one-quarter of the particles is uniformly spread over the computational domain. The other particles are initially located at a fixed separation (about $\frac{5}{2}\eta$, with η the Kolmogorov length scale) in x-, y- z-direction from the reference particles. For two values of the background stratification (cases N0 and N1000) we also studied the influence of the initial particle separation on particle-pair statistics. In these cases 81920 particles are tracked with initial separations of about $\frac{1}{6}, \frac{3}{4}, \frac{3}{2}, 6, 15$ (in units of η) in all three directions. Particles are released when the flow has reached a quasi-stationary state.

3 Results from forced stratified turbulence simulations

3.1 Flow structure and density profile

An important question in this work is to see whether we are able to maintain a state of statistically stationary stratified turbulence by applying artificial forcing. If energy transfer to smaller scales is modified or an inverse cascade is present in the flow, forcing of the large scales could lead to accumulation of energy at these scales and eventually to a collapse of the simulation. We do find that this type of forcing brings us a quasi-stationary state. Checking stationarity is done by looking both at kinetic energy in all three directions and at the velocity derivative skewness. Moreover, the value of the latter quantity for case N0 is consistent with values found in studies of homogeneous isotropic turbulence. Furthermore, (an)isotropy of the forcing is tested, because the resulting flow is anisotropic for cases with $N \neq 0$. Applying equal forcing in all three directions or purely forcing the horizontal wavenumber modes leads to different flow configurations with the main difference the ratio between the vertical and horizontal kinetic energy. Dispersion statistics, however, are very similar despite differences in the large-scale flow structures.

To get an idea of the structures in the flow, in figure 1 we have shown vertical cross-sections of the density profile for two different values of the background stratification. The plot on the right in figure 1 shows an almost linear density profile. The stable stratification is so strong that turbulent fluctuations are

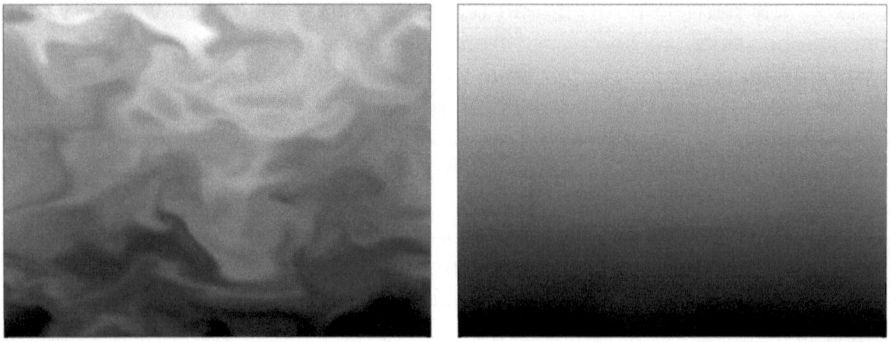

Fig. 1. Vertical cross-sections of the density profile. For case N1 (left) mixing is clearly visible, whereas case N100 (right) only shows slight fluctuations upon the linear background profile.

hardly visible. For moderate stratification (left) much more overturning and mixing between different layers can be seen. The structure of the flow becomes also clear from the isovorticity plots in figure 2. The strong stratification shows the 'well-known' pancake-like structures, whereas the weaker stratification shows traces of small-scale structures like in 3D isotropic turbulence. Wave-

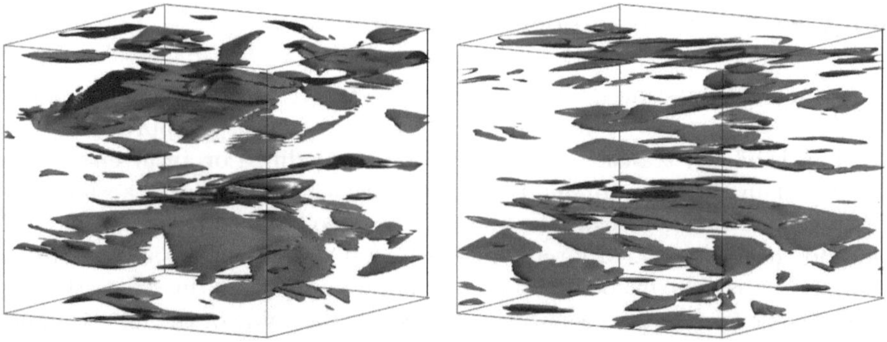

Fig. 2. Isovorticity surfaces of cases N10 (left) and N100 (right) at $1.7\omega_{rms}$. With increasing stratification the structures become more horizontal and 'pancake'-like and less small-scale structures are visible.

like motion is present in both the velocity and the scalar field. It can be seen in most quantities (Eulerian and Lagrangian, mainly in the vertical) of which the evolution in time is studied, like kinetic and potential energy, length-scales, particle dispersion and autocorrelation functions. The frequency of these waves can clearly be linked to the buoyancy frequency and not to the frequency of the forcing.

3.2 Single-particle statistics

The dispersion of particles is given by Taylor's equation

$$\overline{(X(t) - X(0))^2} = 2\overline{u_{rms}^2} \int_0^t (t-\tau) R_L(\tau) d\tau \tag{6}$$

under the assumption that the Lagrangian and Eulerian rms-velocity are the same and that the flow is homogeneous and stationary. $R_L(\tau)$ is the Lagrangian velocity autocorrelation function which is only a function of the time separation $\tau = t - t'$:

$$R_L(\tau) = \frac{\overline{u_p'(t')u_p'(t)}}{\overline{u_{rms}^2}}. \tag{7}$$

Using some known properties of the autocorrelation, the following scaling can be derived for dispersion in homogeneous isotropic turbulence

$$\overline{(X(t) - X(0))^2} \approx \overline{u_{rms}^2} t^2 \qquad t \to 0, \tag{8}$$

$$\overline{(X(t) - X(0))^2} \approx 2\overline{u_{rms}^2} T_L t \qquad t \to \infty. \tag{9}$$

For stratified turbulence, several authors (see for example [3], [4] and [14]) provided evidence of a plateau for vertical dispersion for $t \gtrsim \frac{2\pi}{N}$, though based on different arguments. This plateau scales as

$$\overline{(Z(t) - Z(0))^2} \approx \frac{\overline{w'^2}}{N^2}. \tag{10}$$

Furthermore, the choice of the rms-velocity in equation 10 differs, some take the overall value as defined above [3], while others use only the component in vertical direction [4]. We use the vertical component given by $\overline{w'^2} = 2E_z$ with E_z the kinetic energy in vertical direction.

Single-particle dispersion in horizontal and vertical direction is shown in figure 3 for five different values of the density stratification. For isotropic turbulence (case N0) we do retrieve the classical regimes of equations 8 and 9, as well as for horizontal dispersion in relative weak stratification. For strong stratification the plateau with its accompanying scaling is found. When the horizontal and vertical axes of the right plot in figure 3 are rescaled to $t\frac{N}{2\pi}$ and $\overline{(Z(t) - Z(0))^2} N^2/\overline{w'^2}$, respectively, the graphs nicely collapse. For times up to about $T_E = L_x/u_{rms}$ ($O(50(s))$) these plots resemble the results of [1] and [3]. For longer times a new regime can be identified, which becomes available by tracking the particles for sufficiently long times (possible due to forcing of the flow). Dispersion in vertical direction starts to increase again and is proportional to t, which is a clear indication of a diffusion process.

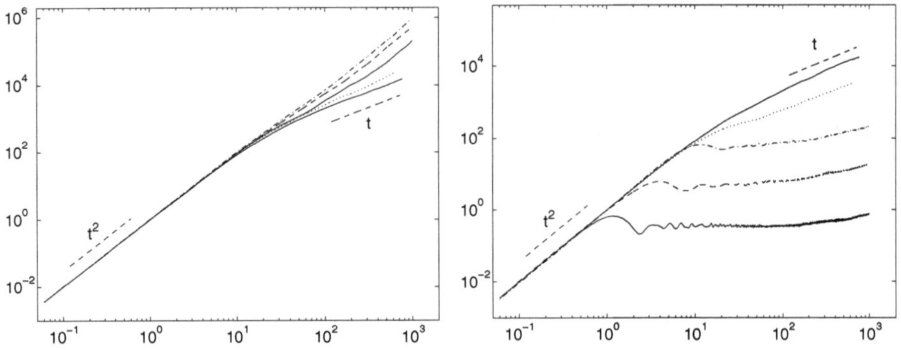

Fig. 3. Horizontal (left) and vertical (right) single-particle dispersion as function of time t for five different cases N0: ——, N1: · · · · ·, N10: — · — · —, N100: - - - -, N1000: —— . Horizontal dispersion is averaged over x- and y-direction, $\frac{1}{2}\left(\overline{(X(t)-X(0))^2}+\overline{(Y(t)-Y(0))^2}\right)$. Vertical dispersion is calculated as $\overline{(Z(t)-Z(0))^2}$. The plotted results are scaled with $\frac{1}{2}(\overline{u'^2}+\overline{v'^2})$ and $\overline{w'^2}$ respectively. The plateau starts around $t=\frac{2\pi}{N}$.

This diffusion of fluid particles away from the original equilibrium position is caused by molecular diffusion of the active scalar (density), what we checked by changing the Schmidt number. This regime was already predicted by the models of Pearson et al. [4] and Das & Durbin [12]. However, to our knowledge this is the first time that it has been observed in numerical studies.

What seems to happen with vertical dispersion is the following. In the early stages particles move away from their initial position with their local velocity. This results in a t^2-regime. Next their vertical movement is trapped within wave-like motions. The amplitude of this wave is determined by the scale of the 'pancake'-like vortex structures. The wave-like motion is evident from the continuous increase and decrease around an averaged plateau level. In case the particle positions are initialized as a point or line source instead of a homogeneous random distribution, dispersion denotes the growth of the particle cloud. A separate simulation in which the particles started in a thin horizontal plane indeed showed that the vertical width of the cloud was oscillating. For long times finally, particles completely forget their initial positions and the resulting behavior is thus diffusive.

Horizontal dispersion in strongly stratified turbulence shows an increase for long times with a much larger slope than linear, viz. $\propto t^2 - t^3$. The cause of this effect is not clear yet, further study is necessary to explain this scaling behavior. It might be related to the same effects that cause a t^2-regime in rotating turbulence [2].

3.3 Particle-pair statistics

In view of future applications, in which cluster formation and aggregation (relevant for dispersion of micro-organisms) might play a role, not only single-particle statistics but also particle-pair statistics are of importance. For homogeneous isotropic turbulence a couple of regimes can be identified [16]. For short times, the mean squared separation between two fluid elements grows either exponentially or like t^2. When the inter-particle separation distance falls in the inertial range this growth behaves as t^3 according to Richardson. For long times a diffusion limit exists similar to single-particle dispersion. Which regimes are passed through depends on the initial separation of the particles, only the final t-regime is universal since at long times particles become uncorrelated, independent of their initial separation Δ_0. The t^3 regime is derived theoretically for high Reynolds number flows, with a clear inertial range.
The behavior for homogeneous isotropic turbulence (case N0) can be seen in figure 4 (left) for an initial separation of about $\frac{3}{2}\eta$. A t^2-regime is expected initially and indeed visible. Next follows a region which is close to t^3, though because of our relative low Reynolds number this is most likely a necessary transition from the t^2-regime to the t-regime of which the levels are determined by the flow. Finally the long-term diffusion limit is clearly visible. Figure

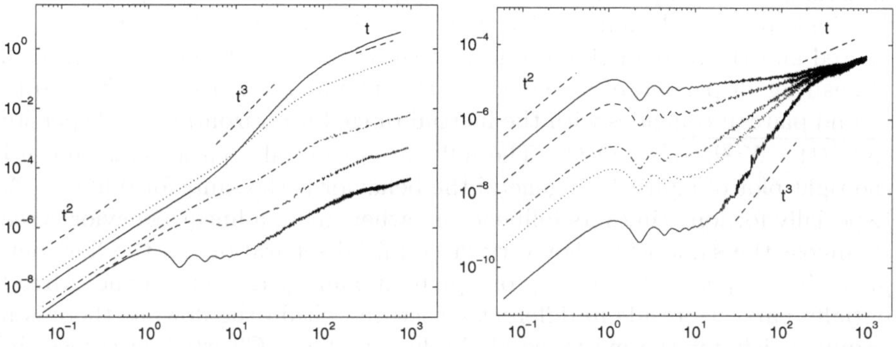

Fig. 4. Second order moment of the vertical separation between two particles $\overline{(Z^{(1)}(t) - Z^{(2)}(t))^2}$ as a function of time t. The left plot shows results for different levels of the background stratification, cases N0: ——, N1: $\cdots\cdots$, N10: $-\cdot-\cdot-$, N100: - - - -, N1000: —— . The initial separations are in horizontal direction, $\frac{3}{2}\eta$ for cases N0 and N1000 and $\frac{5}{2}\eta$ for cases N1, N10 and N100. The first plateau starts around $t = \frac{2\pi}{N}$ whereas this value for the second plateau is $t \sim O(T_E) = \frac{L_x}{u_{rms}}$ and thus not explicitly dependent on the stratification. On the right case N1000 is shown for five different initial separations; $\frac{1}{6}\eta$: ——, $\frac{3}{4}\eta$: $\cdots\cdots$, $\frac{3}{2}\eta$: $-\cdot-\cdot-$, 6η: - - - -, 15η: —— .

4 shows $\overline{(Z^{(1)}(t) - Z^{(2)}(t))^2}$ as a function of time, where superscripts 1 and 2

denote the two particles of a pair. These results are derived from pairs with an initial separation in the horizontal plane. Results of pairs with an initial separation in vertical direction are similar, except for cases where Δ_0 is larger than the levels of the first plateaus. The results of horizontal dispersion are not plotted here. For all values of N they behave similar to the non-stratified case N0 in figure 4. Only for long times and for strong stratification different behavior is seen, which is similar to the long-term behavior for single-particles (see figure 3). The left plot of figure 4 shows that vertical separation starts with the classical t^2-regime. Next a plateau is found, starting around $t = \frac{2\pi}{N}$. At later times an increase to a second plateau (around $t = 400(s)$) can be seen, at least for the strongest stratifications. The time at which the increase to this second plateau starts is of the order of the largest time-scales in the flow (T_E). The final regime is again the linear diffusion limit. Nicolleau & Vassilicos already predicted the existence of the second plateau, although their kinematic simulation was not carried out long enough to see more than just the beginning of this plateau. The most probable explanation of the course of the separation is the following. Initially particles start separating, but they are trapped within a vertical flow structure. This results in a first plateau. Trapping occurs only in vertical direction, they can still freely disperse in the horizontal direction. After some time the horizontal separation is large enough for them to reside in different structures, where again they are able to disperse also in vertical direction. A second plateau is reached that can be linked to the plateau found for single-particle dispersion. By that time the two particles become uncorrelated and the level of the second plateau is therefore twice the level of the plateau for dispersion of a single particle. The increase from the first to the second plateau coincides with the inertial range for horizontal pair dispersion, $\overline{(X^{(1)}(t) - X^{(2)}(t))^2} \propto O(t^3)$. The influence of initial separation is shown in the right plot of figure 4. In general the behavior is the same for different Δ_0. Especially for long times its influence is rather small, a linear behavior is seen at almost the same level. For very large initial separations in the horizontal plane, it is expected to see only one plateau, namely the second one. Already initially particles reside in different structures within the flow, so there is no argument left for the existence of the first plateau. Clearly, our largest initial separation (15η) is smaller than the horizontal size of the flow structures, because the first plateau is still present.

3.4 Modeling

From the particle velocities the Lagrangian autocorrelation as defined in equation 7 can be calculated. The results are shown in figure 5 (left). For case N0 the autocorrelation function can be approximated with a decaying exponential function (except for very short times), which is a general result for homogeneous isotropic turbulence (see for example [15]).

When a background stratification is present, the autocorrelation function in vertical direction can be approximated by

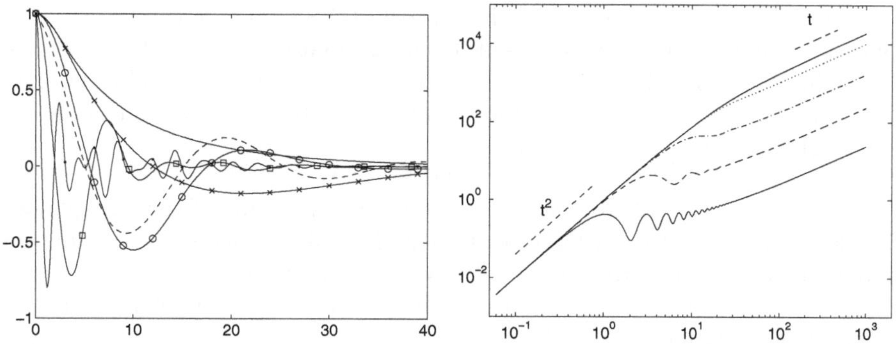

Fig. 5. Left: Lagrangian velocity autocorrelation function ($R_{zz}(t)$) as a function of time t for cases N0: —, N1: ×, N10: ○, N100: □, N1000: •. The dashed line shows equation 11 with values N = N10 and $T_L = 11.9$. Right: single-particle dispersion calculated with use of equation 13, for cases N0: ——, N1: ·····, N10: −·−·−, N100: - - - -, N1000: —— . The results are scaled with $\overline{w'^2}$ for comparison with figure 3 and values for T_L are derived from fitting equation 11 with the autocorrelations in the left figure.

$$R_{zz}(t) \sim \exp(-t/T_L)\cos(Nt), \tag{11}$$

where a wave-like motion is added to the autocorrelation function for pure isotropic turbulence. For isotropic turbulence the Lagrangian time-scale T_L can be derived from

$$T_L = \int_0^\infty R_{zz}(\tau)d\tau. \tag{12}$$

However, with increasing N the area under the R_{zz} graph goes to zero. Here, we determine a value for T_L from fitting equation 11 with the results of figure 5 (left).

Combining equation 11 with the general expression for single particle dispersion (eq. 6) gives a model for the vertical dispersion of particles:

$$\begin{aligned}\overline{Z^2}(t) = \frac{2\overline{w'^2}T_L^2}{1+N^2T_L^2}\Bigg[&\frac{1-N^2T_L^2}{1+N^2T_L^2}\cos(Nt)\exp(-t/T_L)\\ &-\frac{2NT_L}{1+N^2T_L^2}\sin(Nt)\exp(-t/T_L)+\frac{t}{T_L}-\frac{1-N^2T_L^2}{1+N^2T_L^2}\Bigg]\end{aligned} \tag{13}$$

The vertical dispersion according to this model is shown in figure 5 (right) which is plotted using the values of $\overline{w'^2}$, N and T_L from the DNS. The three regimes (t^2, plateau and t) are clearly visible and they start around similar times as in the DNS results in figure 3. The linear diffusion limit is easily identified as the second last term of equation 13. The other regimes are more difficult to trace in that equation, but they can be found under the assumption that $N^2T_L^2 > NT_L > 1$ which holds for strong stratification. The main

difference between the model results and the DNS results is the length of the plateau, which most likely is caused by the estimated value for T_L. Increasing the value of T_L results in a longer plateau. Moreover, the amplitude of the autocorrelation function decays more quickly for the DNS results than for the artificial function of equation 11.
The model described above is only valid for stationary flows. However, if you make $\overline{w'^2}$ a function of time $\overline{w'^2}(t) \sim t^{-p}$ with p the slope of the energy decay for decaying stratified turbulence, the model only displays a t^2-regime and a plateau for vertical dispersion. There seems to be a connection with the decaying simulations of, for example, Kimura & Herring [1], who do not see a final linear regime.

4 Concluding remarks

Stratification suppresses vertical motion in forced turbulent flows, as shown in this work. For both single-particle and particle-pair dispersion in vertical direction this leads to the occurrence of plateaus in dispersion plots. We found a linear diffusion regime for long times, which was not actually seen before. Both the suppression of vertical dispersion and final diffusion might be important for practical purposes. The first might enhance clustering and aggregation for interacting particles, because particles remain close together for long times, whereas the second makes it possible for particles to reach everywhere in the flow for long enough times. We have derived a model for single-particle vertical dispersion in stratified turbulence based on an Ansatz for the velocity autocorrelation function. Although a fit in the autocorrelation function is far from perfect for strong stratification, the model is able to describe the three different dispersion regimes rather well. Finding a similar model for particle-pair dispersion is a next topic to be addressed.

Acknowledgments

This program is funded by the Netherlands Organization for Scientific Research (NWO) and Technology Foundation (STW) under the Innovational Research Incentives Scheme grant ESF.6239. The use of supercomputer facilities for this work is sponsored by NCF (National Computing Facilities). We would like to thank Kraig Winters for kindly providing the (original version) of the numerical code.

References

[1] Kimura Y, Herring JR (1996) J Fluid Mech 328:253–269

[2] Liechtenstein L, Godeferd FS, Cambon C (2005) J Turb 6(24):1–18
[3] Nicolleau F, Vassilicos JC (2000) J Fluid Mech 410:123–146
[4] Pearson HJ, Puttock JS, Hunt JCR (1983) J Fluid Mech 129:219–249
[5] Riley JJ, Lelong MP (2000) Annu Rev Fluid Mech 32:613–657
[6] Herring JR, Métais O (1989) J Fluid Mech 202:97–115
[7] Waite ML, Bartello P (2004) J Fluid Mech 517:281–308
[8] Lindborg E (2006) J Fluid Mech 550:207–242
[9] Mazzitelli IM, Lohse D (2004) New J Phys 6:203
[10] Bourgoin M, Ouellette NT, Xu H, Berg J, Bodenschatz E (2006) Science 311:835–838
[11] Csanady GT (1964) J Atm Sci 21:439–447
[12] Das SK, Durbin PA (2005) Phys Fluids 17:025109
[13] Winters KB, MacKinnon JA, and Mills B (2004) J Atm Oc Tech 21:69
[14] Kaneda Y, Ishida T (2000) J Fluid Mech 402:311–327
[15] Yeung PK (2002) Annu Rev Fluid Mech 34:115–142
[16] Sawford B (2001) Annu Rev Fluid Mech 33:289–317

DNS of particle-laden flow over a backward facing step at a moderate Reynolds number

A. Kubik and L. Kleiser

Institute of Fluid Dynamics, ETH Zürich, Switzerland
`kubik@ifd.mavt.ethz.ch`

Summary. The present study investigates turbulence modification by particles in a backward-facing step flow with fully developed channel flow at the inlet. This flow configuration provides a range of flow regimes, such as wall turbulence, free shear layer and separation, in which to compare turbulence modification. Fluid-phase velocities in the presence of different mass loadings of particles with a Stokes number of $St = 3.0$ are studied. Local enhancement and attenuation of the streamwise component of the fluid turbulence of up to 27% is observed in the channel extension region for a mass loading of $\phi = 0.2$. The amount of modification decreases with decreasing mass loading. No modification of the turbulence is found in the separated shear layer or in the re-development region behind the re-attachment, although there were significant particle loadings in these regions.

1 Background

The use of Direct Numerical Simulations (DNS) to predict particle-laden flows is appealing as it promises to provide accurate results and a detailed insight into flow and particle characteristics that are not always, or not easily, accessible to experimental investigations. In the present study, a vertical turbulent flow over a backward-facing step (with gravity pointing in the mean flow direction) at moderate Reynolds number $Re_\tau \approx 210$ (based on friction velocity u_τ and inflow channel half-width h) is investigated by means of DNS. The main focus is directed towards particle statistics and turbulence modification.

Fessler and Eaton [9] reported the results of experiments on particle-laden flows in a backward-facing step configuration. Like in our simulations, in this work the bulk flow rate was fixed. The corresponding Reynolds number was approximately $Re_\tau \approx 644$. The experiments were performed with glass and copper particles of different diameters in downward air flows. The particles used in our simulations were chosen to match those in experiments and our previous studies. (Due to limited space results for only one particle species are presented here.)

Bernard J. Geurts et al. (eds), Particle Laden Flow: From Geophysical to Kolmogorov Scales, 165–177.

Our previous studies, e.g. [13], concentrated on particle-laden flows in a vertical channel down-flow at the above-mentioned Reynolds number. It was confirmed that particle feedback causes the turbulence intensities to become more non-isotropic as the particle loading is increased. The particles tended to increase the characteristic length scales of the fluctuations in the streamwise velocity, which reduces the transfer of energy between the streamwise and the transverse velocity components.

2 Methodology, numerical approach and parameters

The Eulerian-Lagrangian approach is adopted for the calculations of the fluid flow and the particle trajectories. The two phases are coupled, as the fluid phase exerts forces on the particles and experiences a feedback force from the dispersed phase.

Fluid phase

The fluid phase is described by the 3D time-dependent modified Navier-Stokes equations in which the feedback force on the fluid is added as an effective body force. Additionally, the incompressibility constraint must be satisfied.

$$\frac{\mathrm{D}\underline{u}^f}{\mathrm{D}t} = -\nabla p + \frac{1}{Re}\Delta\underline{u}^f + \underline{f}^g + \underline{f}^r \quad , \quad \nabla \cdot \underline{u}^f = 0 \tag{1}$$

The symbols $\underline{u}^f$, t, p, Re, $\underline{f}^g$ and $\underline{f}^r$ denote the fluid velocity, time, pressure, Reynolds number, gravity force and feedback force per unit mass, respectively. To compute the feedback force the sum of the drag and lift forces acting on a particle is redistributed to the nearest grid points, summed up with feedback forces from other particles and divided by the mass of fluid contained in the volume surrounding the grid point. [11]

The geometry and dimensions of the backward-facing step domain are shown in fig. 1. The Reynolds number of the inlet channel flow in the present simulation is chosen to be $Re_\tau \approx 210$, as in our previous work [12],[13]. This is a moderate number, still manageable in terms of computational costs but securely located in the range of flows considered turbulent. Based on the bulk velocity of the fluid, the Reynolds number is around 3333. This translates to a Reynolds number of the back-step, based on bulk velocity and step height H of $Re_H \approx 6666$.

The equations are solved using a spectral–spectral-element Fourier–Legendre code [20] with no-slip boundary conditions on the walls and periodic boundary conditions in the spanwise direction. Fully developed turbulent channel flow from a separate calculation is applied at the inlet, whereas a convective boundary condition is imposed at the outlet.

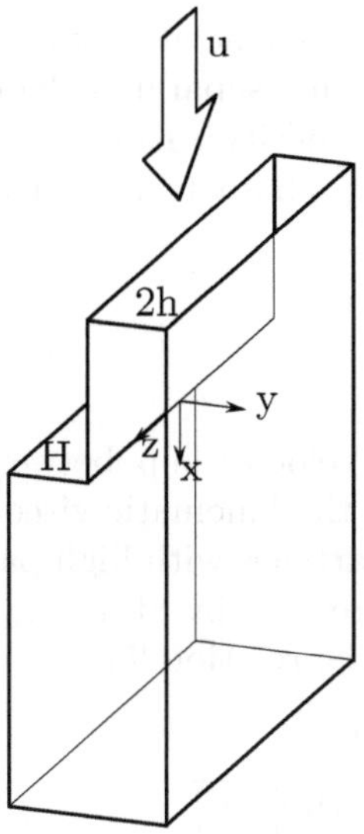

Inlet channel	
Channel half-width	h
Channel span	$3.2h$
Channel length	$5H$
Backward-facing step domain	
Step height	$H = 2h$
Expansion ratio	$H : 2h = 1$
Domain behind step	$52H$

Fig. 1. The geometry of the backward-facing step domain.

Dispersed phase

The particles are tracked individually. Their trajectories are calculated simultaneously in time with the fluid phase equations by integrating the equation of motion for each particle. This is done by solving the equations for the particle velocity and position vectors as given by Maxey and Riley [15]. Several modifications were necessary: A lift force was supplemented ad hoc as described in [16],[18]. Empirical and analytical corrections for the drag and lift were necessary to accommodate moderate Reynolds numbers [5],[17] and the proximity of walls [1],[8],[19]. Only the effects of drag, gravity and lift are taken into account [11]. The equations for the particle velocity and position are thus

$$\frac{\mathrm{d}\underline{u}^{p(k)}}{\mathrm{d}t} = \underline{F}_{drag} + \underline{F}_g + \underline{F}_{lift} \quad , \quad \frac{\mathrm{d}\underline{x}^{p(k)}}{\mathrm{d}t} = \underline{u}^{p(k)} \tag{2}$$

where $\underline{u}^{p(k)}$ denotes the velocity and $\underline{x}^{p(k)}$ the position of the particle k. $\underline{F}_{drag}$, $\underline{F}_g$ and $\underline{F}_{lift}$ represent the drag, gravity and lift force per particle mass, respectively. Equations (2) were discretized in time and solved using a third-order Adams-Bashforth scheme. Particle-wall collisions are modeled taking into account the elasticity of the impact and particle deposition for low-velocity particles in regions of low shear [12],[11]. Particle-particle collisions are omitted in this study and the parameter range for the calculations is restricted such as to keep this assumption valid. At the initial time the backward-facing step computational domain contains no particles. They are introduced via the inlet channel flow, in which they have reached a statistically s spatial distribution (starting with a random field) in a separate calculation.

Monodisperse particles with a particle-to-fluid density ratio of $\rho_p/\rho_f = 7458$ are used. The particle Reynolds number Re_p characterizing the flow around the particle is defined as

$$Re_p = \frac{d_p \ |u^f - u^p|}{\nu} \tag{3}$$

where d_p is the particle diameter, $|u^f - u^p|$ the velocity slip between the particle and the fluid at the particle position and ν the kinematic viscosity of the fluid. The particles response time τ_p for small particles with high particle-to-fluid density ratios can be derived from the expression by Stokes $\tau_{p,Stokes}$ corrected for non-negligible Reynolds numbers by the relation [5]

$$\tau_p = \frac{\tau_{p,Stokes}}{[1 + 0.15Re_p^{0.687}]} \approx \frac{\rho_p d_p^2}{18\mu[1 + 0.15Re_p^{0.687}]} \tag{4}$$

where μ the fluid dynamic viscosity. The Stokes number is the ratio of the particle response time to a representative time scale of the flow, $St = \tau_p/\tau_f$. There are several fluid time scales appropriate for analyzing the backward-facing step flow, such as the approximate large-eddy passing frequency in the separated shear layer $5H/u_{cl}$ [9] or the local turbulence time-scale k/ϵ. (u_{cl} is the fluid velocity at the centerline.) In the present study the nominal Stokes number was chosen to be $St = 3.66$, based on $\tau_{p,Stokes}$ and the large-eddy time scale. This corresponds to a Stokes number of $St = 3.0$ based on τ_p and turbulence time-scale k/ϵ at the inlet channel centerline.

3 Results and Discussion

Figure 2 shows a contour plot of the mean fluid velocity with superimposed streamlines. The flow topology includes the recirculation region behind the step, an enlarged boundary layer at the step-opposite wall (due to the pressure gradient), the re-attachment point at $x/H = 7.4$, a deceleration of the flow behind the step, and a re-development toward an symmetric channel flow at approximately $x/H = 20$. It should be noted that the mean velocity profile is unchanged by the presence of particles, as constant fluid mass flow was enforced in the simulation. (Additionally, relatively low mass loadings of the particles combined with the high particle-to-fluid density ratio result in very low volume loadings of the particles.)

Figure 3 shows a contour plot of the mean particle number density c divided by the particle number density averaged over the inlet c. Very few particles are found in the recirculation region directly behind the step. After the re-attachment point an increasing number of particles can be found below

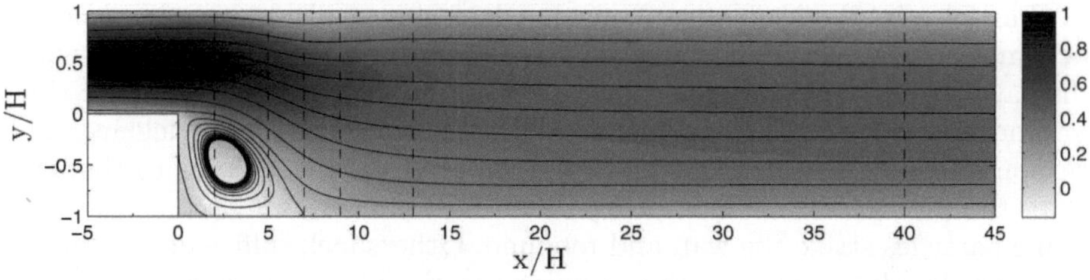

Fig. 2. Contour plot of the mean fluid velocity u/u_{cl} with superimposed streamlines. (Note the strongly enlarged vertical scale.)

$y/H = 0$ until at $x/H \approx 13$ the number density across the section is becoming more uniform. These dispersion field characteristics were also found in the experiments [9]. The lack of particles in the recirculation region is not surprising. Previous studies [10] have found that particles will be dispersed into the recirculation region only if their large-eddy Stokes numbers are less than one. In this study the Stokes numbers of the particles based on the large-eddy time scale, $\tau_f = 5H/u_{cl}$, are significantly larger than unity ($St = 3.66$). Furthermore, heavy particles ($\rho_p/\rho_f \gg 1$) like those in this simulation tend to migrate out of eddies and toward the fringes [6]. Also, particles whose response time is larger than the relevant fluid time scale (i.e. $St > 1$, as in the present case) do not respond quickly to the vortical structures and are ejected from these structures soon after being injected [7]. Another parameter which is important in this vertical downward flow is the ratio of the particle's terminal settling velocity to the maximum velocity of reverse flow, u_T/u_{rev}. The terminal settling velocity for a particle can be calculated from basic principles [5], and is in the range of $0.12u_{cl}$ for the particles considered here. The maximum reverse flow velocity found behind the step is approximately $0.2u_{cl}$ (see fig. 4). The resulting ratio is $u_T/u_{rev} = 0.6$. This is large enough that particles would experience difficulty moving upstream (vertically upward) in the recirculation region.

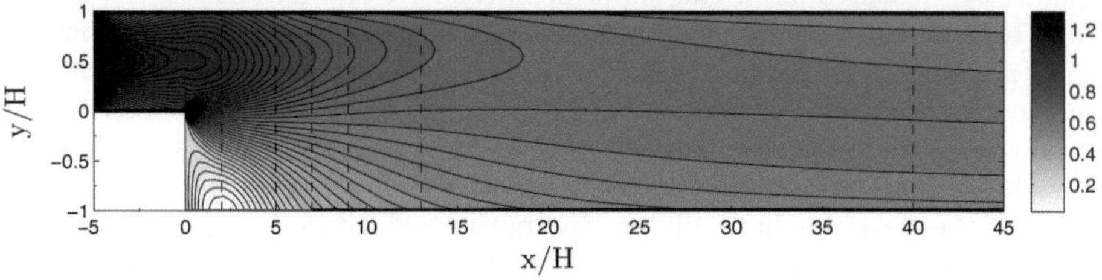

Fig. 3. Contour plot of the mean particle number density c/c_0 distribution.

Particles in the size range considered in this study have a tendency to accumulate near the channel walls. (This is also apparent in fig. 9 below which displays the particle concentration as a function of y/H, averaged over time, and normalized by the initial mean particle concentration.) Particle inertia is responsible for this phenomenon. Particles tend to travel closer to the walls than the fluid elements that bring them into or near the viscous sub-layer. Some particles strike the wall and rebound. Others lack sufficient momentum to reach the closest wall and are confined to the viscous wall region for long periods of time. Therefore, the particles tend to have a higher residence time in regions close to the wall than in the channel core. Several other numerical studies, e.g. [2], [14], report this accumulation of particles near the walls of a vertical channel for a broad range of particle characteristics. The slight increase of particle concentration in the middle of the channel can be partly explained by the turbophoresis phenomenon [4]. Turbophoresis results from small random steps taken by a particle in response to the surrounding fluid turbulence. If there is a gradient in the intensity of the turbulence, the particles will tend to migrate to regions of lower turbulence intensity since they have a longer residence time in those regions. However, the particles in our study have large values of St which limits their response to local turbulence and causes them to move along roughly straight lines over relatively large distances.

In fig. 4, the mean streamwise particle velocities are plotted at different distances from the step (indicated in fig. 2 by dashed lines) for particle mass loadings (ratio of total particle mass to fluid mass in the computational domain) = 0.1 and 0.2. The mean fluid velocity of the unladen flow ($\phi = 0$) is shown for comparison. (The mean fluid velocity profile is unchanged by the presence of particles.) At $x/H = 2$, the particle velocities in the bulk of the channel are lower than the fluid velocities. This is a remnant of a phenomenon observed for the channel flow [13] where the particles show a negative slip velocity due to cross-stream particle movement. Further downstream, the particles are faster than the fluid due to the deceleration of the fluid by the sudden channel expansion. Mean particle velocities in the wall-normal direction (not shown) are generally similar to or slightly smaller than the corresponding fluid velocities. The mean streamwise particle and fluid velocities exhibit the same qualitative features as in the experiments [9]. At the location $x/H = 40$ it may be seen that the symmetric character of the ordinary channel flow has been regained. The particles lag the fluid in the core of the channel but lead in the near-wall region, resulting in profiles that are flatter than those of the fluid phase (see also [13]).

The streamwise velocity fluctuations of the particles are shown in fig. 5. Mostly, the particles have higher fluctuating velocities ($u^p_{rms} = (\overline{u'_p u'_p})^{1/2}$) than the unladen flow ($u^f_{rms} = (\overline{u'_f u'_f})^{1/2}$). In the near-wall region the difference is most eminent. This phenomenon, also found in previous studies (e.g. [9]), is a result of transport of inertial particles out of regions with mean shear. The particle velocity fluctuations in the wall-normal direction (fig. 6)

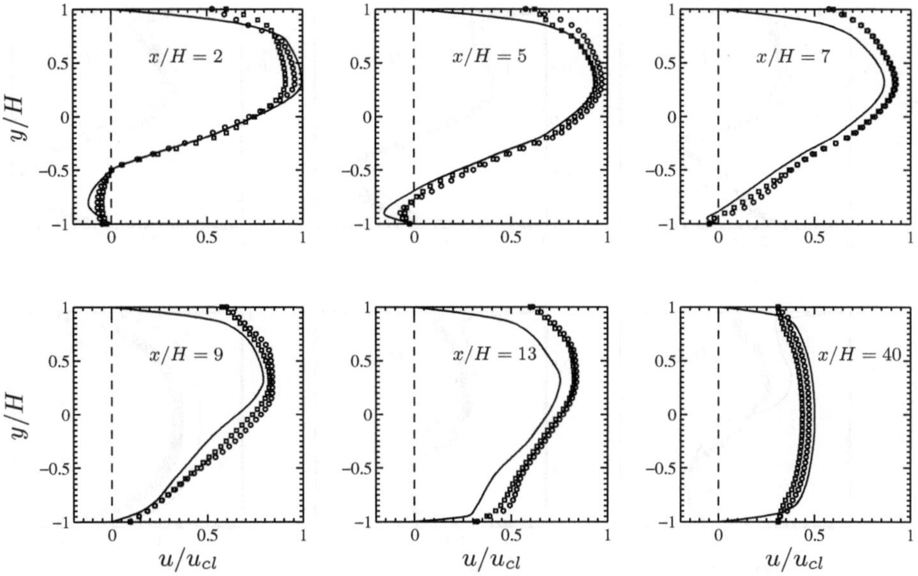

Fig. 4. Mean particle velocity profiles and mean fluid velocity profile for the unladen flow. ——— $(\cdot)_f, \phi = 0$; ○ $(\cdot)_p, \phi = 0.1$; □ $(\cdot)_p, \phi = 0.2$.

are consistently lower than the flow r.m.s. fluctuations. However, further downstream the fluid fluctuation intensities abate significantly whereas those of the particles do not. So the wall-normal fluctuating velocity of the particles near the wall is relatively large compared to that of the fluid. This provides partial explanation for the divergence of the particle velocity from the decreasing fluid velocity in this region and is consistent with the particle trajectory statistics in the y-z-plane, where high-speed particles move towards the wall and rebound, still carrying much of their streamwise momentum.

In fig. 7 the streamwise flow r.m.s. fluctuations for particle-laden flow are compared to those of the unladen flow. At locations directly behind the step ($x/H = 2, 5, 7$) only a small turbulence modification is obtained for $y/H > -0.25$. The turbulence in the shear layer and the recirculation zone is relatively unaffected by the particles. The particle number density in this region indicates that there are few particles in the area $y/H < -0.25$ before the re-attachment point but significant dispersion of the particles has occurred at the locations further downstream. Thus the extremely low level of turbulence modification is not simply a result of an absence of particles in the shear layer, but rather a difference in the response of the turbulence in that region to the presence of particles must be assumed. Further downstream a slight disparity between the fluctuations of laden to unladen flow develops. The effect is nonhomogeneous as the maxima at the walls become higher and broader whereas the intensity in the channel core decreases as with increasing particle mass

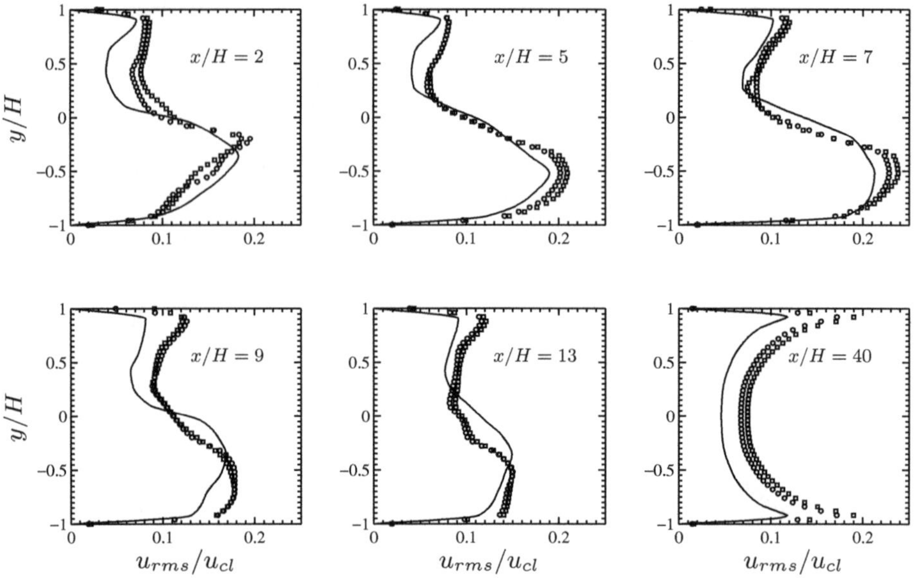

Fig. 5. Streamwise fluctuation intensities of the particles and r.m.s. fluctuations of the unladen flow. —— $(\cdot)_f, \phi = 0$; $\circ\, (\cdot)_p, \phi = 0.1$; $\square (\cdot)_p, \phi = 0.2$.

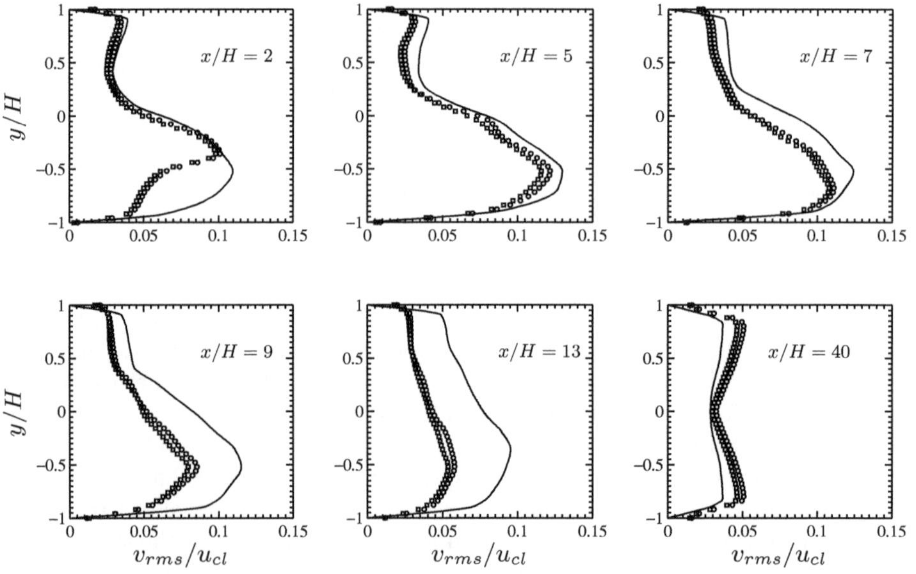

Fig. 6. Wall-normal fluctuation intensities of the particles and r.m.s. fluctuations of the unladen flow. —— $(\cdot)_f, \phi = 0$; $\circ\, (\cdot)_p, \phi = 0.1$; $\square (\cdot)_p, \phi = 0.2$.

loading ϕ. Figure 8 shows the profiles of the wall-normal fluctuating velocities. They display the same trends as the streamwise data, with very slight modification of the turbulence only for $y/H > -0.25$ directly behind the step but with increasing effect of the particle loading further downstream. This findings are confirmed by the results of [9].

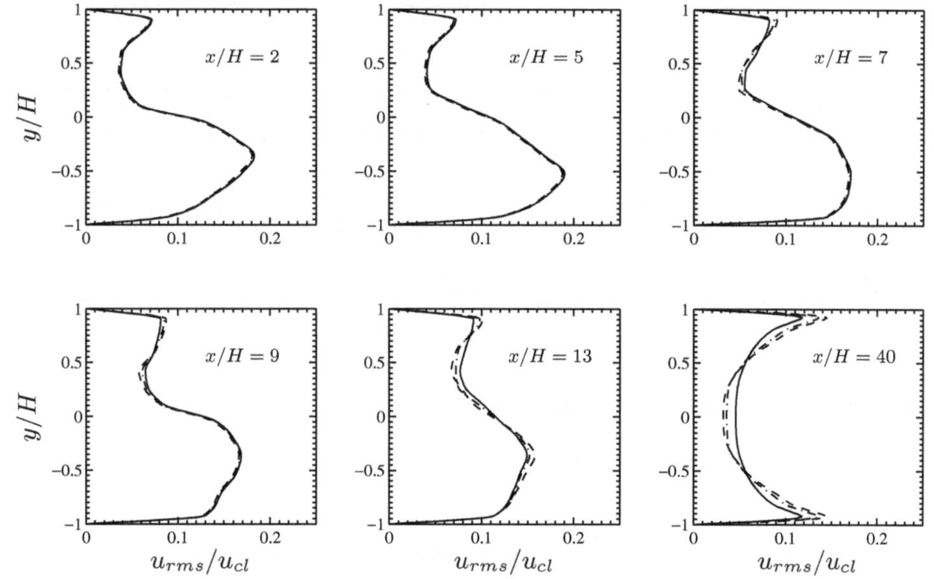

Fig. 7. Streamwise flow r.m.s. fluctuations for different mass loadings. ——— $(\cdot)_f, \phi = 0$; —·— $(\cdot)_f, \phi = 0.1$; - - - - $(\cdot)_f, \phi = 0.2$.

To analyze the differences in turbulence modification at various streamwise locations behind the step ($x/H = 2, 13, 40$) the ratio of the laden to unladen wall-normal r.m.s. fluctuating velocity was calculated for a mass loading of $\phi = 0.2$. Any change in the turbulence due to the presence of particles will appear as a departure of the ratio from unity. (Wall-normal r.m.s. fluctuating velocities were deemed more appropriate for this analysis than the streamwise ones since they do not exhibit the non-homogeneous behavior.) Figure 9 shows these ratios along profiles of the particle number density. Local turbulence attenuation of up to 27% is evident. At x/H locations of 2 and 13 there are still considerably more particles in the region of $y/H > 0$ but at $x/H = 13$ the particles have begun to spread to the $y/H < 0$ region. Despite this fact, the turbulence attenuation is still small for $y/H < 0$. At $x/H = 40$ the turbulence modification is roughly proportional to particle number density. (Near-wall regions are an exception, as the particle number rises while turbulence modification decreases. This is due to the fact that for laden and unladen flow

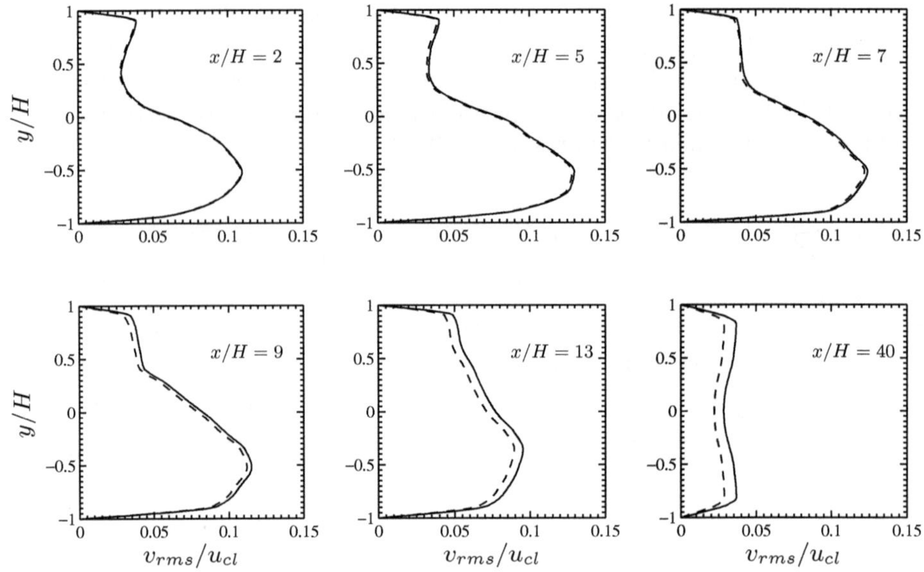

Fig. 8. Wall-normal flow r.m.s. fluctuations for different mass loadings. ——— $(\cdot)_f, \phi = 0$; - - - - $(\cdot)_f, \phi = 0.2$.

alike the no-slip condition must be observed. Furthermore, particles which accumulate close to the wall have rather low momentum.)

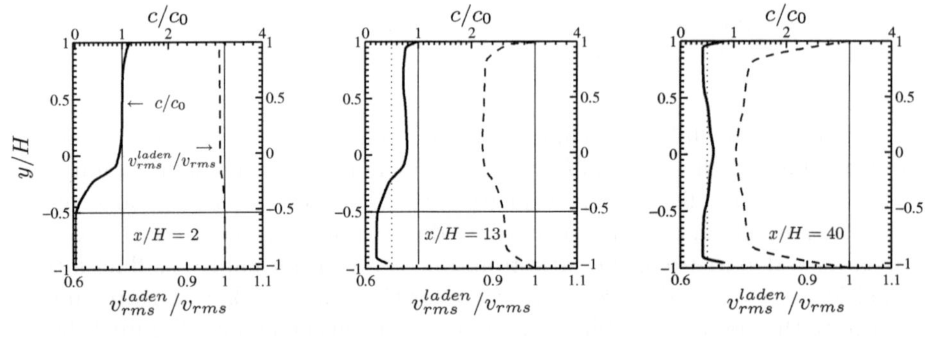

Fig. 9. Spatial development of wall-normal turbulence modification and particle concentration. ——— c/c_0; - - - - $v_{rms}^{laden}/v_{rms}, \phi = 0.2$.

Apparently, the particles' effectiveness in influencing the surrounding fluid depends on the flow structure. For flows with a large ratio of particle to fluid density and particle diameters smaller than the Kolmogorov scale, the particle path, the relative velocity, and the feedback force applied to the fluid should be uniquely determined by the Stokes number. To investigate if the particle

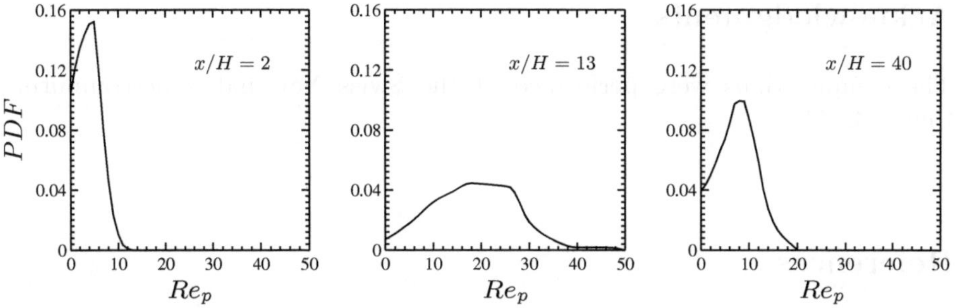

Fig. 10. PDF of particle Reynolds numbers at different locations downstream from the step for $\phi = 0.2$.

Reynolds number may also be a factor in determining the degree of turbulence modification the probability density function (PDF) of this quantity at different locations downstream of the step has been calculated for $\phi = 0.2$. It is obvious from fig. 10 that the mean values as well as the distribution of Re_p vary strongly with the location. The fact that particles with the same Stokes number but different particle Reynolds numbers produce different turbulence modification indicates that the latter is not solely determined by the net inter-phase momentum transfer. The change in particle wake structure and details of the flow distortion associated with different particle Reynolds numbers could affect the level of turbulence response to the presence of particles. Recent calculations [3] have shown that the flow distortion around the particle is a strong function of the particle Reynolds number.

4 Conclusions

The current DNS study investigated turbulence modification in a particle-laden flow over a backward-facing step. The results show that the degree of modification to the fluid phase depends on the particle Reynolds number as well as on the Stokes number. In addition to these parameters, the flow regime was found to strongly affect the degree of turbulence modification. After the re-attachment point in the region behind the step, very little turbulence modification was observed, although the particle number density was the same as that in the channel flow extension where significant modification was found.

Acknowledgments

The computations were performed at the Swiss National Supercomputing Center CSCS.

References

[1] H. Brenner. The slow motion of a sphere through a viscous fluid towards a plane surface. Chemical Engineering Science, 16:242-251, 1961.
[2] J. W. Brooke, K. Kontomaris, T. J. Hanratty, and J. B. McLaughlin. Turbulent deposition and trapping of aerosols at a wall. Phys. Fluids A, 4(4):825-834, 1992.
[3] T. M. Burton and J. K. Eaton. Fully resolved simulations of particle-turbulence interaction. J. Fluid Mech., 545:67-111, 2005.
[4] M. Caporaloni, F. Tampieri, F. Trombetti, and O. Vittori. Transfer of particles in nonisotropic air turbulence. J. Aerosol. Sci., 32:565, 1975.
[5] R. Clift, J. R. Grace, and M. E.Weber. Bubbles, Drops, and Particles. Academic Press, New York, 1978.
[6] S. Elghobashi. On predicting particle-laden turbulent flows. Appl. Sci. Res., 52:309-329, 1994.
[7] S. Elghobashi and G. C. Truesdell. On the two-way interaction between homogeneous turbulence and dispersed solid particles. I: Turbulence modification. Phys. Fluids A, 5(7):1790-1801, 1993.
[8] H. Faxen. Die Bewegung einer starren Kugel laengs der Achse eines mit zaeher Fluessigkeit gefuellten Rohres. Arkiv Mat. Astron. Fys., 17(27):1-28, 1923.
[9] J. R. Fessler and J. K. Eaton. Turbulence modification by particles in a backward-facing step flow. J. Fluid Mech., 394:97-117, 1999.
[10] Y. Hardalupas, A. M. K. P. Taylor, and J. H. Whitelaw. Particle dispersion in a vertical round sudden expansion flow. Proc. R. Soc. Lond. A, 341:411-442, 1992.
[11] A. Kubik. Numerical simulation of particle-laden, wall-bounded attached and separated flows. PhD thesis, ETH Zuerich. In preparation. 2006.
[12] A. Kubik and L. Kleiser. Direct numerical simulation of particle statistics and turbulence modification in vertical turbulent channel flow. In Direct and Large- Eddy Simulation VI. Springer, Dordrecht, The Netherlands, 2005. To appear.
[13] A. Kubik and L. Kleiser. Particle-laden turbulent channel flow and particle-wall interactions. Proc. Appl. Math. Mech., 5:597-598, 2005.
[14] Y. Li, J. B. McLaughlin, K. Kontomaris, and L. Portela. Numerical simulation of particle-laden turbulent channel flow. Phys. Fluids, 13(10):2957-2967, 2001.

[15] M. R. Maxey and J. J. Riley. Equation of motion for a small rigid sphere in nonuniform flow. Phys. Fluids, 26(4):883-889, 1983.

[16] E. E. Michaelides. Review - The transient equation of motion for particles, bubbles, and droplets. Trans. ASME, J. Fluids Eng., 119(2):233-247, 1997.

[17] F. Odar and W. S. Hamilton. Forces on a sphere accelerating in viscous fluid. J. Fluid Mech., 18:302-314, 1964.

[18] P. G. Saffmann. The lift on a small sphere in a slow shear flow. J. Fluid Mech., 22:385-400, 1965.

[19] Q. Wang and K. D. Squires. Large eddy simulation of particle-laden turbulent channel flow. Phys. Fluids, 8(5):1207-1223, 1996.

[20] D. Wilhelm, C. Haertel, and L. Kleiser. Computational analysis of the twodimensional-three-dimensional transition in forward-facing step flow. J. Fluid Mech., 489:1-27, 2003.

Stochastic modeling of fluid velocity seen by heavy particles for two-phase LES of non-homogeneous and anisotropic turbulent flows

Abdallah S. Berrouk[1], Dominique Laurence[1,2], James J. Riley[3] and David E. Stock[4]

[1] MACE/The University of Manchester. PO Box 88 Manchester.M60 1QD. United Kingdom. a.berrouk@postgrad.manchester.ac.uk
[2] Electricite de France, MFTT. 6 Quai Watier 78400. Paris. France. Dominique.laurence@edf.fr
[3] Mechanical Engineering Department. University of Washington. Seattle, Washington. 98116 rileyj@u.washington.edu
[4] School of Mechanical and Materials Engineering.Washington State University. Pullman, Washington. 99164 stock@wsu.edu

Summary. The neglect of the effects of sub-filter scale velocities often remains a source of error in LES predictions of particle dispersion and deposition. Indeed, sub-filter fluctuations should be expected to be more significant for particles with smaller relaxation times compared to the LES-resolved turbulence time scales. In this work, a stochastic diffusion process is used to include the sub-filter scale transport when tracking a dilute suspension of heavy particles (glass beads in air with different Stokes' numbers, namely 0.022 and 2.8) in a high Reynolds number, equilibrium turbulent shear flow ($Re_\tau = 2,200$ based on the friction velocity and the pipe diameter). A Langevin-type equation is proposed to model the Lagrangian fluid velocity seen by solid particles, taking into account inertia and cross-trajectory effects. LES predictions are compared to RANS results and experimental observations. It is shown that the RANS approach is unable to predict particle dispersion statistics as accurately as LES does, especially for inertial particles characterized by a Stokes number smaller than one. For particles with Stokes number higher than one, both LES and RANS predictions compare reasonably well with the experimental results. More importantly, the use of a stochastic approach to model the sub-filter scale fluctuations has proven crucial for results concerning the small-Stokes-number particles. The model requires additional validation for non-equilibrium turbulent flows.

1 Introduction

Understanding the dispersion of heavy particles from a source point in turbulent flows is a domain of research of utmost practical interest. Heavy particle

Bernard J. Geurts et al. (eds), Particle Laden Flow: From Geophysical to Kolmogorov Scales, 179–192.

transport and dispersion are encountered in a wide range of flow configurations, whether they are of industrial or environmental character.

For many environmental applications, the Reynolds-Averaged Navier-Stokes approach has been proven inherently ill-posed [1]. Thus, the use of Large-eddy simulation (LES) has increased over the years as a promising tool to address these types of problems. The modeling of the residual, or sub-filter scale, velocity field becomes especially important as smaller Stokes-number inertial particles are tracked within higher Reynolds-number turbulent flows.

The large-scale velocity field provided by LES can be assumed to mimic the large-scale fluid velocity field seen by inertial particles. This is the simplest approach since it consists of neglecting the effects of the sub-filter fluctuating velocity field on the particle trajectories. This should be a justifiable assumption in most applications in which large Stokes-number, heavy particles are numerically tracked on very fine LES grid using small filters. In this case, the inertial particles do not sense turbulent fluctuations associated with the sub-filter scales and less turbulent kinetic energy of the larger scales is filtered out. For high Reynolds number wall-bounded turbulent flows, the LES grid is often not fine enough, in particular near the walls, where the resolution is dictated by restrictions in computing resources. In this case, discarding the sub-filter fluctuations can be a major source of error in LES predictions of heavy particle statistics, especially for the ones having very small Stokes numbers [2]. For these cases, the instantaneous velocity can be synthetically derived from the LES velocity field. Solving an additional transport equation for the residual or sub-filter kinetic energy is often the approach used by meteorologists [3]. Another approach consists of de-filtering the LES velocities to generate the fluid phase instantaneous field to use to track inertial particles [4]. Stochastic modeling using the Langevin equation has been extensively used in the framework of RANS to construct total turbulence fluctuations based on the mean flow statistics [5]. This approach can be extended for the Lagrangian modeling of the fluid velocity seen by heavy particles for LES of particle-laden turbulent flows. Shotorban *et al.* [6] obtained promising results for particles with very small response times when they used a Langevin equation to account for the sub-filter effects in their LES of particle-laden decaying isotropic turbulent flow. They neglected particle inertia and cross trajectory effects, however, which is valid only for inertial particles with vanishing response times.

In this paper, the effect of the sub-filter scales on the heavy particle dispersion in an equilibrium non-homogeneous and anisotropic turbulent flow are taken into account. The modeling of the fluid velocity seen by solid particles is carried out using a Langevin-type equation allowing for both inertia and cross-trajectory effects. LES predictions are compared to RANS results and experimental observations [7].

2 Governing equations of solid-gas turbulent flow

2.1 Equation of Fluid Flow

In order to simulate the experiment of Arnason [7], a turbulent pipe flow is studied in a Cartesian framework. An unstructured grid consisting of 740,000 cells is used to avoid having too many grid points in the core region of the pipe and in order to properly resolve the near-wall region. A polar grid is used for the first three layers with non-conforming embedded refinement as shown in Tab. (1). Then the polar grid is made to match an octahedral bloc for the core region of the pipe (Fig. 1). The 2D grid is then extruded in the streamwise direction by 192 nodes. The first grid point near the pipe wall at which the axial velocity is computed is located at $y^+ = 1.3$, with 2 grid points placed within the viscous sub-layer, the depth of which approximately equals 5 wall units. A non-uniform grid is employed in the normal-to-the-wall direction within the circular part. This is done in order to locate more grid points in the near-wall region that is characterized by steep gradients and small energy-containing eddies. The Reynolds number of the simulation based on the pipe diameter D and on the centerline velocity u_c equals approximately 50,000 (based on mean velocity u_b and shear stress velocity u_τ, it is 42,000 and 2,200 respectively).

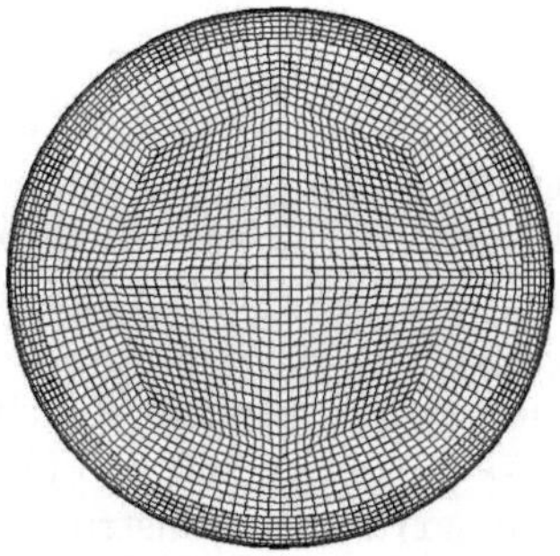

Fig. 1. Unstructured grid used for the simulations.

Table 1. Hanging nodes in the polar part of the unstructured grid

$y^+ = u_\tau . y/\nu$	Rad. direction	Circumf. direction
$0 < y^+ < 30$	4 cells	256 cells
$30 < y^+ < 100$	4 cells	192 cells
$100 < y^+ <\approx 360(r=(2/3)R)$	8 cells	128 cells

A flow solver from the R&D section of Electricité de France named Code_Saturne was used as starting point of the present development. It is based on a collocated unstructured finite volume method, and has been extensively tested for LES of single-phase flows [8] as well as its Lagrangian module for particle tracking based on RANS/Stochastic modeling [9].

The filtered spatial and temporal evolution of an incompressible Newtonian fluid flow can be determined from the following equations :

$$\frac{\partial \overline{u}_i}{\partial x_i} = 0 \tag{1}$$

$$\frac{\partial \overline{u}_i}{\partial t} + \overline{u}_j \frac{\partial \overline{u}_i}{\partial x_j} = -\frac{\partial \overline{p}}{\partial x_i} + \frac{1}{Re}\frac{\partial^2 \overline{u}_i}{\partial x_j \partial x_j} + \frac{\partial \tau_{ij}}{\partial x_j} \tag{2}$$

where

$$\tau_{ij} = \overline{u_i u_j} - \overline{u}_i \overline{u}_j \, . \tag{3}$$

Here τ_{ij} is the sub-grid scale (SGS) stress tensor and is modeled using the algebraic eddy-viscosity model proposed by Smagorinsky [10]:

$$\tau_{ij} - \frac{1}{3}\delta_{ij}\tau_{kk} = -2\nu_{SGS}\overline{S}_{ij} \, , \tag{4}$$

with ν_{SGS} is the subgrid-scale viscosity

$$\nu_{SGS} = C_s \Delta^2 \mid \overline{S} \mid \tag{5}$$

Here C_s is a constant, and $|\overline{S}| = |2\overline{S}_{ij}\overline{S}_{ij}|^{1/2}$, where $\overline{S}_{ij} = \frac{1}{2}(\partial_j \overline{u}_i + \partial_i \overline{u}_j)$ is the resolved rate-of-strain tensor. The length scale Δ is taken equal to $2h$, where h is the grid spacing, and the value of the constant C_s is taken equal to 0.065.

LES calculations provide the large-scale velocity field and allow, under certain assumptions, the evaluation of the sub-filter kinetic energy and its dissipation rate that are necessary for the construction of the velocity field seen by inertial particles, as we shall see in the following sections. Mean velocity and rms turbulent fluctuation profiles are computed from the LES of the single phase flow and they were found to compare very well with previous LES results and experimental observations, and are not presented in this paper.

For RANS calculations, the second-moment model ($R_{ij} - \epsilon$) is used to close the time-averaged Navier-Stokes equations. Mean fluid velocity, turbulent kinetic energy and its mean dissipation rate are computed and used later to model the fluid velocity seen by the inertial particles. The single-phase velocity fields of both RANS and LES will be used to track inertial particles originating from a point source in a turbulent, vertically downward pipe flow.

2.2 Equation of Particle Motion

From a point source located at the center of the pipe, solid particles are released and tracked into the turbulent flow described in the previous section. The physical properties of these solid particles are summarized in Tab.(2). The simplest way to characterize the dynamics of particle motion is by means of its Stokesian relaxation or response time and the corresponding Stokes number given by:

$$St = \frac{\tau_p}{T_E}, \quad \text{where} \quad \tau_p = \frac{\rho_p d_p^2}{18\mu}. \tag{6}$$

Here ρ_p is the mass density of the particles, d_p is the diameter of the particles, μ is the dynamic viscosity of the fluid, and T_E is the Eulerian time scale of the fluid phase. Since large particles are also considered in this study, the particle Reynolds number is expected to often exceed unity. A non-linear drag coefficient, taking into account the high particle Reynolds number, is therefore more appropriate. As a consequence the actual particle response time will, at high particle Reynolds numbers, be smaller than the one defined by Eqn. (6). As a result of the high density ratio between particle and fluid densities, the equation describing particle motion becomes reasonably simple and only the drag and gravity forces will be retained since other forces are in this case negligible [11].

Table 2. Physical characteristics of inertial particles used in the simulations

Mean diameter d_p^μ (μm)	5	57
Standard deviation σ_p (μm)	1	11
clipping (μm)	$2 < d_p^\mu < 10$	$32 < d_p^\mu < 101$
Density (kg/m^3)	2475	2420
Response Time (s)	0.0002	0.025

The tracking of the solid particles within the turbulent flow obeys the following system of equations:

$$dx_{p,i} = u_{p,i}dt \tag{7}$$

$$du_{p,i} = \frac{u_{s,i} - u_{p,i}}{\tau_p}dt + gdt \tag{8}$$

$$\tau_p = \frac{\rho_p}{\rho_f}\frac{4d_p}{3C_D|u_s - u_p|} \tag{9}$$

$$C_D = \begin{cases} \frac{24}{Re_p}(1 + 0.15Re_p^{0.687}) & \text{if } \mathrm{Re_p} < 1000 \\ \\ 0.44 & \text{if } \mathrm{Re_p} > 1000 \end{cases} \tag{10}$$

Here x_p and u_p are the particle position and velocity, u_s is the fluid velocity seen by a solid particle along its trajectory, g is the gravity force by unit of mass, C_D is the drag coefficient and Re_p is the particle Reynolds number, $Re_p = d_p|u_s - u_p|/\nu$.

A tri-linear interpolation scheme is used to obtain the velocities between the grid points. The interaction of solid particles with the wall is considered elastic. According to the Sommerfeld criteria [12], the turbulence level and the size of inertial particles considered in Arnason experiment and in the numerical simulations do not give rise to a wall-collision dominated flow. Also, neither two-way coupling nor particle-particle collision is taken into account, since a dilute concentration of particles is considered for all the simulations (volume fraction $\alpha_p < 10^{-6}$).

The system of Eqns (7-10) can be now used to track inertial particles in a Lagrangian framework as they move down the pipe. The only unknown in this system of equations is the fluid velocity u_s seen by these inertial particles along their trajectories as they move through the turbulent field.

In the next section, a stochastic model is proposed to reconstruct the Lagrangian instantaneous fluid velocity seen by heavy particles from the filtered Navier-Stokes equations.

2.3 Modeling of seen fluid velocity

Langevin models [13] have been attractive stochastic diffusion models developed for fluid particle turbulent velocities [14], and they have been extended for the generation of the fluid turbulent field seen by inertial particles. The general form of the Langevin model chosen for the velocity of the fluid seen by particles is:

$$du_{s,i} = A_{s,i}(t, x_p, u_p, u_s)dt + B_{s,ij}(t, x_p, u_p, u_s)dW_j \, , \tag{11}$$

where the drift vector A and the diffusion matrix B have to be modeled. Each component of the vector dW is a Wiener process (white noise); it is a stochastic process of zero mean, $\langle dW \rangle = 0$, a variance equal to the time interval, $\langle (dW)^2 \rangle = dt$, and delta-correlated in the time domain [15]. This formulation, along with Eqns. (7) and (8), is equivalent to a Fokker-Planck equation for the corresponding filtered density function (fdf) and can be used in a Monte Carlo simulation of the underlying fdf [16].

The theoretical and numerical formulations of the Langevin model have been extensively discussed in the framework of particle-laden RANS [17, 18, 19] and its use is extended herein with the necessary modifications for the modeling of the fluid velocity seen by particles in LES framework. It is:

$$du_{s,i} = dt(-\frac{1}{\rho_f}\frac{\partial \overline{p}}{\partial x_i} + \frac{1}{Re}\frac{\partial^2 \overline{u}_i}{\partial x_j \partial x_j}) - (\frac{dt(u_{s,i} - \overline{u}_{f,i})}{T^*}) + \sqrt{C_0^* \langle \epsilon_r \rangle} dW_i \tag{12}$$

Here ϵ_r is the dissipation rate of the residual or sub-filter turbulent kinetic energy k_r, and C_0^* is the diffusion constant. The fluid Lagrangian time scale seen by heavy particles T^* is T_E in the limit of very large inertia, since in this situation the heavy particles respond slowly to the fluid turbulence. On the other hand, $T^* = T_L$ (the fluid Lagrangian time scale) if $St \to 0$ since in this case the inertial particles reduce to fluid elements. Thus, in general T^* is a function of St and varies between T_L and T_E as it is portrayed by the following equation [20]:

$$T^* = \frac{T_L}{\beta}(1 - (1-\beta)(1+St)^{-0.4(1+0.01St)}), \tag{13}$$

where β is defined below.

This equation, though developed for homogeneous and isotropic turbulence, can also be used for shear flow to compute the fluid time scale seen by inertial particles [21]. To account for the crossing trajectory and the continuity effects, Csanady's expressions [22] can be used to compute the fluid Lagrangian time scale in the direction of the mean drift ($\|$) and the transverse directions ($\perp$):

$$T^*_{\|} = \frac{T^*}{\sqrt{1+\beta^2|\langle u\rangle|^2/(2k_r/3)}}$$

$$T^*_{\perp} = \frac{T^*}{\sqrt{1+4\beta^2|\langle u\rangle|^2/(2k_r/3)}} \tag{15}$$

Here $\langle u_r\rangle$ is the mean slip velocity between fluid and inertial particles.

The diffusion coefficient C_0^* is evaluated according to the following formulation:

$$C_0^* = C_0 b_i \tilde{k_r}/k_r + \frac{2}{3}(b_i\tilde{k_r}/k_r - 1), \tag{16}$$

where C_0 is the Kolmogorov constant, and $\tilde{k_r}$ is a modified kinetic energy:

$$\tilde{k_r} = \frac{3}{2}\frac{\sum_{i=1}^{3} b_i\langle u_i'^2\rangle}{\sum_{i=1}^{3} b_i} \tag{17}$$

Here u_i' is the fluid fluctuating velocity and $b_i = T^*/T_i^*(i = \perp \; or \; \|)$.

For the ratio of the Lagrangian timescale to the Eulerian time scale β, it was shown that its value is Reynolds number dependent [23] and varies considerably in the literature. For this study, it is expected that its influence on the model predictions is very small, since small universal scales are modeled unlike in RANS, where modeling of turbulent fluctuations linked to large scales is sought. When formula (13) is used to take into account the inertia effect, β is chosen to be 0.356 [20]. Two other values are tested: $\beta = 1.3$ [24] and $\beta = 0.8$ [23]. Simulation results showed an insignificant influence of β on the model predictions. These results are not presented in this paper.

For LES, the Lagrangian time scale for the sub-filter fluctuations T_L is assumed to evolve according Eqn. (17) [25], using the sub-filter kinetic energy

k_r and its dissipation rate ϵ_r that have to be evaluated according to the SGS model used to take into account subgrid effects on the large scales.

$$T_L = (\frac{1}{2} + \frac{3}{4}C_0)^{-1}\frac{k_r}{\langle \epsilon_r \rangle} \tag{18}$$

Indeed, in the case of the Smagorinsky model, if equilibrium is assumed at the cutoff, the dissipation rate and then the residual kinetic energy can be evaluated as:

$$\epsilon_r = -\tau_{ij}.\frac{d\overline{u}_i}{dx_j} = (C_s.\Delta)^2|\overline{S}|^3 \tag{19}$$

$$k_r = C_\epsilon(\Delta.\epsilon_r)^{2/3} \tag{20}$$

Typically $C_\epsilon \approx 1$ and $C_0 \approx 2.1$ [16].

At each time step Eqns. (13) to (19) are evaluated. The stochastic differential equation (SDE) system that comprises Eqns. (7), (8) and (12) is integrated at each time step using an appropriate weak second-order integration scheme [26] that account for the nature of the problem characterized by the presence of different time scales; this can lead to stiff equations when the smallest time-scale is significantly less than the time-step of the simulation. As it was highlighted by Maclnnes *et al.* [27] and Minier *et al.* [9], the Langevin model under the formulation (12) does not suffer any spurious drift in the non-homogeneous case.

3 Results and discussions

The numerical predictions for Arnason's experiment using RANS and LES were collected at three sections downstream the seeding point (Z=0.318m, 0.502m and 0.679m) and consist of particle concentration and radial velocity profiles for two different particle diameters, namely 5μm and 57μm. Using these results, the dispersion coefficient ϵ_p for both types of particles is evaluated and then compared to the experimental measurements.

Figures 2 and 3 display numerical predictions of particle concentration and radial velocity profiles for 5μm particles, for RANS, LES with and without taking into account the sub-filter effects on particles dispersion. As expected, a significant improvement in the results were made possible by introducing the effects of the sub-filter scales on the motion of the 5μm particles; these have Stokes numbers less than one (evaluation of the Stokes numbers for the 5μm particles shows that they vary between 0.022 and 0.46). These particles sense the whole spectrum of turbulent fluctuations present in the flow with different time scales. It is clear that LES predictions closely match the experimental results, though slightly underestimating the concentration profiles near the wall. RANS predictions increasingly overestimate the particle concentration spread as we move downstream. The LES without sub-filter model

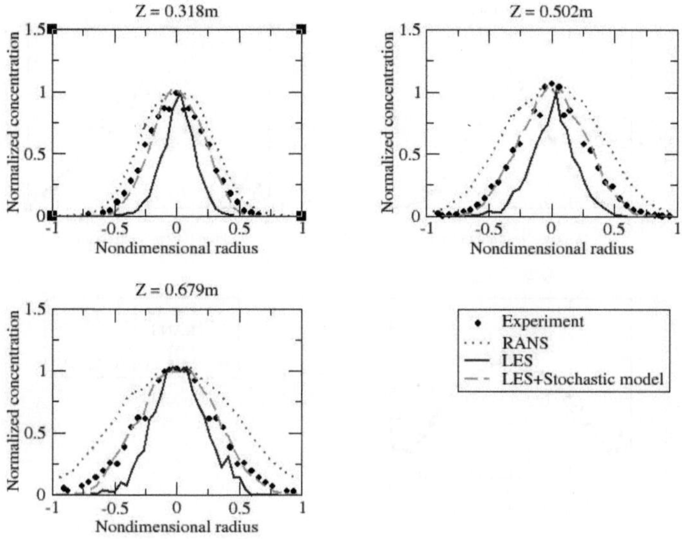

Fig. 2. Concentration profiles for 5μm particles.

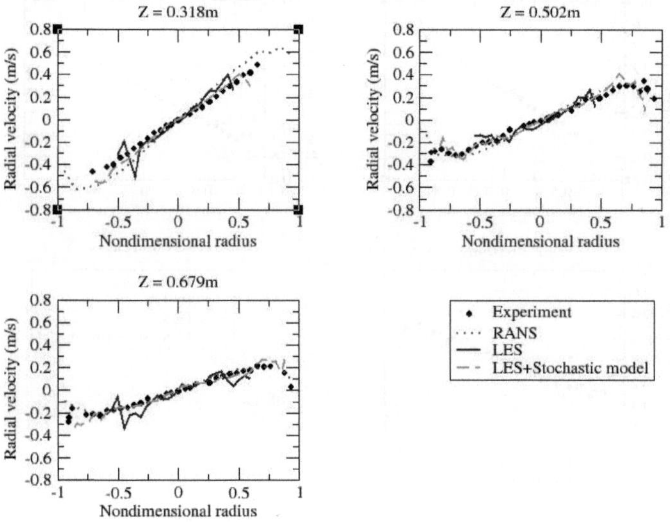

Fig. 3. Radial velocity profiles for 5μm particles.

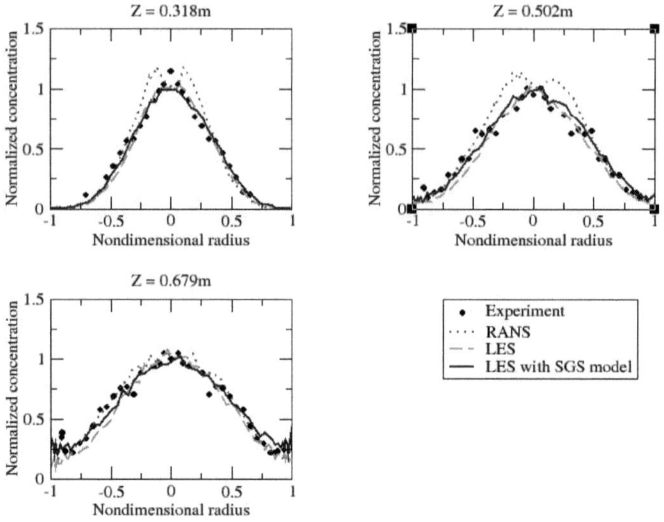

Fig. 4. Concentration profiles for $57\mu m$ particles.

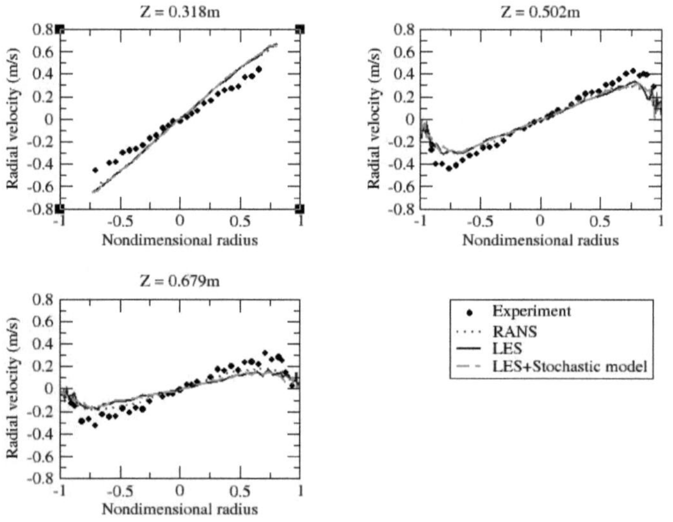

Fig. 5. Radial velocity profiles for $57\mu m$ particles.

shows an initial underestimation of the spread, and this lag remains constant downstream. With the sub-filter model incorporated in the LES, this initial defect is now well corrected. It should be mentioned that the injection is represented by only very a few cells, and this is why the enhanced dispersion by the sub-filter model is more important at the initial stages.

For the 57μ particles, Fig. 4 and 5 show that the effect of the sub-filter scales on their motion is clearly very small; this is because their Stokes numbers based on the sub-filter scale time scale exceed one (they are between 2.8 and 58 throughout the computational domain), and therefore they do not respond significantly to the higher frequency sub-filter-scale motions. It was expected that there would be little effect of the sub-filter scales on 57μm particles, which is verified on particle radial velocity profiles. However, a very small difference has been noticed on the particle concentration profiles, which is probably due to an excessive filtering out of the kinetic energy that is not well captured by the model. Both RANS and LES predict reasonably well the experimental results, with the LES results being marginally closer to the experiment for the concentration profiles.

Figure 6 shows, for both numerical simulations, the prediction of the particle dispersion coefficient along with the experimental observations. Previous dispersion coefficient estimates were all based on observations of number density profiles of particles dispersing from a line or point source [28]. Then the dispersion coefficient or turbulent diffusivity of particles was related to the time rate of change of the mean square dispersion y^2 following [29, 30], and is given as:

$$\epsilon_p = \frac{1}{2}\frac{d}{dt}y^2 . \tag{21}$$

This method is only correct if ϵ_p is constant throughout the flow field and the convective velocity is uniform, as is the case for homogeneous and isotropic turbulence. It is less meaningful for turbulent shear flows, however, which is the case of turbulent pipe flow.

A method is developed to allow local estimates for the diffusivity to be obtained when the turbulence is neither homogeneous nor isotropic [31]. If the flux is considered to be caused only by gradient diffusion (Fick's law), the particle dispersion coefficient at every radial location can be computed according to the following formula, except in regions where $\partial C/\partial r$ is near 0:

$$\epsilon_p = -\langle v_p \cdot C\rangle(\frac{\partial C}{\partial r})^{-1} . \tag{22}$$

This allows the particle diffusivity to be computed locally at all measuring points. The required measurements are the average particle velocity v_p in the radial direction and the concentration profile C.

As it is depicted by Fig. 6., numerical predictions for the 57μm particle diffusivity are about the same, slightly under-predicting the experimental results near the wall. For the 5μm particles, it is clear that LES results more closely match the experimental results than the RANS predictions. It is shown that

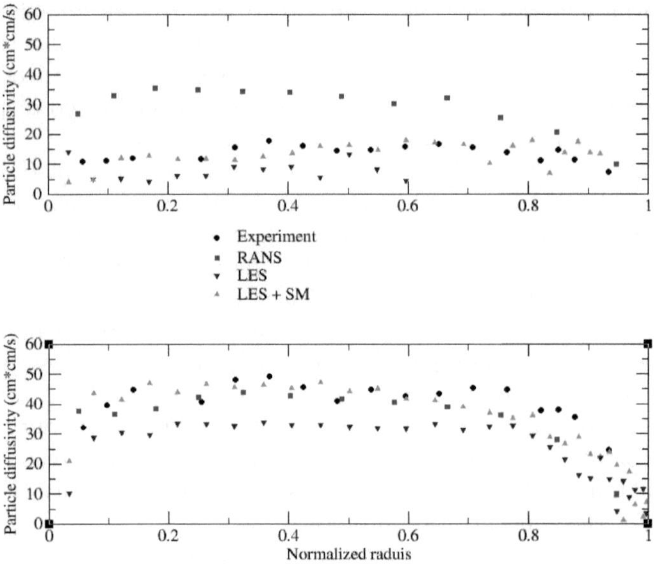

Fig. 6. Particle dispersion coefficient. top:5μm; bottom:57μm.

both RANS and LES predict reasonably well the particle dispersion coefficient of 57μm particles, whereas RANS predictions failed to match the experimental results of the 5μm particles. LES predictions for the 5μm particles are excellent and highlight once again the importance of including the sub-filter effects on small particles dispersion.

4 Concluding remarks

Numerical simulations using two approaches, namely RANS and LES, were conducted to compute inertial particle dispersion in a turbulent gas flow in a pipe at a high Reynolds number. Numerical predictions were compared to the experimental observations of Arnason and Stock.

Stochastic modeling of the turbulent fluctuations seen by inertial particles along their trajectories has been used in the framework of RANS, and is extended to model turbulent sub-filter scale fluctuations seen by the inertial particles in an LES velocity field. Particle dispersion statistics such as particle concentration, radial velocity and the dispersion coefficient were computed for two types of solid particles that have different inertia and drift.

The use of a Langevin-type stochastic approach to model the sub-filter fluctuations has proven crucial for results concerning the small-Stokes-number particles. The stochastic model used was of Langevin type that has been extensively used in the framework of RANS. Its simplistic extension to predict the sub-filter fluctuations for LES gave satisfactory results.

Comparison of RANS and LES results have shown that the RANS approach is unable to predict particle dispersion statistics as accurately as the LES in particular for inertial particles characterized by a Stokes number smaller than one. For particles with Stokes number higher than one, both LES and RANS predictions compare reasonably well with the experimental results.

One possible extension of this study can be the further testing of the ability of the present stochastic sub-filter model, in the framework of LES of non-equilibrium, non-homogeneous and anisotropic turbulent flows, such as flows characterized by the presence of zones of recirculation for instance. It is expected that the assumptions made to implement the model may cause it to fail in the prediction of this type of flow (assumptions of an equilibrium at the LES cutoff in modeling sub-filter time scales). Solving an extra transport equation for the residual kinetic energy may be needed to replace the crude Smagorinsky scaling estimates.

Acknowledgment

This work was funded by the Algerian Government and the Worldwide University Network at Manchester University. The authors would like to thank EDF R&D for making the "Code_Saturne" software available in source code.

References

[1] Lakehal D. (2002) Int. J.Multiphase Flows, 28:823-863
[2] Armenio V., Piomelli U., Fiorotto V. (1999) Phys. Fluids. A(11), pp. 3030-3042.
[3] Schumann U. (1975) J. Comp. Phys. 18,376.
[4] Shotorban B., Mashayek, F. (2005) Phys. Fluids 17, 081701.
[5] Pope S.B (2000) Cambridge University Press
[6] Shotorban B., Mashayek, F. (2006) J.of Turbulence. Vol 7, No. 18.
[7] Arnason G. (1982) Ph.D. dissertation, Washington State University, Pullman
[8] Archambau F., Mehitoua N., Sakiz M. (2004) International Journal on Finite Volume.
[9] Minier J-P., Peirano E. (2001) Physics Reports, 352:1-214.
[10] Smagorinsky J. (1963) Mon. Weather Rev. 91-99.
[11] Maxey M.R., Riley J.J. (1983) Phys. Fluids., v 26, n4, pp. 883-9.
[12] Sommerfeld M.(1992) Int. J. Multiphase Flow, 18: pp.905-926.
[13] Langevin P. (1908) Comptes Rendus Acad. Sci., Paris 146, 530-533.
[14] Haworth. D.C., Pope S. (1986) Phys. Fluids. 29(2) pp. 387-405
[15] Kloeden P.E., Platen E., Schurz H. (1994) Springer-Verlag.
[16] Gicquel L.Y.M., Givi P., Jaberi F.A, Pope S.B. (2002) Phys. Fluids. A(14)3.
[17] Pozorski J., Minier J-P. (1998) Int. J. Multiphase Flow, 24: 913-945
[18] Minier J-P., Peirano E., Chibarro S. (2004) Phys. Fluids, 18:302-14.

[19] Minier J. P. (2000) Monte Carlo Methods and Appl., Vol.7, No. 3, pp. 295-310.
[20] Wang L.P., Stock D.E. (1993) J. Atmos Sci. Vol.50, No 13, pp. 1897-1913.
[21] Carlier J. Ph., Khalij M., Oesterle B. (2005) Aerosol Sci. and Tech. 39:196-205
[22] Csanady G.T. : (1963) J. Atmos. Sci. Vol.20, pp 201-208.
[23] Sato Y., Yamamoto K. (1987) J. Fluid Mech. 175, 183.
[24] Riley J.J., (1971) Ph.D dissertation, The Johns Hopkins University, Baltimore.
[25] Heinz S.(2003) Springer-Verlag. Berlin.
[26] Minier J-P., Peirano E., Chibarro S. (2003) Monte Carlo Methods and Appl., Vol.9, No. 2, pp. 93-133.
[27] Maclnnes J. M., Bracco F. V. (1992) Phys. Fluids A4(12). pp.2809-2824
[28] Arnason G., Stock D.E. (1983) American Society of Mechanical Engineers. Fluid Engineering Division. v10, pp. 25-29.
[29] Taylor G.I. (1921) Proc. Roy. Soc. A 151, pp. 421-478.
[30] Hinze J.O. (1975) McGraw-Hill, 253,460-471.
[31] Arnason G., Stock D.E. (1984) Experiments in Fluids. v2, n2, pp. 89-93.

DNS study of local-equilibrium models in dilute particle-laden turbulent pipe flows

A.M.P. Boelens and L.M. Portela

Department of Multi-Scale Physics, Delft University of Technology, Prins Bernhardlaan 6, 2628 BW Delft, The Netherlands L.Portela@tudelft.nl

Summary. A two-fluid closure model, commonly used in engineering simulations, is the drift-velocity model of Simonin. In this work, this model is evaluated for dilute particle-laden pipe flows, using Direct Numerical Simulations (DNS) with particle tracking. The simulations were performed with both heavy and light particles (particle relaxation-times of $\tau_2^+ = 100$ and $\tau_2^+ = 10$, respectively) and with reflecting and absorbing walls as the boundary conditions for the particles, resulting in four different cases. For all the cases, except the combination of heavy particles and absorbing walls, from a pragmatic point of view, the assumptions of local-equilibrium and homogeneous turbulence seem to hold. For the drift-velocity two models were evaluated: (i) a simple Schmidt-number model, and (ii) a more advanced drift-tensor model. From an engineering perspective, for light particles with reflecting walls the Schmidt-number model appears to be the best choice. However, when the particles are heavier or when the walls are absorbing, the more advanced drift-tensor model gives better results. For these cases, provided that there exists good closure models for the time-scales and particle-fluid velocity correlations, a drift-tensor model could be a better option.

1 Introduction

Particle-laden turbulent pipe-flows are an important area of interest within the broader field of multiphase flows. The applications are numerous: pneumatic conveying of granular material, annular flow, catalytic crackers, to name just a few examples. For engineering purposes, a method often used for numerical simulations of particle-laden turbulent gas flows is the two-fluid model. In this model, the particle-phase as well as the continuous-phase, is described by a set of Reynolds-averaged Navier-Stokes equations. Although this has the advantage of not requiring a large amount of computational resources, it has

Bernard J. Geurts et al. (eds), Particle Laden Flow: From Geophysical to Kolmogorov Scales, 193–206.

the disadvantage that modeling is required for both the Reynolds-stresses and the interfacial momentum-transfer.

Over the years, quite some research has been done in particle-laden flows. A possible modeling candidate to close the dispersed-phase Reynolds-stresses, the local-equilibrium model of Tchen and Hinze [3], is found in the classic book on turbulence from Hinze [3]. Some more recent work comes from Portela et al. [6], who have done a numerical study of equilibrium models in turbulent pipe flows near the wall. Portela et al. evaluated the model of Tchen and Hinze, and, from an engineering perspective, found it is able to accurately predict the dispersed-phase Reynolds-stresses, except very close to the wall.

For dilute systems with heavy small particles and conservation of mass for each individual phase, the factor dominating the interfacial momentum transfer is the drag force, exerted by the continuous-phase on the dispersed-phase. This drag force is proportional to the difference between the average velocity of the particles and the average velocity of the continuous-phase as seen by the particles. The average continuous-phase velocity as seen by the particles needs to be closed, and a method that can be used for this, is turbulent diffusion modeling.

An example of a simple turbulent-diffusion model, is the Schmidt-number model. This model is used by Young and Leeming [10], for their evaluation of the influence of the gradient of the Reynolds stresses on particle deposition for different particle relaxation-times. In their calculations of the turbulent diffusion, Young and Leeming assume the Schmidt number to have a value equal to one. Both Reeks [8] and Simonin [9] have developed more-advanced turbulent-diffusion models, based on the local-equilibrium assumption. The objective of this work is to evaluate the turbulent diffusion model of Simonin, the drift-velocity model [9], using DNS with particle tracking, and to compare it with the Schmidt-number model, which can be seen as a particular case of Simonin's model.

In the next section the basic, two-fluid equations are presented, together with the models for the closure of the Reynolds stresses and the drift-velocity. Then, in the subsequent section, the results are presented, followed by the conclusions.

2 Theory of two-fluid modeling

2.1 General equations

The general two-fluid mass and momentum balances can be obtained in various ways. One method, is by deriving them from the single-phase Navier-

Stokes equation or the Maxey and Riley equation [4] with a phase-operator [2]. This method is similar to Reynolds-averaging, used in RANS modeling. Another approach, used by Simonin in the derivation of his drift-velocity model, is the statistical approach with a probability density function [5]. This probability density function (PDF) gives a Lagrangian description of both the particles an the fluid. Using the Fokker-Plank equation, which describes the evolution of system considered, the Lagrangian PDF can be rewritten into the two-fluid mass and momentum balances. When using the statistical approach, also the transport equation for the average continuous-phase velocity as seen by the particles can be obtained. From this transport equation, the drift-velocity model is derived.

The focus of this work is on the drift-velocity modeling. In order not to mix different influences, a few assumptions have been made: (i) the system is assumed to be in a statistical steady state, (ii) there is no evaporation nor chemical reaction taking place, so there is conservation of mass for each individual phase, (iii) the system is very dilute, so it is valid to assume one-way coupling and neglect collision terms, (iv) there is no gravity, and (v) there are only small heavy rigid particles. These assumptions result in the Maxey and Riley equation being reduced to only the drag term.

Applying the above assumptions to the general two-fluid equations of the dispersed-phase, results in:

$$\nabla \cdot (\alpha_2 \mathbf{U}_2) = 0 \tag{1}$$

for the average mass balance and:

$$\nabla \cdot (\alpha_2 \mathbf{U}_2 \mathbf{U}_2) = -\nabla \cdot \left(\alpha_2 \langle \mathbf{u}_2' \mathbf{u}_2' \rangle\right) - \frac{\alpha_2}{\langle \tau_2 \rangle} \mathbf{U}_{\text{rel}} \tag{2}$$

for the momentum-balance. Here, α_2, is the dispersed-phase volume-fraction, $\mathbf{U}_2$, the average dispersed-phase velocity, $\langle \mathbf{u}_2' \mathbf{u}_2' \rangle$, the dispersed-phase Reynolds stresses, $\langle \tau_2 \rangle$, the average particle relaxation-time, and $\mathbf{U}_{\text{rel}}$, the average relative velocity between the continuous and dispersed-phase. Equation 2 shows that the dispersed-phase is only influenced by advection, divergence of its Reynolds stresses, and drag. Closure models are needed for the dispersed-phase Reynolds-stresses and the average relative velocity.

2.2 Reynolds stresses

A model often used for the Reynolds stresses, is the local-equilibrium model of Tchen and Hinze [3]. This model assumes that the dispersed-phase Reynolds-stresses and the continuous-phase Reynolds-stresses are directly proportional to each other:

$$\langle \mathbf{u}_2'\mathbf{u}_2' \rangle = \Gamma \langle \mathbf{u}_1'\mathbf{u}_1' \rangle \tag{3}$$

where Γ is given by:

$$\Gamma = \frac{\tau_1}{\tau_1 + \tau_2} \tag{4}$$

Here τ_1 is the continuous-phase Lagrangian integral time-scale. Although not shown here, the local-equilibrium model Tchen and Hinze was evaluated [1], and previous results, from Portela et al. [6], were confirmed. From an engineering perspective, Γ is a useful definition for modeling the dispersed-phase Reynolds-stresses in a turbulent pipe flow. Albeit the model assumes homogeneous isotropic turbulence, as long as both the dispersed-phase average radial-velocity and the continuous-phase average radial-velocity are close to zero, the particle-fluid interaction is similar to the interaction found in homogeneous isotropic turbulence, and from a pragmatic point of view, the model gives good results, except very close to the wall.

2.3 Interfacial momentum transfer

Besides the dispersed-phase Reynolds-stresses, the average relative velocity, $\mathbf{U}_{\mathrm{rel}}$, needs to be closed too. The definition of $\mathbf{U}_{\mathrm{rel}}$ is:

$$\mathbf{U}_{\mathrm{rel}} = \mathbf{U}_2 - \tilde{\mathbf{U}}_s \tag{5}$$

with $\tilde{\mathbf{U}}_s$ being the average continuous-phase velocity, as seen by the particles. This average velocity, $\tilde{\mathbf{U}}_s$, can also be written as:

$$\tilde{\mathbf{U}}_s = \mathbf{U}_1 + \langle \mathbf{u}_1' \rangle_2 \tag{6}$$

Here, $\mathbf{U}_1$, is the average continuous-phase velocity and, $\langle \mathbf{u}_1' \rangle_2$, is the average continuous-phase velocity fluctuation as seen by the particles, also known as the drift-velocity, $\mathbf{U}_d$. Assuming homogeneous turbulence and steady state, the transport equation of the continuous-phase velocity, as seen by the particles, reduces to [5]:

$$\mathbf{U}_d = -\mathsf{D}_{12} \frac{1}{\alpha_2} \nabla \alpha_2 \tag{7}$$

The drift-velocity, $\mathbf{U}_d$, is assumed to be in local equilibrium with the normalized gradient of the dispersed-phase volume-fraction, $\frac{1}{\alpha_2}\nabla\alpha_2$. They are proportional to each other with, D_{12} being the diffusion tensor. In this work, two different models for D_{12} were evaluated: (i) a simple Schmidt-number model, which assumes D_{12} to be a scalar, and (ii) a more advanced tensor-diffusion model, where D_{12} is a tensor.

The Schmidt-number model is the limiting case of the tensor-diffusion model for small particles and homogeneous isotropic turbulence. It assumes the diffusion of the drift-velocity, D_{12}, to be directly proportional to the eddy viscosity. The ratio between the two is defined as the Schmidt number, Sc, resulting in the following equation for, D_{12}:

$$\mathsf{D}_{12} = \frac{\nu_t}{\mathrm{Sc}} \tag{8}$$

The more advanced diffusion-tensor model [9] is given by:

$$\mathsf{D}_{12} = -\mathsf{G}_{12}^{-1} \langle \tilde{\mathbf{u}}_1' \mathbf{u}_2' \rangle^T \tag{9}$$

where $\langle \tilde{\mathbf{u}}_1' \mathbf{u}_2' \rangle^T$ is the transpose of the correlation tensor between the velocities of the continuous and dispersed phases. G_{12}^{-1}, is the inverse tensor of G_{12} and has the dimension of a time-scale. G_{12} represents crossing-trajectory effects, and is modeled by Simonin [9] as:

$$G_{12,ij} = -\frac{1}{\tau_{12,\perp}^t}\delta_{ij} - \left[\frac{1}{\tau_{12,\parallel}^t} - \frac{1}{\tau_{12,\perp}^t}\right]\frac{U_{\mathrm{rel},i}U_{\mathrm{rel},j}}{|\mathbf{U}_{\mathrm{rel}}|^2} \tag{10}$$

Here $\tau_{12,\parallel}^t$ and $\tau_{12,\perp}^t$ are the Lagrangian integral time-scales, measured along the particle paths, of the fluid velocity fluctuation, in the directions parallel and orthogonal to the mean relative velocity, respectively. These time-scales need to be modeled, and Simonin follows Csanady's approach (with $C_\beta = 0.45$) to obtain them from the standard Lagrangian integral time-scale of the fluid, τ_1^t. This gives the following set of equations to calculate the parallel and orthogonal Lagrangian time-scales, $\tau_{12,\parallel}^t$ and $\tau_{12,\perp}^t$:

$$\begin{aligned}
\tau_{12,\parallel}^t &= \tau_1^t\left[1 + C_\beta\zeta_r^2\right]^{-1/2} & \tau_{12,\perp}^t &= \tau_1^t\left[1 + 4C_\beta\zeta_r^2\right]^{-1/2} \\
\zeta_r^2 &= \frac{3}{2}\frac{|\mathbf{U}_{\mathrm{rel}}|^2}{k_1} & \tau_1^t &= \frac{1}{\beta_1}\frac{k_1}{\epsilon_1} \\
\beta_1 &= \frac{1}{2} + \frac{3}{4}C_0 & k_1 &= \frac{1}{2}\langle u_{1,i}' u_{1,i}' \rangle_1
\end{aligned} \tag{11}$$

Here k_1 is the continuous-phase turbulence kinetic energy and ϵ is the dissipation. For the standard Lagrangian integral time scale ,τ_1^t, Simonin uses a model proposed by Haworth and Pope with $C_0 = 2.1$ [9].

For the evaluation of the Schmidt-number model, the Schmidt number is calculated using the formula:

$$\mathrm{Sc} = -\frac{\nu_t}{\mathbf{U}_d}\frac{1}{\alpha_2}\nabla\alpha_2 \tag{12}$$

with ν_t, the eddy viscosity, defined as:

$$\nu_t = -\frac{\langle u'_{1,r}u'_{1,z}\rangle}{\frac{\partial U_{1,z}}{\partial r}} \tag{13}$$

In both equations 12 and 13 all variables on the right-hand-side are obtained from the data obtained from the DNS with particle tracking. The drift-tensor model is evaluated using equation 9 to calculate the theoretically predicted drift-velocity. This theoretically-predicted drift-velocity is then compared to the drift-velocity obtained from the DNS.

3 Results

We performed standard DNS with particle tracking on a non-uniform staggered grid [7]. At both ends of the pipe, which has a length of five diameters, periodic boundary conditions are applied. At the wall, the continuous phase has a no-slip boundary condition, and for the particles two different boundary conditions are implemented: (i) reflecting walls, which is akin to the situation encountered when a solid particle hits the wall and bounces back elastically, and (ii) absorbing walls, which is akin to the situation encountered when a liquid droplet in annular flow hits the liquid film at the wall and disappears. For each individual particle it is counted how often it has crossed the five diameter long computational domain of the pipe, and thus how far it has traveled in total. In the code, when a particle hits the wall and is absorbed, its counter for the total distance it traveled is reset to zero. It is reinserted at the very beginning of the pipe with an uniform random distribution over the cross-section, and with the instantaneous velocity equal to the local velocity of the continuous-phase. After having traveled approximately twenty-five pipe diameters down-stream, a reinserted particle is uncorrelated with its initial state, and is accounted for again in the particle statistics.

In the work of Young and Leeming [10], a classification is made which distinguishes three different particle deposition regimes, depending on the particle relaxation-time. (i) a regime where particle deposition is dominated by Brownian motion, (ii) a regime where particle deposition is dominated by the inertia of the particles, and (iii) an intermediate regime. In our simulations two different particles are used: (i) "light" particles with a non-dimensional relaxation-time of $\tau_2^+ = 10$, which are in the intermediate regime and (ii) "heavy" particles with a relaxation-time of $\tau_2^+ = 100$, which are in the inertia-dominated regime. Together with the two different boundary conditions, these

Table 1. An overview of the values of the different parameters used in the DNS.

all:	One-Way Coupling			
	Grid points (radial, tangential and streamwise)	64x128x128		
	Radial grid stretching	C_g	$=$	1.1745
	Friction Reynolds number	Re_τ	$=$	360
	Bulk Reynolds number	Re_B	$\approx$	5300
	Density ratio	ρ_2/ρ_1	$=$	1000
	Kinematic viscosity	ν_1	$=$	$1.5 \cdot 10^{-5}$
	Kolmogorov length-scale	η_k^+	$\approx$	2
	Kolmogorov time-scale	τ_k^+	$\approx$	4
case 1:	Reflecting walls			
	Particle relaxation-time	τ_2^+	$=$	10
	Particle diameter	d_2^+	$=$	0.42
case 2:	Reflecting walls			
	Particle relaxation-time	τ_2^+	$=$	100
	Particle diameter	d_2^+	$=$	1.34
case 3:	Absorbing walls			
	Particle relaxation-time	τ_2^+	$=$	10
	Particle diameter	d_2^+	$=$	0.42
case 4:	Absorbing walls			
	Particle relaxation-time	τ_2^+	$=$	100
	Particle diameter	d_2^+	$=$	1.34

two particles result in a total of four different simulations. An overview of the different parameters used in each simulation is given in table 1.

All parameters have been non-dimensionalized using the friction velocity, $\mathbf{u}_\tau$, and the viscosity, ν_1, denoted by the superscript $+$.

Figure 1 shows the normalized concentration profiles for all four cases. For all of them, except heavy particles combined with absorbing walls, a huge increase of the particle concentration towards the wall can be seen. The increase in concentration in the near-wall region compared to the center of the pipe can be as big as four decades. For reflecting walls, the two terms of the momentum balance (equation 2) that determine the particle-concentration profile are (i) the gradient of the radial Reynolds-stresses, driving the particles towards the wall, and (ii) the drag in radial direction, caused by the drift-velocity, pushing the particles away from the wall. Particles constantly move towards the wall. However, once they arrive there, they tend to stay there. Close to the wall, the turbulence kinetic energy of the continuous phase is small, therefore it is hard for a particle to gain enough kinetic energy to go back to the center of the pipe. Due to their bigger inertia, for heavy particles it is even more difficult to escape from the near-wall region. This explains the higher concen-

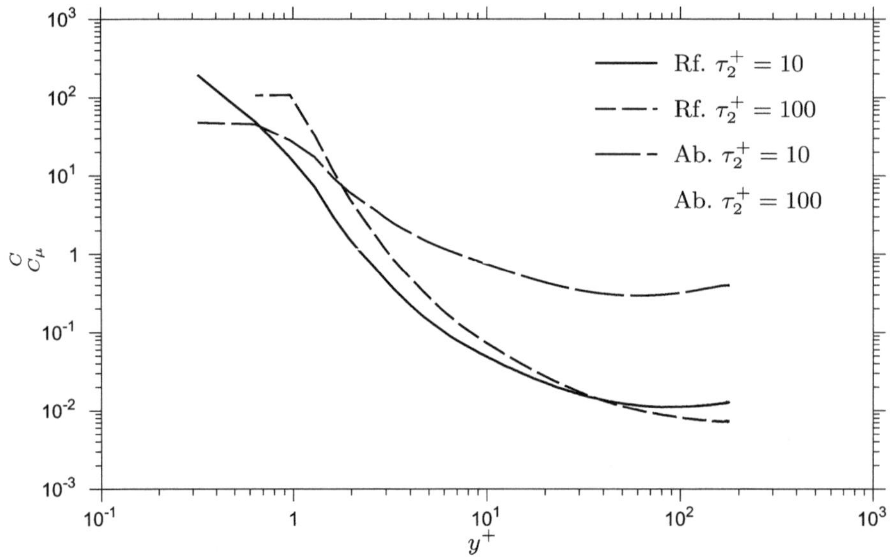

Fig. 1. Normalized particle-concentration as a function of the distance from the wall, for reflecting (Rf) and absorbing (Ab) walls, and non-dimensional particle relaxation-times of $\tau_2^+ = 10$ and $\tau_2^+ = 100$.

tration of heavy particles towards the wall for reflecting walls, compared to the concentration of the light particles.

For absorbing walls, an interesting difference between the light and heavy particles can be seen. Although all the particles which hit the wall are absorbed, the light particles are still able to pile-up near the wall. The light particles are slowed-down a lot in the near-wall region, and are not able to move straight to the wall. Their deposition rate is dominated by the small-scale near-wall turbulence. However, the heavy particles, due to their inertia, are accelerated by the large scale eddies in the center region of the pipe, and then have enough kinetic energy to move straight through the near-wall turbulence and hit the wall. Contrary to the light particles the deposition rate of the heavy particles is not dominated by small-scale near-wall turbulence, but by the large-scale center-region turbulence. As will be shown below, for heavy particles combined with absorbing walls, the relatively flat concentration profile makes the use of drift-velocity models quite problematic.

Figure 2 shows the radial, tangential and streamwise drift-velocities for all the cases. Due to symmetry arguments, the tangential component of the drift-velocity, $U_{d,\theta}^+$, should be equal to zero, as indeed is the case.

There exist two limits for the drift-velocity as a function of the particle relaxation-time. Particles with a relaxation-time approaching zero behave as

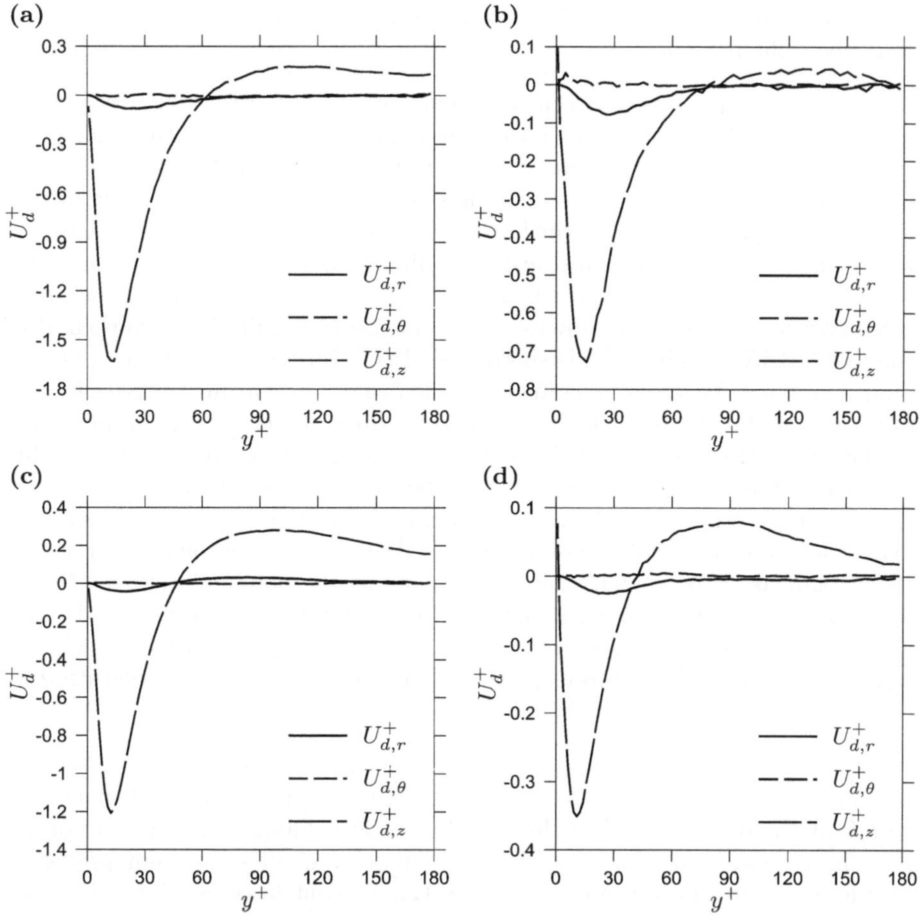

Fig. 2. Drift-velocities in radial, $U_{d,r}^+$, tangential, $U_{d,\theta}^+$, and streamwise direction, $U_{d,z}^+$, for (a) reflecting walls and a particle relaxation-time of $\tau_2^+ = 10$, (b) reflecting walls and a particle relaxation-time of $\tau_2^+ = 100$, (c) absorbing walls and a particle relaxation-time of $\tau_2^+ = 10$, and (d) absorbing walls and a particle relaxation-time of $\tau_2^+ = 100$.

tracers and there is no difference between the continuous-phase velocity they see and the "true" continuous-phase velocity, hence there is no drift-velocity for those particles. In the other limit, very-heavy particles are not influenced by the turbulence, they stay at their place (in the case of zero gravity) and also do not see another continuous-phase velocity. Therefore, for very-heavy particles, the drift-velocity is also equal to zero. Somewhere in between, there exists a particle relaxation-time for which a maximum drift-velocity occurs. This can also be seen in figure 2. The heavy particles are less correlated to the continuous-phase fluctuations, and consistently show lower drift velocities than the light particles.

Also interesting to see are the different streamwise drift-velocities. Firstly, in all graphs of figure 2, except for graph (d), at approximately the same place that the radial drift-velocity goes to zero, the streamwise drift-velocity changes sign. Due to preferential-concentration effects, in the large-scale turbulence in the center region of the pipe, the particles see an average continuous-phase velocity higher than the actual continuous-phase velocity, whereas in the small-scale turbulence in the near-wall region, they see a lower continuous-phase velocity. Secondly, the streamwise drift-velocities are quite large compared to the radial drift-velocities. This shows one of the shortcomings of the Schmidt-number model. As can be seen in equations 7 and 8, according to the Schmidt-number model, the drift-velocity in a certain direction is directly proportional to the concentration gradient in that direction. However, the streamwise direction is a homogeneous direction and there is no concentration gradient in this direction. Therefore, the Schmidt-number model wrongly predicts a zero drift-velocity in the streamwise direction. Although this wrong prediction of the streamwise drift-velocity can be a problem when one is interested in the residence time of the particles, for a lot of engineering purposes, one is only interested in the drift-velocity in the radial direction. The drift-velocity in radial direction is the component that determines the particle deposition at the wall, and thus the particle distribution.

Another case where the Schmidt-number model has problems, is light particles combined with absorbing walls. In graph (c), of figure 2, it can be seen that the drift-velocity in the radial direction changes sign, instead of approaching zero, as in the other figures. This change of sign is not predicted by the Schmidt-number model, because the concentration gradient does not change sign at that position.

In figure 3, the Schmidt number is shown for all cases. Since for this a-priory evaluation of the Schmidt number equation 12 is used, in the center region, a very small particle-concentration gradient is divided by a very small drift-velocity. This results in a lot of statistical-noise in the center region, and this region is left out in figure 3. However, when actually using the Schmidt-number model, equation 8 is used, and the almost zero particle-concentration gradient in the center region correctly results in a very small drift-velocity.

As explained above, the Schmidt number is not a very useful definition for absorbing walls. For light particles, the Schmidt number, calculated with equation 12, shows a strange change of sign, due to the change of sign of the radial drift-velocity. For heavy particles and absorbing walls, the Schmidt number is far from constant, because of the almost uniform concentration profile. However, for reflecting walls the situation is better. Except very close to the wall, the Schmidt number seems to be quite constant. For heavy particles

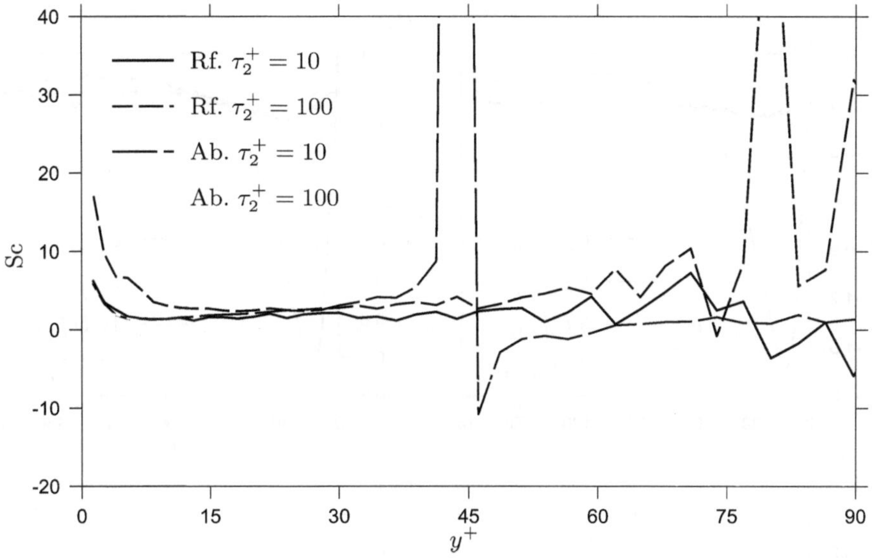

Fig. 3. Schmidt number in the radial direction for reflecting and absorbing walls and non-dimensional particle relaxation-times of $\tau_2^+ = 10$ and $\tau_2^+ = 100$, as a function of the distance from the wall.

an increase in the Schmidt number can be seen from $y^+ \approx 75$ on, but for light particles its value is almost constant. Also, its value is of the order of one, as was assumed by Young and Leeming [10]. From an engineering perspective, the Schmidt-number model is a good modeling candidate for light particles combined with reflecting walls. Especially, since using the Schmidt number to calculate the drift-velocity only involves taking the right value of the Schmidt number and modeling the eddy viscosity, whereas for the diffusion-tensor model a much more complicated modeling is needed.

Figure 4, shows both the drift-velocity obtained directly from the DNS, $U^+_{d,\mathrm{sim}}$, and the drift-velocity calculated using the diffusion-tensor model, $U^+_{d,\mathrm{D}}$, which was introduced in section 2. When using the model presented above for the two time-scales, the drift-tensor model gives quite a wrong prediction of the magnitude of the drift velocities. However, when introducing a new fitting constant, whose value is simply obtained by visually fitting the maximum value of $U^+_{d,\mathrm{sim}}$ to $U^+_{d,\mathrm{D}}$, the drift-velocities turn out to fit quite well for most cases. Graphs (a) and (b), in figure 4, show the drift-velocities for reflecting walls combined with light and heavy particles, respectively. In both graphs, especially for the light particles, the shape of the radial drift-velocity is predicted well. Also, a drift-velocity in the streamwise direction is predicted, although not its change of sign. Graphs (c) and (d), in figure 4, show the drift-velocities for absorbing walls. As was also seen in the concentration profiles (figure 1) the

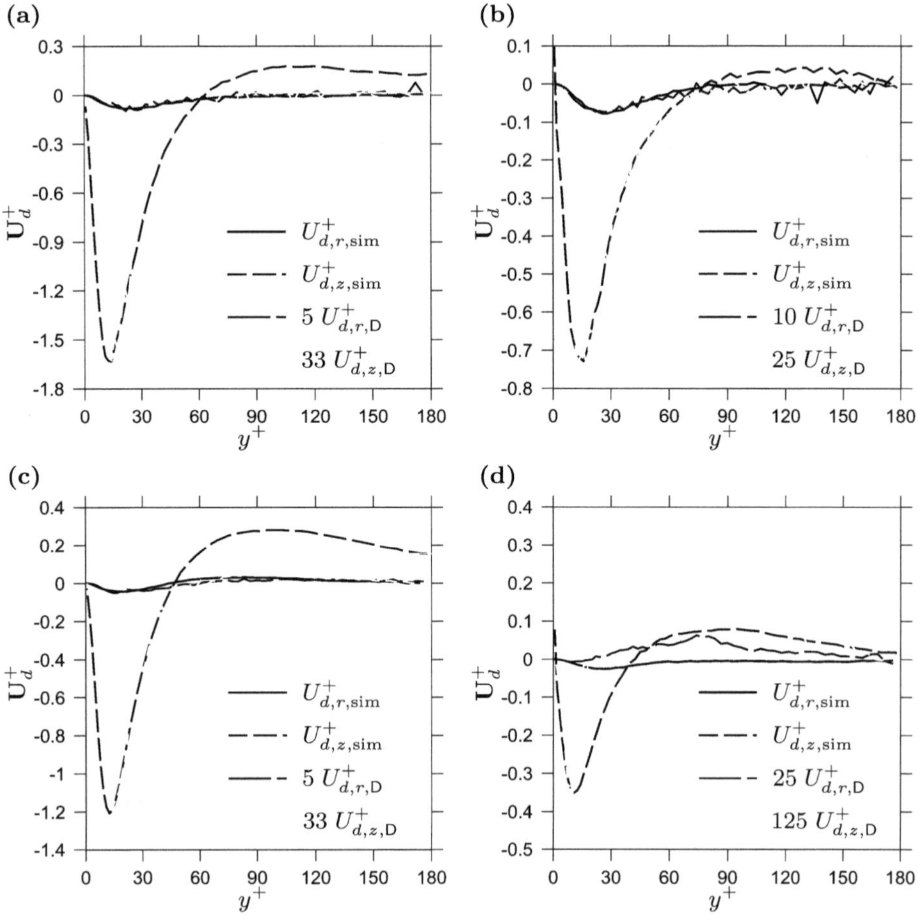

Fig. 4. Theoretically-predicted drift-velocities and drift-velocities obtained from DNS as a function of the wall-distance, for (a) reflecting walls and a particle relaxation-time of $\tau_2^+ = 10$, (b) reflecting walls and a particle relaxation-time of $\tau_2^+ = 100$, (c) absorbing walls and a particle relaxation-time of $\tau_2^+ = 10$, and (d) absorbing walls and a particle relaxation-time of $\tau_2^+ = 100$.

light particles in combination with absorbing walls behave very similarly to the case of light particles and reflecting walls. The drift-velocity is quite well predicted. However, for the heavy particles neither the radial drift-velocity nor the streamwise drift-velocity is predicted correctly. Only very close to the wall, the theoretically-predicted radial and streamwise drift-velocities correspond to the actual drift-velocities. Due to the fact, that the heavy particles move straight through the near-wall turbulence, the concentration profile is too flat, and both the radial and the streamwise drift-velocities are not well predicted anymore. Due to the flux of particles towards the wall, the basic assumption of the drift-velocity model, local equilibrium, is not valid anymore.

4 Conclusions

For all cases, except very close to the wall, for both light and heavy particles in combination with reflecting walls, and light particles combined with absorbing walls, the tensor-diffusion model holds. From a pragmatic perspective, local equilibrium and neglecting non-homogeneity of the flow, which are the basic assumptions of the drift-velocity model, seem to be valid for these cases.

The Schmidt-number model for interfacial momentum transfer, has some problems. It wrongly predicts a zero drift-velocity in streamwise direction and it does not predict the drift-velocities for absorbing walls correctly. However, for light particles with reflecting walls the radial drift-velocity is predicted correctly. Because for engineering purposes the radial direction is usually the only important direction, the Schmidt number is a simple way to predict the drift-velocity for this case.

The more advanced drift-tensor model predicts both the radial and streamwise drift-velocity with the wrong value, when one uses the "standard" time-scales proposed by Simonin. However, after multiplying the predicted drift-velocities with a fitting constant, the model did predict the right shape of the drift velocities for all cases, except heavy particles with absorbing walls. With a better model for the time-scales, and when modeling the particle-fluid velocity correlations correctly, this model is more flexible than the Schmidt-number model. For light particles and reflecting walls the Schmidt number seems better: there is less modeling involved and the results work equally well.

References

[1] A.M.P. Boelens. DNS study of local-equilibrium models in dilute particle-laden turbulent pipe flows. M.Sc. Thesis, Delft University of Technology, Juli 2006.

[2] H. Enwald, E. Peirano, and A.-E. Almstedt. Eulerian two-phase flow applied to fluidization. International Journal of Multiphase Flow, 22:21-66, 1996.

[3] J.O. Hinze. Turbulence. McGraw-Hill series in mechanical engineering. McGraw-Hill Book Company, second edition, 1975.

[4] M.R.Maxey and J.J. Riley. Equation of motion for a small rigid sphere in a nonuniform flow. Physics of Fluids, 26(4):883-889, April 1983.

[5] J.-P. Minier and E. Peirano. The pdf approach to turbulent polydispersed two-phase flow. Physics Reports, 352:1-214, 2001.
[6] L.M. Portela, P. Cota, and R.V.A. Oliemans. Numerical study of the near-wall behaviour of particles in turbulent pipe flows. Powder technology, 125:149-157, 2002.
[7] L.M. Portela and R.V.A. Oliemans. Eulerian-lagrangian DNS/LES of particle-turbulence interactions in wall-bounded flows. International Journal for Numerical Methods in Fluids, 43(9):1045-1065, 2003.
[8] M.W. Reeks. On the constitutive relations for dispersed particles in nonuniform flows – dispersion in a simple shear flow. Physics of Fluids A, 5(3):750-761, March 1993.
[9] O. Simonin. Two-fluid model approach for turbulent reactive two-phase flows. ERCOFTAC Summerschool, June 1999.
[10] J. Young and A. Leeming. A theory of particle deposition in turbulent pipe flow. Journal of Fluid Mechanics, 340:129-159, 1997.

Numerical particle tracking studies in a turbulent round jet

Giordano Lipari, David D. Apsley and Peter K. Stansby

School of Mechanical Aerospace and Civil Engineering - University of Manchester - Manchester - M60 1QD - UK g.lipari@manchester.ac.uk

1 Overview

This paper discusses numerical particle tracking of a 3D cloud of monodisperse particles injected within a steady incompressible free round turbulent jet. With regard to particle-turbulence interaction, the presented modeling is adequate for dilute suspensions [7], as the carrier and dispersed phase's solutions are worked out in two separate steps.

Section 2 describes the solution of the carrier fluid's Reynolds-averaged flow. The Reynolds numbers of environmental concern are generally high, and here the turbulence closure is a traditional k-ϵ model *á la* Launder and Spalding [2] with an *ad hoc* correction of Pope's to the ϵ equation to account for circumferential vortex stretching in a round jet [21]. The resulting mean-flow and Reynolds-stress fields are discussed in the light of the LDA measurements by Hussein *et al.* (1994) with $Re \sim 10^5$ [11].

Section 3 deals with the solution of the dispersed phase. The carrier fluid's unresolved turbulence is modeled as a Markovian process. We particularly refer to the reviews of Wilson, Legg and Thomson (1983) [30] and McInnes and Bracco (1992) [18]. Clouds of marked fluid particles, rather than trajectories, are used for visualizing the dispersing power of fluctuations. As fluctuations in inhomogeneous turbulence are known to entail sizeable spurious effects, the consistency of the Eulerian and Lagrangian statistics are checked by comparing the first- and second-order moments of the particle velocity with the mean flow and Reynolds stresses of the Eulerian solution, as well as the concentration fields from either solution.

Surprisingly, our tests failed to confirm the full effectiveness of the corrections proposed in either model. The particle spurious mean-velocity vanishes towards the jet edge, thus abating the unphysical migration towards low-turbulence regions. However, because of a residual disagreement between the Lagrangian and Eulerian mean velocities, mass conservation entails concentration profiles that do not follow the anticipated scaling. Possible reasons for this are discussed in the closing section.

Bernard J. Geurts et al. (eds), Particle Laden Flow: From Geophysical to Kolmogorov Scales, 207–219.

2 Eulerian modeling of the Reynolds-averaged jet flow

The symbolism for the Reynolds averaging is $u_i = \overline{u}_i + u_i'$ $(i = 1 \ldots 3)$.

The Eulerian governing equations are the elliptic Reynolds-averaged momentum and continuity equations and the transport equations for the turbulence scalars k and ϵ. In the Cartesian space and at the steady state, they read

$$\overline{u}_j \frac{\partial \overline{u}_i}{\partial x_j} = -\frac{1}{\rho}\frac{\partial p}{\partial x_i} - \frac{\partial R_{ij}}{\partial x_j} + \nu_t \frac{\partial^2 \overline{u}_i}{\partial x_i x_j}; \quad \frac{\partial \overline{u}_j}{\partial x_j} = 0.$$

The Reynolds-stress tensor R_{ij} is modeled with the Boussinesq approximation:

$$R_{ij} = -\nu_t \left(\frac{\partial \overline{u}_i}{\partial x_j} + \frac{\partial \overline{u}_j}{\partial x_i} \right) + \frac{2}{3} k \delta_{ij}.$$

The eddy viscosity is based on the scaling $\nu_t = C_\mu k^2/\epsilon$, where the fields of the k and ϵ turbulent scalars are computed with

$$\overline{u}_j \frac{\partial k}{\partial x_j} = -R_{ij} \frac{\partial \overline{u}_i}{\partial x_j} - \epsilon + \frac{\partial}{\partial x_j} \left[\left(\nu + \frac{\nu_t}{\sigma_k} \right) \frac{\partial k}{\partial x_j} \right],$$

$$\overline{u}_j \frac{\partial \epsilon}{\partial x_j} = -C_{\epsilon 1} \frac{\epsilon}{k} R_{ij} \frac{\partial \overline{u}_i}{\partial x_j} - (C_{\epsilon 2} - C_{\epsilon 3}\chi) \frac{\epsilon^2}{k} + \frac{\partial}{\partial x_j} \left[\left(\nu + \frac{\nu_t}{\sigma_\epsilon} \right) \frac{\partial \epsilon}{\partial x_j} \right]. \quad (1)$$

The constants C_μ through $C_{\epsilon 2}$ and σ_ϵ take the classic values of Launder and Spalding [2]. The extra term having $C_{\epsilon 3}$ in Eq.(1) depends on the vortex-stretching invariant χ:

$$\chi = \left(\frac{k}{2\epsilon} \right)^3 \left(\frac{\partial \overline{u}_i}{\partial x_j} - \frac{\partial \overline{u}_j}{\partial x_i} \right) \left(\frac{\partial \overline{u}_j}{\partial x_k} - \frac{\partial \overline{u}_k}{\partial x_j} \right) \left(\frac{\partial \overline{u}_k}{\partial x_i} - \frac{\partial \overline{u}_i}{\partial x_k} \right),$$

put forward by Pope (1978) to reconcile the spreading rate of the axial velocity profile [21], for which the uncorrected model would yield 0.11 against the measured 0.094-0.096. We used a value $C_{\epsilon 3} = 0.5$, lower than 0.7, to match more recent measurements than those originally used by Pope – Fig. 1*a*. Benefits and limitations of this correction are also discussed in [25].

The equations are solved in dimensionless form by normalization with the nozzle diameter D and jet exit velocity $\overline{u}_{(0,0)}$ and, exploiting axi-symmetry, in the polar-cylindrical space (x, r, θ). The origins of both frames are placed at the jet exit centerline.

The physical domain is the flow's symmetry half-plane 100×20 diameter long and wide respectively. A pipe protrudes into the domain for 8 diameters. A structured grid of 200×90 suitably clustered, orthogonal cells is more than adequate to resolve the expected gradients accurately. A plug-flow profile is assigned as inflow condition.

The general-purpose in-house research code STREAM, thoroughly described in [16], has been used to solve the above equations with a finite-volume method. Suffice it here to say that the norms of the algebraic-equation residuals could be brought down below the order of 10^{-13} routinely.

2.1 Results

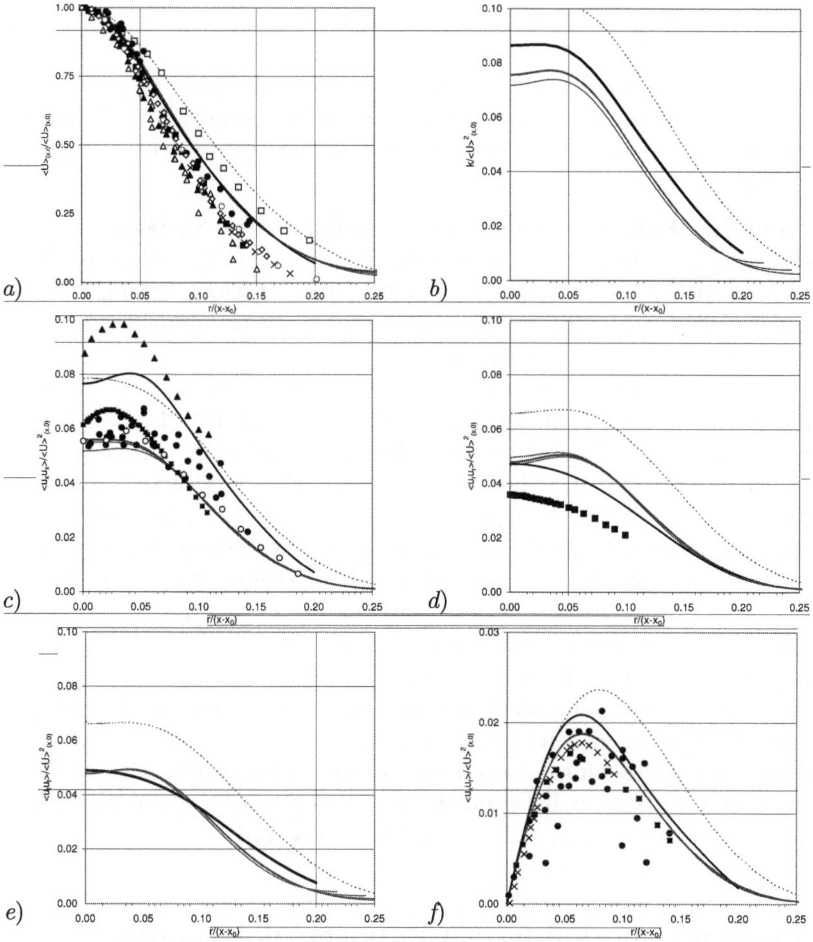

Fig. 1. Radial profiles in self-similarity variables of: *a)* $\overline{u}_x$; *b)* k; *c)* R_{xx}; *d)* R_{rr}; *e)* $R_{\theta\theta}$; *f)* R_{xr}. Thin lines: k-ϵ results at transects $x/D = 55, 74, 83, 92$. Bold line: measured data fit by Hussein *et al.* [11]. Dashed line: selected profile at $x/D = 74$ without Pope's correction. Symbols: measurements from [23] (◇), [15] (○), [19] (■),[9] (×), [28] (▲), [10] (•), [24] (△) and [8] (□).

Centerline values (not shown here) . The normalized axial velocity is expected to decay as x^{-1}. The inverse quantity $\overline{u}_{(0.0)}\overline{u}^{-1}_{(x.0)}$ increases linearly with a slope of 0.1564 very close to 0.1538 as measured. The virtual origin at $x_0 = 1.07D$, less than $4D$ as measured, implies a shorter zone of flow establishment.

The turbulent kinetic energy k is to decay as x^{-2} [2], and the quadratic fit of the inverse quantity is excellent from $x = 10D$ onwards. Similarly, the rate of turbulent energy dissipation ϵ should decay as x^{-4}, which is well reproduced by computation; the fourth-order polynomial fit to the inverse quantity has a leading-order coefficient of 0.0188 against 0.0208 as measured by Antonia *et al.* for $Re = 1.5 \cdot 10^5$ [5].

The turbulence timescale $T' = k/\epsilon$, therefore, increases as x^2, *e.g.* in accordance with Batchelor's analysis [6], with values ranging from 5 to 50 time units. This derived quantity is central to modeling the autocorrelated part in the fluctuation velocity – Eq. (2).

Radial profiles (**Fig. 1**). All plots are in self-similarity variables. Bold lines represent the data fits of the benchmark experiment [11]. Continuous lines show the computed quantities at selected far-field stations, which do collapse on a single curve, achieving self-similarity. Dashed lines indicate the k-ϵ performance without Pope's correction.

Symbols are used to report the LDA measurements of high-Re single-phase jets made available by some authors – Popper *et al.* (1974) [23], Levy and Lockwood (1981) [15], Modarres *et al.* (1984) [19], Fleckhaus *et al.* (1987) [9], Tsuji *et al.* (1988) [28], Hardalupas *et al.* (1989) [10], Prevost *et al.* (1996) [24] and Fan *et al.* (1997) [8] – prior to studying the two-phase case.

Pope's correction helps reduce to some extent the discrepancy between measured and computed flow quantities. A $C_{\epsilon 3}$-value to match the axial-velocity spreading rate (the point of ordinate 0.5 in Fig. 1*a*) worsens the prediction of the turbulent axial stress R_{xx} only (Fig. 1*c*), while those of R_{rr}, $R_{\theta\theta}$ and R_{xr} improve to match the correct proportion with the scaling variable $\overline{u}^2_{(x,0)}$ (Fig. 1*d*-*f*). The off-axis peaks of R_{rr} and $R_{\theta\theta}$ are not supported by the corresponding measurements though.

Further, the cross-comparison between the experimental data sets reveals a noticeable disagreement between the benchmark and the two-phase studies that, except for Fan *et al.*s, spread less than expected

3 Lagrangian modeling of the particulate cloud

Particles enter the domain at uniformly-distributed random positions on a pipe cross-section with a chosen input rate $\dot{N}$ (equal to 100 particles per unit time as a baseline default). The flow properties at a particles position are worked out by mapping the Cartesian position (x_1, x_2, x_3) into the computational grid (x, r) and, then, working out the Reynolds-averaged dependent variables with a bilinear interpolation. The resulting values are then mapped back into the Cartesian space with the standard vector/tensor rotation operations. The local instantaneous fluid velocity u_i is then created by summing $\overline{u}_i$ and u'_i as obtained from Sec 2 and 3.1 respectively. The Lagrangian equations of motion are finally resolved.

Initial values at the injection point $x_i^{(0)}$ are set as $v_i^{(0)} = \overline{u}_i[x_i^{(0)}]$ and $a_i^{(0)} = 0$, where v_i and a_i are the instantaneous velocity and acceleration to the dispersed phase. Given $x_i^{(n)}$ and $v_i^{(n)}$, the particle acceleration is computed as $a_i^{(n)} = a_i[v_i^{(n)}, u_i(x_i^{(n)})]$. A first-order Euler scheme yields the solution

$$v_i^{(n+1)} = v_i^{(n)} + a_i^{(n)} \Delta t, \quad x_i^{(n+1)} = x_i^{(n)} + v_i^{(n)} \Delta t.$$

When dealing with marked fluid particles, $v_i \equiv u_i$ and only the displacement equation is solved for.

A cloud then progresses and disperses within the previously-computed mean flow. Particles leave the domain if either $x_1 > 40$ or $r > 20$ diameters. The cloud reaches a statistically-steady state when the particles entering the domain equates in mean value to those leaving it, and the co-ordinates statistics start to oscillate closely around steady values. To compare Lagrangian and Eulerian statistics, the particle instantaneous properties are averaged first over the volumes of a monitoring grid and, then, over time.

3.1 Fluctuation velocity field

The results discussed here only regard statistically-independent fluctuation components which, at time $t = n\Delta t$ for $n > 1$, take their values from the Markov sequence:

$$u'^{(n)}_i = \sqrt{\Phi_{ii}^{(n)}}\beta_i^{(n)} + c_i F_{ii}^{(n)} u'^{(n-1)}_i + d_i. \tag{2}$$

For $n = 0$, $c_i = d_i = 0$.

In the first contribution, β_i is a Gaussian random number with zero mean and unit variance generated with the 'polar Marsaglia' method [1], and $\Phi_{ii} = \overline{\phi_i \phi_i}$ are the diagonal components of the 'randomness covariance matrix' defined farther in Eq.(3). (No summation convention on tensor components.)

The 'fluctuation variances', resulting from squaring and averaging (2), follow the sequence:

$$\overline{u'_i u'_i}^{(n)} = \Phi_{ii}^{(n)} + c_i^2 F_{ii}^{2(n)} \overline{u'_i u'_i}^{(n-1)}. \tag{3}$$

Thereby, on requiring $\overline{u'_i u'_i} = R_{ii}$ for consistency between the representations of the same flow viewed either in Eulerian or Lagrangian terms [22], and after little manipulation, the 'randomness variances' read

$$\Phi_{ii}^{(n)} = R_{ii}^{(n)} \left[1 - c_i^2 \left(R_{ii}^{(n-1)}/R_{ii}^{(n)}\right) F_{ii}^{2(n)}\right]. \tag{4}$$

In the second contribution, the c_i coefficients are scaling quantities dependent on modeling choices discussed below. Farther, F_{ii} belongs to the autocorrelation tensor, modeling the 'memory' of the previous value in the present component. We employ the exponential autocorrelation function

$$F_{ii} = \exp(-\Delta t/T_{p,i}), \tag{5}$$

wherein $T_{p,i}$ is a (directional) Lagrangian particle-memory timescale. The condition $\Delta t \ll T_p$ in F is recommended to limit the time-step dependence of the cloud dispersion in homogeneous turbulence, and a stricter limitation is anticipated in inhomogeneous turbulence [17, 31]. On taking $\Delta t < 0.1T_p$, $F > 0.905$ follows, *i.e.* the fluctuation is strongly autocorrelated.

In the third contribution, finally, d_i is a drift-correction term proposed by various authors to remove spurious effects arising from modeled fluctuations in inhomogeneous turbulence.

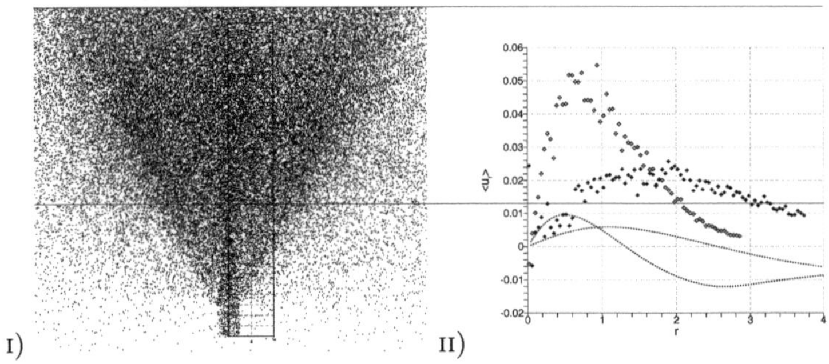

Fig. 2. Baseline Markovian fluctuation model. I) Side view of a 3D cloud of marked fluid particles (distorted scales), $\overline{N} \approx 69,000$; II) Radial profiles of radial mean velocity $\overline{u}_r$ at stations $x/D = 10$ and 20. Lines: $k-\epsilon$ results. Symbols: volume/time-averages of particles. The farther downstream the station, the lower the data set.

Baseline model (**Fig. 2**). The baseline model follows from the choices $T_{p,i} = T_p$ (isotropic timescale), $c_i = 1$ (no rescaling), $d_i = 0$ (no drift correction). The postulate $T_p \propto T'$ is commonly accepted, although there is no consensus on its value even for isotropic homogeneous turbulence. An interesting, direct measurement of this quantity in a jet flow, which effectively controls the cloud spread, was presented at this conference by Bourgoin *et al.* [3]. Reviews report estimates in the range of 0.06-0.63 [18, 26]. $K_{\mathrm{T}} = T_p/T'$ is here taken as 0.2 after Picart *et al.* (1986) [20]. For the resulting range of T_p here, this entails $\Delta t < 0.1$ time units.

Plot 2.I shows the cloud of marked fluid particles. The fluid particles injected from the pipe drift away against the entraining mean flow un-physically. This process, acting like spurious turbophoresis, is expected from stochastic differential equations properties [12, 14] or on statistical [27] and physical [29, 18] grounds.

A spurious velocity component appears in the radial mean-velocity profiles of Plot 2.II, for the curves of the Lagrangian particles (symbols) and Eulerian field (lines) should rather collapse, as the fluctuations (2) are required to have zero mean.

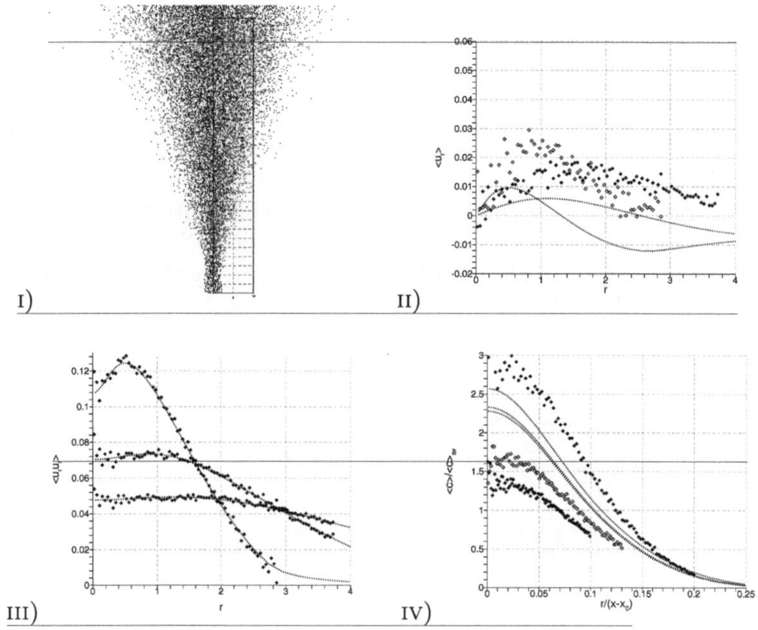

Fig. 3. Wilson-Legg-Thomson model. I) Side view of a 3D cloud of marked fluid particles (distorted scales), $\overline{N} \approx 25,200$; II) Radial profiles of radial mean velocity $\overline{u}_r$ at stations at $x/D = 10$ and 20; III) Radial profiles of radial turbulent stress R_{rr} at $x/D = 10$, 20 and 30. IV) Normalized profiles of concentration $\overline{c}$ at $x/D = 20$, 30 and 40. Same symbols as in Fig. 2.

WLT-1983 model (**Fig. 3**). Wilson, Legg and Thomson (1983) [30] elaborated on the previous works of Wilson *et al.* (1981) and Legg and Raupach (1982) [14] on fundamental atmospheric dispersion problems. Here, the baseline model is modified by assuming

$$a)\ c_i = \sqrt{R_{ii}^{(n)}/R_{ii}^{(n-1)}} \quad b)\ d_i = \frac{1}{2}\frac{\partial R_{ii}}{\partial x_i}(1-F)T_p, \tag{6}$$

Here $\Phi_{ii} > 0$ is guaranteed unconditionally owing to (6a), as $\Phi_{ii} = R_{ii}(1 - F^2)$ from (4). (Legg (1983) [13] and Thomson (1984) [27] presented further analyses.)

The cloud of marked fluid particles (Plot 3.I) now remains neatly confined within an ideal cone as expected [6]. However, the comparison of the Lagrangian and Eulerian radial mean-velocity profiles (Plot 3.II) shows that the drift velocity is reduced, but far from removed.

Plot 3.III shows the corresponding profiles of the radial turbulent stress which are closely collapsing, predominantly as an effect of the c_i rescaling coefficients. This ensures that the particle velocity variance locally corresponds to the turbulent stress field.

Plot 3.IV shows the self-similar concentration profiles with the ordinates normalized with the cross-section average, rather than centerline, value to reduce sensitivity on local scatter. Here, lines represent the solution of the advection-diffusion equation of a passive tracer; symbols represent the volume-time averages of the particle probability density (conditional to being at a given streamwise location). The particle concentration profiles are bell-shaped, but they do not reach a self-similar collapse as the centerline concentration (commented later in Fig. 5) decays faster than x^{-1} [6]. Plausibly, this is a consequence on the particle depletion off the axis caused by the residual spurious radial velocity of Plot 3.II.

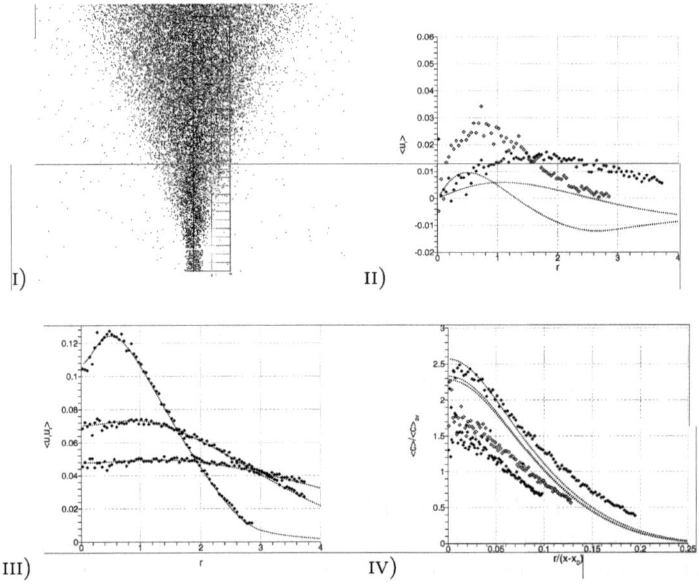

Fig. 4. Zhou-Leschziner/McInnes-Bracco model. Cloud population: $\overline{N} \approx 32,000$. Same symbols as in Fig. 3.

Interestingly, a separate test run with $d_i = 0$ also showed that similar results can be obtained by enforcing (6*a*) alone.

ZL/MB-1992 model (**Fig. 4**). McInnes and Bracco (1992) reviewed a number of random-walk models [18], including Wilson *et al.*'s (1981) [29] and Zhou and Leschziner's (1991) [4], yet leaving out the previous WLT-1983 model and, as a result, the assumption (6a). The baseline model is modified by assuming

$$a)\ T_{p,i} = K_{\mathrm{T}} \sqrt{\frac{2k}{3\overline{u_i^2}}} T' \quad b)\ d_i = \frac{\partial R_{ij}}{\partial x_j} \Delta t, \tag{7}$$

where d_i is the divergence of the stress vector acting on the surface element normal to x_i. The anisotropic Lagrangian time-scales (7a) originate from Zhou and Leschziner, while the drift term (7b) belongs to McInnes and Bracco.

This model predicts a more active dispersion, as the particle cloud now remains confined within a wider cone than Fig. 3 – see Plot 4.I. Again, the drift velocity is reduced, but not entirely removed – Plot 4.II.

Plot 4.III shows three pairs of profiles of the radial turbulent stress, which collapse as closely as in the WLT-1983 model, the less pronounced scatter resulting from a larger number of particles in the cloud. Plot 4.IV finally shows the concentration profiles in self-similar variables.

Overall, those plots make it apparent that the far-field difference between the WLT-1983 and ZL/MB-1992 formulations is one of detail, rather than character. We also recall that this analysis is restricted to independent fluctuations ($i.e.$ $\overline{u_i u_j} = 0$ while in fact $R_{ij} \neq 0$ for $i \neq j$), although separate runs having covariances accounted for in the fluctuations did not effect improvements.

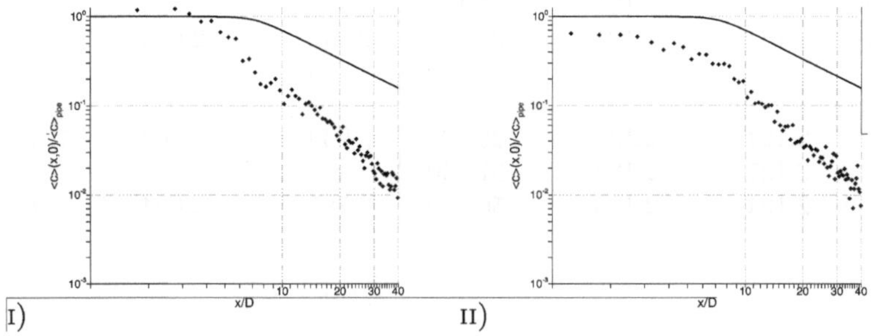

Fig. 5. Centerline concentration decay in the range $x/D = 1 - 40$. Ordinates are normalized with the initial in-pipe concentration. I) Wilson-Legg-Thomson model; II) Zhou-Leschziner/McInnes-Bracco model. Axes in log scale. Line: Eulerian passive tracer. Symbols: volume/time-averages of particles. The sloping line indicates the x^{-1} decay.

Finally, Fig. 5 compares the two fluctuation models and the Eulerian results with regard to the centerline concentration decay. Both models produce a decay faster than x^{-1} (in fact, very close to x^{-2}), while the ZL MB-1992 decay starts from earlier within the unmixed core that predicted by the Eulerian computation.

Sensitivity tests

The cloud radial dispersion has been measured in an aggregated fashion by time-averaging the standard deviations $\overline{\sigma}_2, \overline{\sigma}_3$ of the particle transversal co-ordinates (x_2, x_3) all over the cloud outside the pipe. The time-averaged number of particles, $\overline{N}$, has also been monitored.

Table 1*a* compares the above quantities obtained from either fluctuation model with a given input rate of $\dot{N} = 100$ and different time-steps. The results are sensibly insensitive to the time step-refinement in both models.

Table 1*b* shows the same quantities against the increasing input rate in otherwise identical conditions ($\Delta t = 0.04$ time units). The time-averaged number of particles increases proportionally to the input rate, and dispersion is correctly insensitive to the cloud population.

Table 1. Cloud dispersion sensitivity to fluctuation models and: *a*) time-step refinement; *b*) particle population. Time-averages of the bulk standard deviations of the particle x_i co-ordinates ($\overline{\sigma}_i$) and of particle number $\overline{N}$.

	WLT-1983			ZL/MB-1992		
a) Δt	$\overline{\sigma}_2$	$\overline{\sigma}_3$	$\overline{N}$	$\overline{\sigma}_2$	$\overline{\sigma}_3$	$\overline{N}$
0.0400	2.124	2.123	25,180	2.939	2.702	31,630
0.0100	2.115	2.125	25,120	–	–	–
0.0025	2.124	2.121	25,160	2.945	2.695	31,680
b) $\dot{N}$						
100	2.124	2.123	25,180	2.939	2.702	31,630
150	2.127	2.123	37,740	–	–	–
200	2.124	2.123	50,340	2.787	2.604	61,780

4 Closure

Regarding the carrier flow's mean properties. Pope's round-jet correction to the standard k-ϵ closure does improve the agreement with the benchmark LDA measurements of Hussein *et al.*, though not sufficiently to enable accurate numerical particle tracking (Fig. 1). A new value for Pope's $C_{\epsilon 3}$ constant has been proposed allowing the k-ϵ and experimental axial-velocity profiles to collapse.

Perhaps surprisingly, the benchmark and the published single-phase jet measurements carried out prior to two-phase jet experiments show marked discrepancies. Such basic inconsistencies shall affect a state-of-the-art calibration of particle-laden jet models.

Regarding the marked-fluid particles. Basic Markov fluctuation models are well known to produce a fluctuation field violating the divergence properties of the background flow (Fig. 2). We applied two mainstream modifications thereof, E's. (6) and (7) labeled as WLT-1983 and ZL/MB-1992, to a 3D cloud of marked fluid particles injected within the jet flow.

Both models prescribe a drift-correction term having the gradient of the turbulent normal stress – Eqs. (6*b*) and (7*b*). While WLT-1983 re-scale the autocorrelated fluctuation according to the change of the background turbulence intensity felt by the particle across the time step (6*a*), ZL/MB-1992 bring in the turbulence anisotropy by making the particle memory timescale directional (7*a*). A working assumption of ours regarded modeling independent random numbers, with the neglect of the fluctuation correlation implied in the shear stress, although this appeared not to be a crucial factor. Further, our testing departed from the WLT-1983 specification for being applied to a fully 3D dispersion problem, while the original was conceived for 2D atmospheric dispersion; and from the ZL/MB-1992 model for not having included the time cross-correlation between the fluctuation components.

Both approaches do result in a cloud with a bounded shape, reducing but not correcting the spurious drift. In fact, the mean of the particle radial fluctuations does not collapse onto the Eulerian mean value across the whole cloud radius – plots II in Figs. 3 and 4. The same bias did not affect the axial mean flow, as the axial turbulence inhomogeneity is greatly smaller than the radial one. In consequence, particle mass conservation requires that the Lagrangian concentration field does not follow the anticipated scalings, with a faster streamwise decay (Fig. 5) and larger spread (Plots IV in Figs. 3 and 4, where the maxima approach the average value, the more so the farther downstream, rather than keeping a constant ratio).

Both approaches, further, performed equally well as far as the Lagrangian fluctuation variances and the Eulerian turbulent normal stresses are concerned – plots III in Figs. 3 and 4.

Those results have been shown to be independent of a range of particle input rates and time-step sizes. Therefore, the residual spurious drift, and the consequences thereof, seem to point to modeling insufficiencies in the physical description and/or more sophisticated approaches needed for the numerical solution of stochastic differential equations. It should be stressed, however, that this study is for fluid particles. When particles are solid, inertial effects are likely to override subtle drift inaccuracies, so that the inadequacies highlighted here are relevant when both effects are of the same order.

Acknowledgments

GL is grateful to Mark Muldoon in the School of Mathematics at the University of Manchester for his statistics support. This piece of research was supported by the UK Engineering and Physical Sciences Research Council with grant GR/S25128/01.

References

[1] Kloeden PE Platen E Schurz H (1997) *Numerical Solution of SDE through Computer Experiments*. Springer Verlag.
[2] Wilcox DC (2000) *Turbulence Modeling for CFD*. DCW Industries Inc, La Cañada (US*a*).
[3] Bourgoin M Baudet C Cartellier A Gervais P Gagne Y (2006) In Geurts BJ *et al.* (eds.), *Particle-Laden Flow: from Geophysical to Kolgomorov Scales*. This volume.
[4] Zhou Q Leschziner (1991) In Stock DE *et al.* (eds.), *Gas-solid flows 1991*, FED Vol. 121, ASME. p.255-260
[5] Antonia RA Satyaprakash BR Hussain AKMF (1980) *Phy Fluids* 23(4):695-700
[6] Batchelor GK (1957) *J Fluid Mech* 3:67-80
[7] Elghobashi S (1994) *Appl Sci Res* 52:309-329
[8] Fan J Zhang X Chen L Cen K (1997) *Chem Eng J* 66:207-215
[9] Fleckhaus D Hishida K Maeda M (1987) *Exp Fluids* 5:323-333
[10] Hardalupas Y Taylor AMKP Whitelaw JH (1989) *Proc R Soc London A* 426:31-78
[11] Hussein HJ Capp SP George WK (1994) *J Fluid Mech* 258:31-75
[12] van Kampen NG (1981) *J Stat Phys* 24(1):175-187
[13] Legg BJ (1983) *Quart J R Met Soc* 109:645-660
[14] Legg BJ Raupach MR (1982) *Bound-Lay Meteorol* 24:3-13
[15] Levy Y Lockwood FC (1981) *Combust Flame* 40:333-339
[16] Lien FS Leschziner MA (1994) *Comput Methods Appl Mech Engrg* 114:123-148
[17] Lin C-H Chang L-FW (1996) *J Aerosol Sci* 27(5):681-694
[18] MacInnes JM Bracco FV (1992) *Phys Fluids A* 4(12):2809-2824
[19] Modarres D Tan H Elghobashi S (1984) *AIAA J* 22(5):624-630
[20] Picart A Berlemont A Gouesbet G (1986) *Int J Multiphase Flow* 12(2):237-261
[21] Pope SB (1978) *AIAA J* 16:279-280
[22] Pope SB (1987) *Phys Fluids* 30(8):2374-2379
[23] Popper J Abuaf N Hetsroni G (1974) *Int J Multiphase Flow* 1:715-726
[24] Prevost F Borée J Nuglish HJ Charnay G (1996) *Int J Multiphase Flow* 22(4):685-701
[25] Rubel A (1985) *AIAA J* 23:1129-1130
[26] Shirolkar JS Coimbra CFM Queiroz McQuay M (1996) *Prog Energy Combust Sci* 22:363-399
[27] Thomson DJ (1984) *Quart J R Met Soc* 110:1107-1120

[28] Tsuji Y Morikawa Y Tanaka T Karimine K Nishida S (1988) *Int J Multiphase Flow* 14(5):565-574
[29] Wilson JD Thurtell GW Kidd GE (1981) *Bound-Lay Meteorol* 21:423
[30] Wilson JD Legg BJ Thomson DJ (1983) *Bound-Lay Meteorol* 27:136
[31] Wilson JD Zhuang Y (1989) *Bound-Lay Meteorol* 49:309-316

Acceleration and velocity statistics of Lagrangian particles in turbulence

Guido Boffetta

Dip. Fisica Generale and INFN, via P.Giuria 1, 10125 Torino, Italy

Summary. The statistics of Lagrangian tracers is a fundamental problem in fully developed turbulence. On the basis of high resolution direct numerical simulations, velocity and acceleration statistics will be discussed. The first part will be devoted to ideal fluid tracers, while the second part will consider the more realistic case of finite size particles with inertia.

1 Introduction

The knowledge of the statistical properties of Lagrangian tracers advected by a turbulent flow is not only a fundamental problem in the theory of fully developed turbulence but also a fundamental ingredients for the development of stochastic models for different applications. Despite the importance of this problem, there are still relatively few experimental studies of Lagrangian turbulence [1, 2]. This is mainly due to the intrinsic difficulty to follow tracers for long times at high resolution in a turbulent flow. An alternative approach is given by direct numerical simulations, which have clear advantages in terms of accuracy and possibility to make simultaneous measurement of different statistical quantities albeit at a smaller Reynolds number.

This contribution discusses the statistics of Lagrangian velocity fluctuations and accelerations in turbulent flows on the basis of high resolution direct numerical simulations. Most of the results presented here were published in previous papers [3, 4, 5, 6] where the interested reader can find more details.

2 Numerical method

Direct numerical simulations of turbulent flow were done by using a parallel, fully de-aliased, pseudo-spectral code on an IBM-SP4 parallel computer at Cineca at resolution up to 1024^3. Energy is injected at the average rate ϵ by

Bernard J. Geurts et al. (eds), Particle Laden Flow: From Geophysical to Kolmogorov Scales, 221–228.

keeping constant the total energy in each of the first two wavenumber shells [7] and is dissipated by a normal viscosity operator. In stationary conditions, particles are injected into the flow and their trajectories integrated according to [8]:

$$\begin{aligned}\frac{d\mathbf{x}}{dt} &= \mathbf{v}(t) \\ \frac{d\mathbf{v}}{dt} &= -\frac{\mathbf{v}(t) - \mathbf{u}(\mathbf{X}(t),t)}{\tau_s}\end{aligned} \tag{1}$$

where $\tau_s = r^2/(3\beta\nu)$ is the response time of a particle of radius r and density ρ_p in a fluid of viscosity ν and density ρ_f, and $\beta = 3\rho_f/(\rho_f + 2\rho_p)$. These equations are valid for a dilute suspensions of heavy ($\beta \ll 1$), small, spherical particles. In the limit of $\tau_s \to 0$, equations (1) simplifies to $d\mathbf{x}/dt = \mathbf{u}(\mathbf{x}(t),t)$, i.e. the motion of fluid particles. Lagrangian velocity was calculated using linear interpolation on the Eulerian grid. Particles' positions, velocities and accelerations have been recorded along the particle paths about every $0.1\tau_\eta$.

The range of Stokes number investigated is $0 \leq St \leq 3.31$ with $1.9 \cdot 10^6$ particles at $St = 0$ and $0.5 \cdot 10^6$ for each $St > 0$. Table 1 contains the most important numerical parameters. Details can be found in [5, 6].

N	R_λ	T_E/τ_η	T/T_E	$L/\delta x$	$\eta/\delta x$
512	183	43.8	2.4	523	0.83
1024	284	54.6	2.4	1047	0.83

Table 1. Parameters of the numerical simulations. Resolution N, micro-scale Reynolds number R_λ, large-eddy turnover time $T_E = L/u_{rms}$, Kolmogorov timescale $\tau_\eta = (\nu/\varepsilon)^{1/2}$, total integration time T, box size L, grid spacing δx, Kolmogorov length-scale $\eta = (\nu^3/\varepsilon)^{1/4}$

3 Fluid particle statistics

The simplest statistical object of interest in Lagrangian turbulence is single particle velocity increment $\delta_t \mathbf{v} \equiv \mathbf{v}(t) - \mathbf{v}(0)$ following a Lagrangian trajectory. In homogeneous, isotropic fully developed turbulence, dimensional analysis predicts [9]

$$\langle \delta_t v_i \delta_t v_j \rangle = C_0 \varepsilon t \delta_{ij} \tag{2}$$

where ε is the mean energy dissipation and C_0 is a dimensionless constant. The remarkable coincidence that the variance of $\delta_t v$ grows linearly with time is the physical basis for the development of stochastic models of particle dispersion. It is important to recall that the diffusive nature of (2) is purely incidental, consequence of Kolmogorov scaling in the inertial range of turbulence.

It could be useful to recall the argument leading to the scaling in (2). Consider the velocity $v(t)$ advecting the Lagrangian tracer as the superposition of the different velocity contributions coming from turbulent eddies. After a time lag t the components associated with the smaller (and faster) eddies, below a certain scale r are de-correlated and thus at the leading order one has $\delta_t v = \delta_r v$. Within Kolmogorov scaling, velocity fluctuations at scale r is given by $\delta_r v \sim V(r/L)^{1/3}$ where V represents the typical velocity at the largest scale L. The correlation time of $\delta_r v$ scales as $\tau_r \sim \tau_0(r/L)^{2/3}$ and thus one obtains the scaling in (2) with $\varepsilon = V^2/\tau_0$.

Equation (2) can be generalized to higher-order moments with the introduction of a set of temporal scaling exponents $\xi(p)$:

$$\langle \delta_t v^p \rangle \sim (\varepsilon t)^{\xi(p)} \tag{3}$$

The dimensional estimation sketched above gives the prediction $\xi(p) = p/2$ but one might expect deviations in the presence of intermittency. In this case, a generalization can be easily developed on the basis of the multi-fractal model of turbulence [10, 11, 3]. The above dimensional argument is repeated for the local scaling exponent h, giving $\delta_t v \sim V(t/\tau_0)^{h/(1-h)}$. Integrating over the distribution of h one obtains the prediction [3]:

$$\xi(p) = \min_h \left[\frac{ph - D(h) + 3}{1 - h} \right] \tag{4}$$

The set of fractal dimensions $D(h)$ is related to the Eulerian structure function scaling exponents $\zeta(q)$ by the Legendre transform [10] $\zeta(q) = \min_h[qh - D(h) + 3]$. The standard inequality in the multi-fractal model $D(h) \leq 3h + 2$ implies for (4) that even in presence of intermittency, $\xi(2) = 1$. This is a direct consequence of the fact that energy dissipation enters into (2) at the first power.

Recent experimental results [2] have shown that Lagrangian velocity fluctuations are intermittent and display anomalous scaling exponents, as predicted by the above arguments. We remark that, despite the high Reynolds number of the experiments, the scaling range in temporal domain is very small. This is due to the presence of *trapping events* in which particles are trapped for relatively long times within small-scale vortices thus contaminating the inertial range scaling [5]. Therefore, an estimate of the scaling exponent $\xi(p)$ can be done only relatively to a reference moment (the ESS procedure [12].

Figure 1 shows the Lagrangian structure functions as obtained from our DNS for one component of the velocity. The inset shows that the relative exponents, as obtained from the ESS procedure, are in very well agreement with the multi-fractal prediction (4).

For very small time increments, $\delta_t v$ reproduces the acceleration of transported particles. It is now well known that turbulent acceleration is an extremely intermittent quantity, with a probability density function (pdf) characterized by large tails corresponding to fluctuations up to 80 times the root

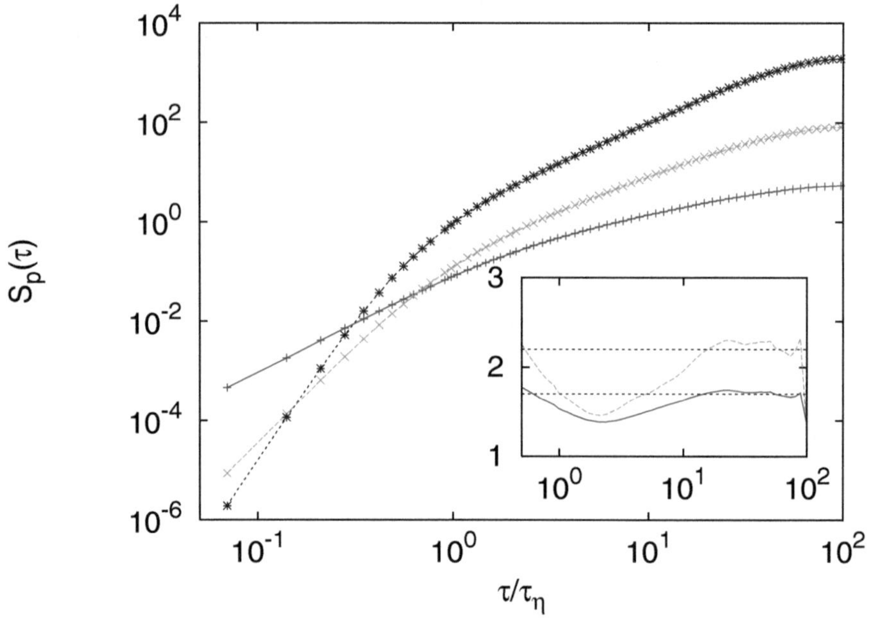

Fig. 1. Lagrangian structure functions of orders $p = 2, 4, 6$ (bottom to top) as a function of τ in log-log coordinates. Inset: ESS local slopes with respect to the second order structure function for $p = 4, 6$ (bottom to top). Straight lines correspond to the Lagrangian multi-fractal prediction (4) with the set of fractal dimensions $D(h)$ obtained from Eulerian velocity structure functions. Data refer to $R_\lambda = 284$.

mean square value a_{rms} [1]. The multi-fractal description of turbulence can be used also for predicting the shape of acceleration pdf. The basic idea [4] is to define the acceleration as the velocity increments at the smallest Kolmogorov scale, $a = \delta_{\tau_\eta} v/\tau_\eta$. Taking into account the fluctuations of the Kolmogorov scale and integrating over the distribution of large-scale velocity fluctuations, one ends with the prediction for the pdf of dimensionless Lagrangian acceleration $\tilde{a} = a/\sigma_a$:

$$\mathcal{P}(\tilde{a}) \sim \int \tilde{a}^{\frac{(h-5+D(h))}{3}} R_\lambda^{y(h)} \exp\left(-\frac{1}{2}\tilde{a}^{\frac{2(1+h)}{3}} R_\lambda^{z(h)}\right) dh \tag{5}$$

where $y(h) = \chi(h-5+D(h))/6+2(2D(h)+2h-7)/3$ and $z(h) = \chi(1+h)/3+4(2h-1)/3$ in which we put $\chi = 2\sup_h(D(h)-4h-1)/(1+h)$. We remark that the above expression contains an unphysical divergence for $a \to 0$ for several models $D(h)$. This is due to the fact that in general multi-fractal model cannot be used to describe small velocity (and acceleration) increments [4]. Therefore, we have to limit $\tilde{a}$ in a range of value above $\tilde{a}_{min} = O(1)$. This is the only free parameter in (5), as the set $D(h)$ is given from Eulerian measurements.

It is simple to recover from (5) the prediction in the case of non-intermittent Kolmogorov scaling. Assuming $h = 1/3$ with $D(h) = 3$ one

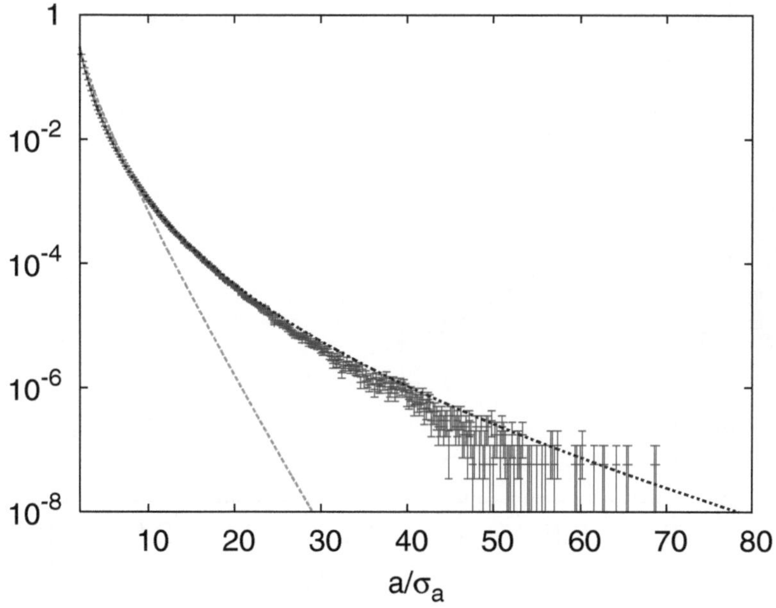

Fig. 2. Lin-log plot of the acceleration pdf. Points are the DNS data at $R_\lambda = 284$, solid line is the multi-fractal prediction and the dashed line if the K41 prediction.

obtains the form

$$\mathcal{P}(\tilde{a}) \sim \tilde{a}^{-5/9} R_\lambda^{-1/2} \exp\left(-\tilde{a}^{8/9}/2\right) \tag{6}$$

Figure 2 shows the pdf obtained from numerical data compared with the theoretical predictions. The agreement with (5) is remarkable, especially considered the range of fluctuations from 1 to $70\sigma_a$.

4 Heavy particle acceleration

We now consider the case of inertial particles with $St > 0$. It is well known that inertial particles spontaneously concentrate on inhomogeneous sets, a phenomenon called *preferential concentration* [13]. The clustering of inertial particles has important physical applications, from rain generation [14] to planet formation [15].

Preferential concentration has dramatic consequences on Lagrangian statistics, in particular on the acceleration as inertial particles sample the turbulent flow in non-homogeneous way. It is relatively simple to predict that in general turbulent acceleration for inertial particles will be reduced with respect to fluid. This is due to two different effects. From one hand, centrifugal forces will expel particles from most intense vortices. Therefore we expect a

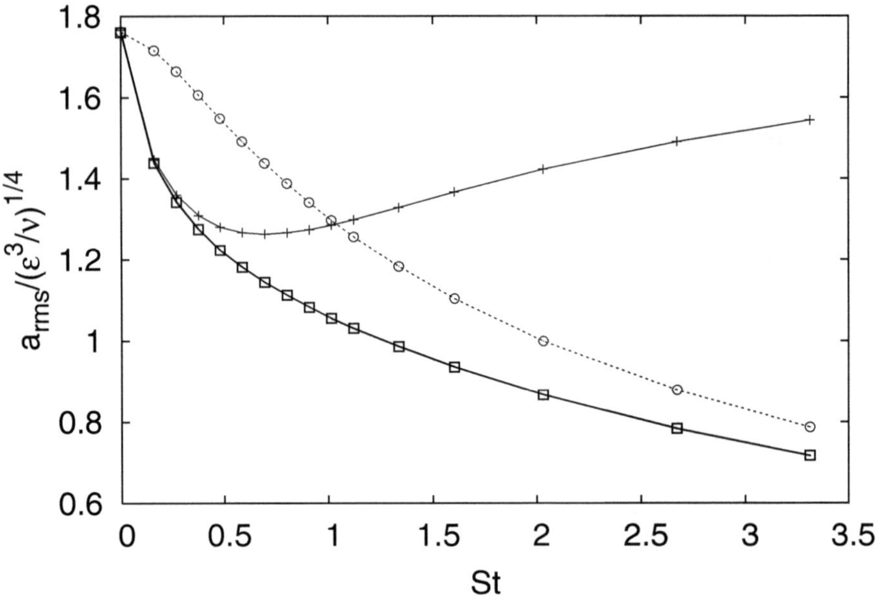

Fig. 3. Normalized acceleration variance $a_{rms}/(\varepsilon^3/\nu)^{1/4}$ as a function of Stokes number for $R_\lambda = 185$ (square). Acceleration of the fluid tracer conditioned to particle positions (crosses) and acceleration obtained from filtered velocity (circles).

preferential concentration on the region of minor pressure gradient (i.e. minor acceleration). On the other hand, the formal solution to (1) yields [6]

$$\mathbf{a}(t) = \frac{1}{\tau_s^2} \int_{-\infty}^{t} e^{-(t-s)/\tau_s} [\mathbf{v}(\mathbf{X}(t), t) - \mathbf{v}(\mathbf{X}(s), s)] \, ds \ . \tag{7}$$

therefore inertial particles acceleration is the result of a low-pass filtering of velocity differences and thus we expect the suppression of fluctuations at higher frequencies. These two mechanisms, preferential concentration and filtering, act in the same direction in a reduction of acceleration fluctuations.

Figure 3 shows the behavior of the acceleration variance as a function of Stokes number. At the maximum $St = 3.3$ the acceleration rms has been reduced by a factor 2.5 with respect the fluid case $St = 0$. In Fig. 3 we also show the two different contributions discussed above. The contribution from preferential concentration has been estimated by computing fluid acceleration conditioned to heavy particle positions. The agreement of this quantity with the inertial particle acceleration indicates that this is the main mechanism for $St < 0.5$. The second contribution has been computed by filtering the Lagrangian velocity with a low-pass filter which suppresses frequencies above τ_s^{-1} and then computing the acceleration as the time derivative of the filtered velocity. Figure 3 shows that filtered acceleration recovers inertial particle accel-

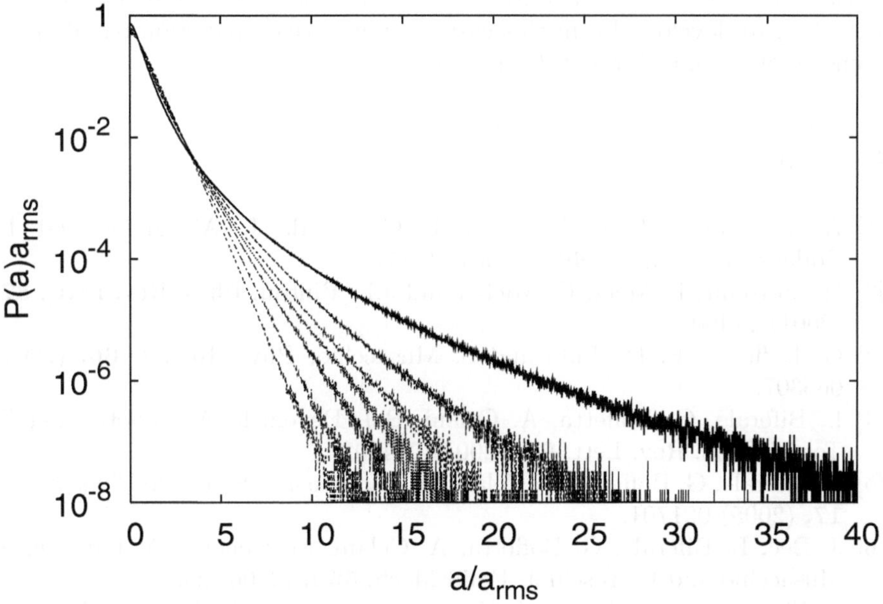

Fig. 4. Acceleration pdf in lin-log plot for inertial particles at $St = 0, 0.16, 0.37, 0.58, 1.01, 2.03, 3.31$ (from top to bottom) for the simulation at $R_\lambda = 185$.

eration for large St. In conclusion, the two described mechanisms, preferential concentration and filtering, are complementary as they become important in two limits of Stokes number.

The effects of inertia on acceleration pdf is shown in Fig. 4. Increasing St the inertial particle acceleration becomes less and less intermittent with a flatness which decreases from $F \simeq 30$ at $St = 0$ to $F \simeq 5$ at $St = 3.31$. The change in the shape of the pdf can be qualitatively captured by an argument similar to the one discussed for a_{rms} [6].

5 Conclusions

In conclusion, we have shown that single particle Lagrangian statistics in turbulence can be described by a simple extension of the multi-fractal formalism. Compared with other existing models, our proposal is very simple as it is based on the assumption that Lagrangian velocity increments are dimensionally related to Eulerian velocity increments.

In the case of inertial particles, we have shown that acceleration statistics is modified by two different mechanisms, namely preferential concentration and filtering and we have discussed which mechanism is dominant in the small and large Stokes number regimes. Of course, our comprehension of the effects of inertia on acceleration is here only at a qualitative level. It would be extremely interesting to develop also in this case a quantitative prediction based on an extension of the multi-fractal formalism.

References

[1] A. La Porta, G.A. Voth, A.M. Crawford, J. Alexander and E. Bodenschatz, Nature **409,** (2001) 1017.
[2] N. Mordant, P. Metz, O. Michel and J.F. Pinton, Phys. Rev. Lett., **87,** (2001) 214501.
[3] G. Boffetta, F. De Lillo and S. Musacchio, Phys. Rev. E **66,** (2002) 066307.
[4] L. Biferale, G. Boffetta, A. Celani, B.J. Devenish, A. Lanotte and F. Toschi Phys. Rev. Lett. **93,** (2004) 064502.
[5] L. Biferale, G. Boffetta, A. Celani, A. Lanotte and F. Toschi Phys. Fluids **17,** (2005) 021701.
[6] J. Bec, L. Biferale, G. Boffetta, A. Celani, M. Cencini, A. Lanotte, S. Musacchio and F. Toschi J. Fluid Mech. **550,** (2006) 349.
[7] S. Chen, G.D. Doolen, R.H. Kraichnan and Z.S. She Phys. Fluids A **5,** (1993) 458.
[8] M.R. Maxey and J. Riley, Phys. Fluids **26,** (1983) 883.
[9] A.S. Monin and A.M. Yaglom, *Statistical Fluid Mechanics*, (MIT Press, Cambridge, MA 1975).
[10] U. Frisch, *Turbulence*, (Cambridge Univ. Press, Cambridge, UK 1995).
[11] M.S. Borgas, Phil. Trans. R. Soc. Lond. A, **342**, (1993) 379.
[12] R. Benzi, S. Ciliberto, R. Tripiccione, C. Baudet, F. Massaioli and S. Succi, Phys. Rev. E **48,** (1993) R29.
[13] J.K. Eaton and J.R. Fessler, J. Multiphase Flow **20,** (1994) 169.
[14] G. Falkovich and A. Fouxon and M.G. Stepanov, Nature **419,** (2002) 151.
[15] A. Bracco, P.H. Chavanis, A. Provenzale and E.A. Spiegel, Phys. Fluids **11,** (1999) 2280.

Numerical studies of viscous effects for particle fluxes to perfectly absorbing spherical surfaces in turbulent environments: biological applications

H. L. Pécseli[1] and J. Trulsen[2]

[1] University of Oslo, Department of Physics, Box 1048 Blindern, N-0316 Oslo, Norway hans.pecseli@fys.uio.no
[2] University of Oslo, Institute of Theoretical Astrophysics, Box 1029 Blindern, N-0315 Oslo, Norway jan.trulsen@astro.uio.no

Summary. The motion of point-particles is studied by numerical simulations. We analyze the turbulent particle fluxes to perfectly absorbing spheres, which are moving with the flow. Particular attention is given to the effect of viscosity for the case where the radius of the sphere is comparable to or smaller than the Kolmogorov length scale. By relatively simple model arguments, we arrive at analytical expressions which give a very good agreement with the observed scalings.

1 Introduction

The problem of turbulent diffusion in neutral turbulent flows is often analyzed in terms of an initial value problem [1, 2], where particles are released at some reference time. However, for many applications, a boundary value problem is more relevant. As such an example we consider the turbulent particle flux to a perfectly absorbing spherical surface, which is a realistic physical model for many practical applications. This formulation of the problem serves, for instance, also as a model for predator-prey encounters in turbulent waters, and seems to be the application of the problem that has received most attention recently [3, 4]. For small predators, fish larvae for instance [5], it can often be assumed that their self-induced motion is small or negligible, and that they are passively convected by the local flow velocity, at least to a good approximation. Similarly, it can be assumed that their food (micro-zooplankton, for instance) is also passively convected by the same flow. The feeding process can be modeled by assuming that any individual prey entering a suitably defined "sphere of interception" is captured. (The surface is thus "virtual" in the sense that it does not disturb the flow.) Assuming that capture is certain, such surfaces can be considered as perfect absorbers.

Bernard J. Geurts et al. (eds), Particle Laden Flow: From Geophysical to Kolmogorov Scales, 229–241.

In turbulent waters, the prey flux to a passively convected predator is related to the problem of relative diffusion, but now considered as a boundary value problem, with the sphere of interception acting as a perfect absorber of prey. This is the standard reference model for this particular problem [6, 7]. We use this as a terminology in the following for simplicity and definiteness. The general interest in the problem arises essentially from the simple observation that the food concentration in the near region of a predator will rapidly be depleted, and without any self-induced motion a predator will be starving, unless the prey within its sphere of interception is replaced by turbulent mixing in the surrounding flow.

For the case where the radii of the absorbing surfaces are significantly larger than the Kolmogorov length-scale, η, the problem has been studied by analytical models as well as model laboratory experiments. Also by simple dimensional arguments, it is possible to obtain a scaling law for the asymptotic flux to a perfectly absorbing co-moving sphere as $\epsilon^{1/3}\mathcal{R}^{7/3}$, in terms of the radius $\mathcal{R}$ of the sphere and viscous dissipation of energy per unit mass of the fluid, ϵ. This scaling has found experimental support [8, 9]. It can be argued that the scaling law is related to the $\sim (r\epsilon)^{2/3}$ variation of the structure function in the universal Kolmogorov-Oubokhov subrange. The two results are thus not independent, and the proposed scaling can be seen also as a *consistency relation* between the saturated turbulent flux and the structure function associated with the universal subrange.

Table 1. The Kolmogorov length scale is here η and the specific energy dissipation (dissipated energy per gram fluid) is ϵ.

Open ocean	$\epsilon \sim 10^{-4} - 1$ mm^{2}s^{-3}	$\eta \sim 10 - 1$ mm
Shelf	$\epsilon \sim 10^{-1} - 1$ mm^{2}s^{-3}	$\eta \sim 2 - 1$ mm
Coastal zone	$\epsilon \sim 10^{-1} - 10^2$ mm^{2}s^{-3}	$\eta \sim 2 - 0.2$ mm
Tidal front	$\epsilon \sim 10$ mm^{2}s^{-3}	$\eta \sim 0.5$ mm

When relating the proposed parameter scaling of the turbulent flux we note that the universal subrange has only limited relevance for aquatic microorganisms, where relevant capture ranges are 1 - 5 mm. For herring larvae an estimated contact radius is ~ 3 mm, for instance [10, 11], while it can be smaller for other species. As indicated by the data summarized in Table 1, we can often find that the capture range of the micro-organisms is comparable to or even smaller than the Kolmogorov scale $\eta \equiv (\nu^3/\epsilon)^{1/4}$, and any result based on the scalings of the universal subrange will fail.

Although viscosity is damping small scale motions, this does not imply that the flow is entirely quiet when analyzed on sub-Kolmogorov scales: any locally linear velocity shear will remain unaffected by a standard viscosity term in the Navier-Stokes equation. If we consider a small surface (spherical or other), the flux through this surface will remain finite for any radius, also when $\mathcal{R} < \eta$. We found it important to extend our previous analysis also to capture ranges smaller than η where no analytical results seem to be available.

A viscous subrange for scales smaller than the Kolmogorov scale is observed in many turbulent flows [12], best when the Reynolds number is large. Unfortunately, in most cases only parts of the power spectrum for the velocity is determined. To our knowledge no particle tracking has yet been carried out in nature, where scales below the Kolmogorov scale could be resolved. The only available data originate from numerical flow simulations and some recent laboratory experiments [13]. We here use two dataset from numerical simulations [14, 15].

Our aim in this study is two-fold. We will analyze data obtained by numerical simulations, in order to determine an empirical relation for the fluxes to a perfectly absorbing spherical surface for those conditions. These results are presented in Section 2. Next, we suggest a model analysis which gives a consistency relation between these observed fluxes and the structure function also obtained in the given simulations. These results are presented in Section 3. Finally, Section 4 contains our conclusions.

2 Numerical results

Our database consists of two very large numerical simulations with parameters summarized in Table 2. The trajectories of point particles are followed, and the positions and velocities of these points are recorded at each time-step. The data are obtained by numerical solutions of the Navier-Stokes equation

$$\frac{\partial}{\partial t}\mathbf{u} + \mathbf{u} \cdot \nabla \mathbf{u} = -\frac{1}{\rho}\nabla p + \nu \nabla^2 \mathbf{u} \tag{1}$$

for incompressible motions, $\nabla \cdot \mathbf{u} = 0$, with $\mathbf{u} = \mathbf{u}(\mathbf{r}, t)$ being the fluid velocity field, p is the pressure, ρ is the mass density of the fluid, assumed constant, while ν is the kinematic fluid viscosity. For water we have $\nu \approx 0.89$ mm^2 s^{-1}. Viscous effects are here included by the physically correct viscous term, and not by some "hyper viscosity" model, which is otherwise often used for numerical modeling.

In figure 1 we show an example for an un-normalized structure function $\langle(\mathbf{u}(\boldsymbol{\xi}, t) - \mathbf{u}(\boldsymbol{\xi} + \mathbf{r}, t))^2\rangle$, obtained on the basis of the positions and velocities of the point-particles followed in simulation II, see Table 2. The structure function for the other dataset is very similar: the two inertial subranges are almost overlapping, with the most conspicuous difference being the somewhat larger Kolmogorov scale in the other case. The power-law variation with the separation distance of the structure functions for the viscous subranges in the two simulations are identical within the uncertainty of the estimates.

The detailed properties of the observed viscous subranges are a feature of the present numerical simulations and can not be claimed to be universally valid. They may be different from those observed in nature or in a laboratory.

Table 2. Simulation I (column 2) lasts 1167 time steps with $dt = 0.005$, simulation II (column 3) lasts 1841 time steps with $dt = 0.0023$. The Kolmogorov length scale is here η and the specific energy dissipation is ϵ, while τ_η is the Kolmogorov time scale. The Reynolds number is Re_λ.

dx	$2\pi/512 = 0.012272$	$2\pi/1024 = 0.0061359$
ν	2.05×10^{-3}	8.8×10^{-4}
ϵ	0.8853212	0.810878
$E = \frac{1}{2}\langle u^2 \rangle = \frac{3}{2}\langle u_{rms}^2 \rangle$	3.01	2.96
u_{rms}	1.42	1.40
λ	0.2642	0.179
Re_λ	183	286
η	$0.00993 = 0.81\ dx$	$0.0054 = 0.88\ dx$
τ_η	0.048	0.033
T	5.84	4.23

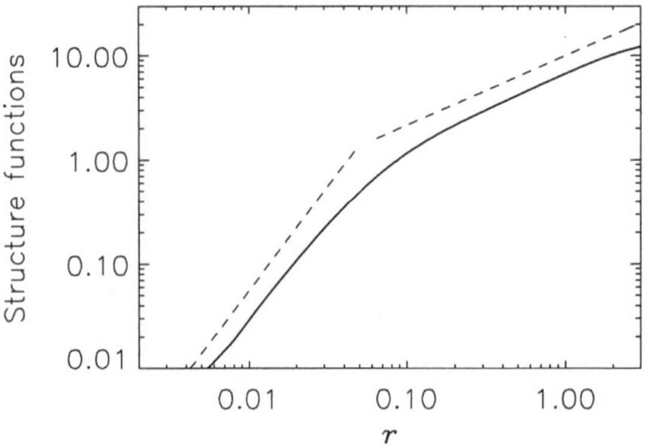

Fig. 1. Structure function, obtained on the basis of point-particle trajectories and velocities from the dataset with the largest Reynolds number, see Table 2. Dashed lines for $r^{2/3}$ and r^2 are inserted as references for the universal and viscous subranges, respectively.

In simulations we have an "artificial" small scale, the minimum length-scale resolved by the algorithm (see Table 2), which does not have a counterpart in natural turbulence. The so-called "bottle-neck" effect, is a feature of many numerical simulations of turbulent flows [16, 17]. Physically, the bottleneck effect arises because of the finite resolution of the sub-Kolmogorov scales, giving a lack of small scale vortices, which makes the energy cascade less effective around the Kolmogorov scale, as compared to the ideal, physical, conditions. Some numerical results may indicate that the bottle-neck effect is a consequence of viscous effects stabilizing small vortex tubes against the kink instability [18]. However, as argued in the following, the basic features of the viscous subrange are well represented by the simulations, and these are the most relevant for the present analysis. The numerical inaccuracies give rise only to slight deviations in the observed power-law subranges in the small scale part of the structure functions as compared to exact theoretical results.

In figure 1 we clearly identify a universal subrange $\sim r^{2/3}$ and a viscous subrange for smaller scales. For scales below the Kolmogorov scale the structure function follows a power-law r^{β} to a very god approximation, with $\beta \approx 2 \pm 0.05$. We note that the "cross-over" from the dissipation to the universal subrange occurs at a scale $r \approx 0.07$ which is approximately 10 times larger than the Kolmogorov scale for these parameters. Similar features are found also in other numerical simulations, and in particular also in the other one of the two datasets in Table 2. By following "virtual" absorbing reference spheres, we are thus able to encompass scales in the viscous as well as the universal subrange of the simulated flow. Due to the spatial periodicity of the simulations, for instance, the numerical results do not give a precise representation of turbulence in nature, but the results can serve as tests for analytical models. If the applicability of such models can be demonstrated here, they might be applied also to viscous subranges observed in nature, for instance.

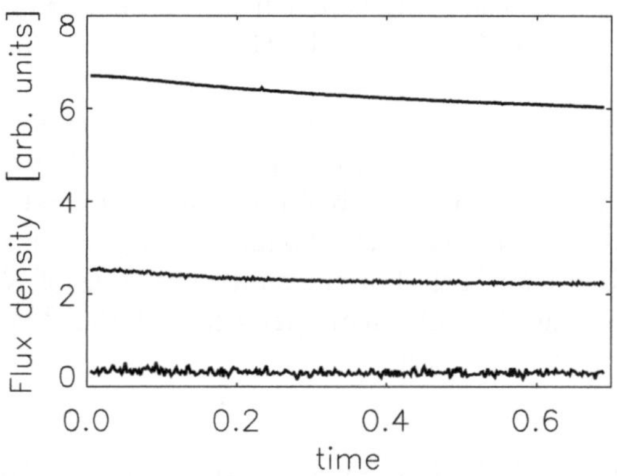

Fig. 2. Examples for turbulent particle fluxes per surface area, as obtained for a spherical absorbing surface. The curves are normalized with the same reference particle density. Radii are $\mathcal{R} = 1, 0.1$ and 0.01. Conditions are as in figure 1.

In figure 2 we show the turbulent particle flux to a perfectly absorbing spherical surface per surface area. The three radii are $\mathcal{R} = 1, 0.1$ and 0.01, i.e. one radius in the universal subrange, one close to the cross-over to the viscous subrange and one radius *in* the viscous subrange. It is readily evident that the fluxes do not scale directly with the surface. If that was the case, the curves should be on the top of each other.

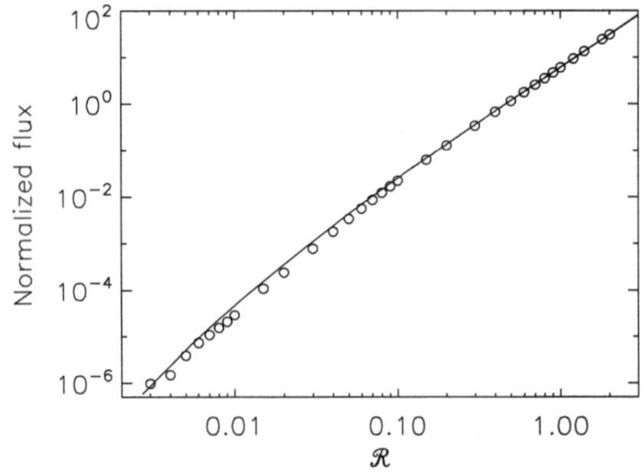

Fig. 3. Normalized turbulent flux $\langle J\rangle/n_0$ for different $\mathcal{R}$, for the case with $Re_\lambda =$ 286, see also Table 2. The full line shows the flux obtained by the simple model (10).

In figures 3 and 4 we show examples of the normalized asymptotic turbulent flux to perfectly absorbing spheres, in the sense discussed before, for different radii $\mathcal{R}$. The two cases shown refer to the two realizations listed in Table 2. The two corresponding structure functions are close to identical in the universal subranges. Consequently, the fluxes are similarly indistinguishable for that interval. As $\mathcal{R}$ becomes smaller, and comparable to η, we see the effect of the viscous subrange, and the flux variation with $\mathcal{R}$ becomes stronger, i.e. the curves become steeper in the double logarithmic presentation of figures 3 and 4. Since the Kolmogorov scale is slightly larger for the smaller Reynolds number case, the transition to the viscous subrange happens for larger $\mathcal{R}$ in that case. For the smallest values of $\mathcal{R}$ we have relatively few particles crossing the spherical surface, and the flux estimates become correspondingly more uncertain.

3 Analytical models for the saturated turbulent flux

We discuss here dimensional arguments for deriving a diffusion equation, which can model the turbulent flux of prey into the capture region of a small predator. Both predator and prey are assumed to be passively convected by the flow. When their separations are small, we can distinguish turbulent eddies which are larger than their separation and eddies which are smaller. The large eddies predominantly move predator and prey together, without changing their relative distance. The time variation of the separation is controlled by turbulent eddies with sizes which are smaller than or equal to the separation. The mean square value of the relative velocities at separation r are given by the second order structure function, which can be modeled by dimensional reasoning. Thus the fluid is characterized by its kinematic viscosity, ν, the

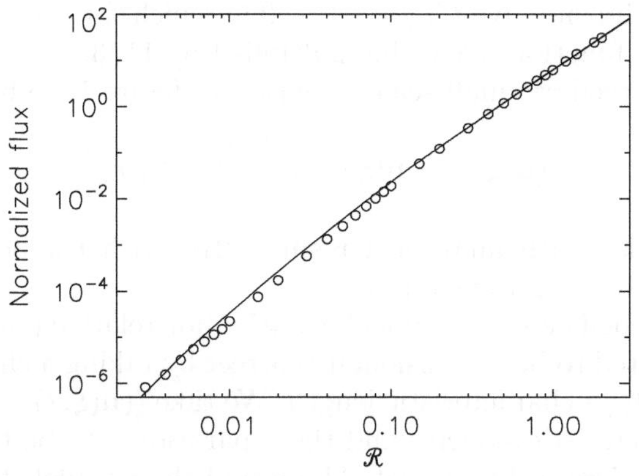

Fig. 4. Normalized asymptotic turbulent flux $\langle J\rangle/n_0$ for different $\mathcal{R}$, for the case with $Re_\lambda = 183$, see also Table 2. The full line shows the flux obtained by the simple model (10).

turbulence by the specific energy dissipation ϵ, and the geometry by the separation r, assuming that all physical scales are much larger than this. It is readily demonstrated [19] that the only possible normalized distance we can obtain by combinations of these quantities is $r\epsilon^{1/4}/\nu^{3/4}$, and the only quantity of dimension "velocity" is $(\nu\epsilon)^{1/4}$. We recognize the Kolmogorov length scale $\eta \equiv (\nu^3/\epsilon)^{1/4}$ as the normalizing length unit. Since the variation of a physical quantity can be written as dimensionless function of a dimensionless variable multiplied with a constant which is giving the correct physical dimension [20, 19], we have the second order structure function in the form

$$\left\langle (\mathbf{u}(\boldsymbol{\xi},t) - \mathbf{u}(\boldsymbol{\xi}+\mathbf{r},t))^2 \right\rangle = F\left(\frac{r\epsilon^{1/4}}{\nu^{3/4}}\right)\sqrt{\nu\epsilon}\,. \tag{2}$$

For homogeneous and isotropic turbulence the structure function introduced in (2) has a simple relation to the longitudinal as well as the transverse structure functions [12].

It is a general experience that viscosity is effective only for the smallest scales in the turbulence, and a subrange may exist where the structure functions are independent of viscosity. In order to have (2) independent of ν we evidently have to require the functional dependence of F to be the power $2/3$ of its argument, which is the only way ν will cancel. The result is the well known Kolmogorov-Oubokhov second order structure function for the universal subrange

$$\left\langle (\mathbf{u}(\boldsymbol{\xi},t) - \mathbf{u}(\boldsymbol{\xi}+\mathbf{r},t))^2 \right\rangle = C_K(r\epsilon)^{2/3}\,, \tag{3}$$

where the Kolmogorov constant C_K has been introduced. For the component structure function (where the velocity component is parallel to the separation vector $\mathbf{r}$), experiments give $C_K \sim 2.1. - 2.5$, which for the present definition of the structure function has to be multiplied by 11/3.

For the observed r^2 small scale variation in figure 1 we have,

$$\left\langle (\mathbf{u}(\boldsymbol{\xi},t) - \mathbf{u}(\boldsymbol{\xi}+\mathbf{r},t))^2 \right\rangle = C_\nu r^2 \frac{\epsilon}{\nu}, \tag{4}$$

in agreement also with analytical results [21], with C_ν being a numerical constant characterizing the subrange.

A diffusion coefficient in a model equation for relative particle separations can be constructed to be dimensionally correct by taking a characteristic velocity multiplied by a characteristic length. We take $\langle (\mathbf{u}(\boldsymbol{\xi},t) - \mathbf{u}(\boldsymbol{\xi}+\mathbf{r},t))^2 \rangle^{1/2}$ to be the characteristic velocity and the separation r to be the characteristic length. Both choices seem natural. The use of the parallel structure function merely implies a change of a numerical constant, which has to be determined by other means anyhow. We then find a diffusion coefficient as

$$K(r) \approx r\,(\nu\epsilon)^{1/4} \sqrt{F\left(\frac{r\epsilon^{1/4}}{\nu^{3/4}}\right)} \tag{5}$$

where F is still an unknown function which contains also a universal numerical constant as a coefficient.

For spherical geometry, we can then argue for a diffusion equation

$$\frac{\partial}{\partial \tau} n(\zeta,\tau) = \frac{1}{\zeta^2}\frac{\partial}{\partial \zeta}\zeta^2 K(\zeta)\frac{\partial}{\partial \zeta} n(\zeta,\tau) \tag{6}$$

where $n(\zeta,\tau)$ is the ensemble averaged particle density, being related to the distance-neighbor function introduced by L.F. Richardson [22]. We introduced normalized variables $\zeta = r\epsilon^{1/4}/\nu^{3/4} \equiv r/\eta$ and $\tau = t\sqrt{\epsilon/\nu}$. We here recognize the Kolmogorov time scale $\tau_\eta \equiv \sqrt{\nu/\epsilon}$ in the normalization of time.

We solve (6) for the case where we have a spherical absorbing boundary at $r = \mathcal{R}$, i.e. with $n(\mathcal{R}/\eta,\tau) = 0$. To have a steady state solution for (6), we obviously require the left hand side to vanish, implying that $\zeta^2 K(\zeta) dn(\zeta)/d\zeta = \text{const}$. The constant can be determined by $n(\zeta \to \infty) = n_0$, which is assumed known. We have

$$n(\infty) - n(\mathcal{R}/\eta) = \int_{\mathcal{R}/\eta}^{\infty} \frac{d}{d\zeta} n(\zeta) d\zeta = \int_{\mathcal{R}/\eta}^{\infty} \frac{\text{const}}{\zeta^2 K(\zeta)} d\zeta,$$

or

$$\text{const} = \frac{n_0}{\displaystyle\int_{\mathcal{R}/\eta}^{\infty} \frac{d\zeta}{\zeta^2 K(\zeta)}}.$$

The time stationary diffusion flux of particles to a perfectly absorbing sphere is in natural units given by $J_0 = 4\pi \mathcal{R}^2 K(\mathcal{R}) dn(r)/dr|_{r=\mathcal{R}}$, in the negative r-direction, towards the spherical surface. In normalized units:

$$\frac{J_0}{n_0} = \frac{4\pi}{\displaystyle\int_{\mathcal{R}/\eta}^{\infty} \frac{d\zeta}{\zeta^2 K(\zeta)}}\,, \tag{7}$$

where J_0 is the number of particles absorbed on the surface per *dimensionless* time units, and also the density is measured with a dimensionless length unit.

Hitherto, we have not specified the diffusion coefficient $K(r)$. We now introduce the estimate (5) or other forms derived from it.

The Richardson model [22] has $K(r) = C\epsilon^{1/3}r^{4/3}$, including a universal constant, C. Using this model for a three dimensional case, we find

$$\frac{J_0}{n_0} = C\frac{4\pi\epsilon^{1/3}}{\displaystyle\int_{\mathcal{R}}^{\infty} \frac{d\zeta}{\zeta^{10/3}}} = \frac{28\pi}{3}\,C\epsilon^{1/3}\mathcal{R}^{7/3}\,, \tag{8}$$

having the well known $\epsilon^{1/3}\mathcal{R}^{7/3}$ scaling, which has been confirmed experimentally [8] and numerically [15]. The model equation (6) together with the present diffusion coefficient $K(r) = C\epsilon^{1/3}r^{4/3}$ has been studied in detail [23].

For a slightly more general case where the second order structure function has a (large) range of power-law variation, $\langle(\mathbf{u}(\boldsymbol{\xi},t) - \mathbf{u}(\boldsymbol{\xi}+\mathbf{r},t))^2\rangle \sim r^{\beta}$, we have $K(\zeta) \sim \zeta^{1+\beta/2}$ and find

$$\frac{J_0}{n_0} = C\,2\pi(4+\beta)\mathcal{R}^3\sqrt{\frac{\epsilon}{\nu}}\left(\left(\frac{\nu^3}{\epsilon}\right)^{1/4}\frac{1}{\mathcal{R}}\right)^{1-\beta/2}\,, \tag{9}$$

which reproduces (8) for $\beta = 2/3$. In particular, for fixed ν and ϵ we have the scaling $J_0 \sim \mathcal{R}^{2+\beta/2}$. If ν and $\mathcal{R}$ are kept constant, we have the scaling $J_0 \sim \epsilon^{(2+\beta)/8}$, while for ϵ and $\mathcal{R}$ constant, we have $J_0 \sim \nu^{(2-3\beta)/8}$. The result (9) can be seen as a predicted consistency relation between a (long) spectral subrange and the associated particle flux to a perfectly absorbing spherical surface having a radius in that range.

We are particularly interested in analyzing the transition from the universal subrange to a viscous subrange. No exact analytical model exists, but assuming that for some length-interval the viscous subrange can be approximated by a power-law r^2, we can postulate an interpolation formula as

$$K(r/\eta_0) \sim \frac{r}{\eta_0}\frac{(r/\eta_0)^{1/3}r/\eta_0}{\sqrt{(r/\eta_0)^{2/3} + (r/\eta_0)^2}} \equiv \frac{(r/\eta_0)^{7/3}}{\sqrt{(r/\eta_0)^{2/3} + (r/\eta_0)^2}}\,, \tag{10}$$

apart from a numerical constant, which has to fitted empirically. It is readily seen that this expression reproduces the previous results for small as well as large normalized separations, r/η_0.

We can estimate the "cross over" scale η_0 between the viscous and the universal subrange by equating (3) and (4) to have $C_K(\eta_0\,\epsilon)^{2/3} = C_\nu\eta_0^2\,\epsilon/\nu$:

$$\eta_0 = \left(\frac{C_K}{C_\nu}\right)^{3/4}\left(\frac{\nu^3}{\epsilon}\right)^{1/4} \equiv \left(\frac{C_K}{C_\nu}\right)^{3/4}\eta\,, \tag{11}$$

in terms of the Kolmogorov length η. We use the generally accepted value for the Kolmogorov constant $C_K \approx 2.1$ for the parallel-velocity component structure function, together with the multiplication factor of 11/3 mentioned before, and the analytical result [21] for $C_\nu = 1/3$, as relevant for the structure function in the form presented in figure 1. We then find $\eta_0 \approx 11\,\eta$. If we choose to use the longitudinal structure functions in the universal and viscous subranges to determine η_0, we will find a slightly different numerical coefficient. The cross over scale is sometimes defined [24] with an "ad hoc" numerical coefficient of 2π for η. This value is, as we have seen, approximately a factor 2 too small.

Using the model (10), we obtain numerical solutions for the normalized flux as shown in figures 3 and 4 by a solid line. The numerical constant is fitted by the flux at the largest $\mathcal{R}$-values. The general agreement is quite good. For the universal subrange the results are close to identical, and indeed the two ϵ–values are very close, see Table 2. For $\mathcal{R} < \eta_0$ the two curves deviate, since the viscous subranges are somewhat different with the present simulation data. The model curves follow the flux-values equally well in both cases.

We note that there is only a minor difference in the variation of the flux with radius $\mathcal{R}$ in the viscous range ($\sim \mathcal{R}^3$) and in the universal range ($\sim \mathcal{R}^{7/3}$), in spite of the pronounced differences in the second order structure function, where we have $\sim r^2$ and $\sim r^{2/3}$, respectively, for the two subranges.

As an approximation we might accept power laws in $K(\xi)$ entering (9), as they are indeed often found to give a good experimental fit in a subrange. For a general case, we believe that (7) can be used for modeling by inserting a phenomenological model for K, and the integral then solved numerically.

4 Conclusions

We have analyzed the turbulent flux to a perfectly absorbing spherical surface embedded in a turbulent flow. Using results from a numerical simulation which resolves also parts of a viscous subrange, we are able to cover cases where the radii of the spheres are in the universal as well as in the viscous subranges.

We have developed a simple model which gives a scaling law for the flux which is in very good agreement with the observations. For the universal subrange we recover the well known $\epsilon^{1/3}\mathcal{R}^{7/3}$ scaling. Since the scaling of the saturated turbulent flux can be interpreted as a universal relation to be derived from the structure functions, we can take the observed viscous subrange, which quite accurately follows r^2 for our case. Applying this variation to our analytical model, we find a flux variation $\sim \mathcal{R}^3$, which is in very good agreement with the observations, see figures 3 and 4. We point out that the good agreement between the model and the simulation results is not automatically given: it is not self evident that a diffusion equation in the form (6) is useful for the turbulent flux. Indeed, if we apply it uncritically to the full time evolution of the flux, we are likely to get results of little or no value [8]. The interesting observation here is that (6) gives a good agreement for the *asymptotic* fluxes, apart from a universal constant which has to be obtained empirically.

It is interesting to note that the beginning of a viscous subrange might have been observed already in the studies of laboratory data [8], where the variation of the turbulent flux with $\mathcal{R}$ had a tendency to become steeper than the predicted $\epsilon^{1/3}\mathcal{R}^{7/3}$-law, when the smallest $\mathcal{R}$-values were considered. In that case, however, the transition from a universal to a viscous subrange in the structure function could not be unambiguously determined.

It is important to note that the scale η_0 in (11) is approximately one order of magnitude larger than the Kolmogorov scale, η, with relevant numerical values summarized in Table 1. It can easily happen that the range of interception for a micro-organism is smaller than η_0, and studies of the viscous subrange may be particularly relevant for accurate models for the predator-prey encounter rate of aquatic micro-organisms.

We expect the results of the present analysis to be relevant for very small scale (sub Kolmogorov scale) transport in turbulent flows. We have already mentioned the possible importance for our understanding of the feeding process of aquatic micro-organisms, but models for the dynamics of coagulation in turbulent environments [25] has also many similarities with that analysis. Our results can be relevant also for such studies.

Acknowledgments

We thank Prof. Guido Boffetta for giving us access to the database from the numerical simulations, and for his interest and comments. We thank also Jakob Mann and Søren Ott for valuable discussions. The authors were in part supported by the "Effects of North Atlantic Climate Variability on the Barents Sea Ecosystem" (ECOBE) project. The present study was initiated while the authors were affiliated with the Norwegian Center for Advanced Studies.

References

[1] Batchelor, G. K. 1952, Diffusion in a field of homogeneous turbulence II. The relative motion of particles. *Proc. Cambridge Philos. Soc. (London)*, **48**, 345–362.

[2] Roberts, P. H. 1961, Analytical theory of turbulent diffusion. *J. Fluid Mech.*, **11**, 257–283.

[3] Sundby, S. and Fossum, P. 1990, Feeding conditions of arcto-norwegian cod larvae compared with the Rothschild-Osborn theory on small-scale turbulence and plankton contact rates. *J. Plankton Res.*, **12**, 1153–1162.

[4] Kiørboe, T. and Saiz, E. 1995, Planktivorous feeding in calm and turbulent environments, with emphasis on copepods. *Mar. Ecol. Prog. Ser.*, **122**, 135–145.

[5] Muelbert, J. H., Lewis, M. R. and Kelley, D. E. 1994, The importance of small-scale turbulence in the feeding of herring larvae. *J. Plankton Res.*, **16**, 927–944.

[6] Rothschild, B. J. and Osborn, T. R. 1988, Small-scale turbulence and plankton contact rates. *J. Plankton Res.*, **10**, 465–474.

[7] Osborn, T. 1996, The role of turbulent diffusion for copepods with feeding currents. *J. Plankton Res.*, **18**, 185–195.

[8] Mann, J., Ott, S., Pécseli, H. L., and Trulsen, J. 2005, Turbulent particle flux to a perfectly absorbing surface. *J. Fluid Mech.*, **534**, 1–21.

[9] Mann, J., Ott, S., Pécseli, H. L. and Trulsen, J. 2006, Laboratory studies of predator-prey encounters in turbulent environments: effects of changes in orientation and field of view. *J. Plankton Res.*, **28**, 509–522.

[10] Blaxter, J. H. S. and Staines, M. E. 1971, Food searching potential in marine fish larvae. In D. J. Crisp, editor, *Proceedings of the 4th European Marine Biology Symposium*, pages 467–485. Cambridge University Press.

[11] Lewis, D. M. and Pedley, T. J. 2000, Planktonic contact rates in homogeneous isotropic turbulence: theoretical predictions and kinematic simulations. *J. Theor. Biol.*, **205**, 377–408.

[12] Hinze, J. O. 1975, *Turbulence.* McGraw-Hill, New York, 2 edition.

[13] Mann, J. and Ott, S. 2006, Private communications: Recent laboratory experiments at the Risø National Laboratory has also demonstrated an r^2-variation for the viscous subrange of the velocity structure function.

[14] Biferale, L., Boffetta, G., Celani, A., Lanotte, A., and Toschi, F. 2005, Particle trapping in three-dimensional fully developed turbulence. *Phys. Fluids*, **17**, 021701.

[15] Boffetta, G., Pécseli, H. L., and Trulsen, J. 2006, Numerical studies of turbulent particle fluxes into perfectly absorbing spherical surfaces. *J. Turbulence*, **7** (22), 1–16. doi:10.1080/14685240600573138.

[16] Lohse, D. and Müller-Groeling, A. 1996, Anisotropy and scaling corrections in turbulence. *Phys. Rev. E*, **54**, 395–405.

[17] Dobler, W., Haugen, N. E. L., Yousef, T. A., and Brandenburg, A. 2003, Bottleneck effect in three-dimensional turbulence simulations. *Phys. Rev. E*, **68**, 026304.

[18] Woodward, P. R., Porter, D. H., Edgar, B. K., Anderson, S., and Bassett, G. 1995, Parallel computation of turbulent fluid-flow. *Comp. Appl. Math.*, **14**, 97–105.

[19] Chandrasekhar, S. 1957, The theory of turbulence. *J. Madras Univ. B*, **27**, 251–275.

[20] Buckingham, E. 1914, On physically similar systems; illustrations of the use of dimensional equations. *Phys. Rev.*, **4**, 345–376.

[21] Davidson, P. A. *Turbulence. An introduction for scientists and engineers.* Cambridge University Press, UK, 2004.

[22] Richardson, L. F. 1926, Atmospheric diffusion shown as a distance-neighbour graph. *Proc. Roy. Soc. London, Ser. A*, **6**, 709–737.

[23] Pigolotti, S., Jensen, M. H., and Vulpiani, A. 2006, Absorbing processes in Richardson diffusion: Analytical results. *Phys. Fluids*, **18**, 048104.
[24] Mann, K. H. and Lazier, J. R. N. 1996. *Dynamics of Marine Ecosystems.* Blackwell Science, UK, 2. edition.
[25] Delichatsios, M. A. and Probstein, R. F. 1975, Coagulation in turbulent flow - theory and experiment. *J. Colloid Interface Sci.*, **51**, 394–405.

3D acoustic Lagrangian velocimetry

M. Bourgoin[1], P. Gervais[2], A. Cartellier[1], Y. Gagne[1], C. Baudet[1]

[1] [1]L.E.G.I. - U.M.R. 5519 - C.N.R.S. / I.N.P.G. / U.J.F. , BP53 - 38041 Grenoble Cedex 09 - France
[2] [2]L.E.A. - U.M.R. 6609 - C.N.R.S. / Universit de Poitiers, BP 30179 - 86962 Futuroscope Chasseneuil Cedex - France

Summary. We report Lagrangian measurements obtained with an acoustic Doppler velocimetry technique. From the Doppler frequency shift of acoustic waves scattered by tracer particles in a turbulent flow, we are able to measure the full three-component velocity of the particles. As a first application, we have studied velocity statistics of Lagrangian tracers in a turbulent air jet at $R_\lambda \sim 320$ and at various distances from the nozzle. The choice of an air jet is motivated by the fact that jets produce a well characterized high level turbulence and open air flows are well suited to simultaneously achieve classical hot wire Eulerian measurements. Therefore, we are also able to explicitly address the question of the differences between Eulerian and Lagrangian statistics. As Lagrangian tracers we use soap bubbles inflated with Helium which are neutrally buoyant in air and can be assimilated to fluid particles. Velocity statistics are analyzed. We show that the Lagrangian autocorrelation decays faster in time than its Eulerian counterpart.

1 Introduction

Experimental and theoretical studies of turbulence has longly been dominated by the Eulerian approach, where a given quantity in the fluid is measured with a probe located at a fixed point in space and is represented as a continuous spatial field. An alternative point of view is given by the Lagrangian approach, where the same quantity is measured along the trajectory of a fluid particle and is represented as a function of time parameterized by the initial position of the considered fluid particle. One reason for the domination of Eulerian approaches in turbulence is probably due to technical difficulties inherent to Lagrangian experiments, which requires the tracking of particles in strongly fluctuating flows. This has remained out of reach of experimentalist until very recently, due to lagging technological advances in fast imaging, and ultrasonic techniques. Simultaneously theoretical advances of stochastic models for turbulence gave a renewed interest to Lagrangian experiments. Moreover many practical situations are naturally described in the Lagrangian framework. This is particularly the case of dispersion in particle laden flows.

Bernard J. Geurts et al. (eds), Particle Laden Flow: From Geophysical to Kolmogorov Scales, 243–256.

An important advance in Lagrangian measurements have been done in the nineties by Virant and Dracos [1] who developed a 3D-Particle Tracking Velocimetry (PTV) technique. They used simultaneously 4 video cameras at a frame rate of 25 fps to access the 3D trajectories of several hundreds of particles at once. Ott and Mann [2] developed a similar technique to study relative dispersion of fluid particles. Because of the low frame rate, particle dynamics could be resolved only for moderate Reynolds numbers, typically $R_\lambda < 100$, where R_λ is defined based on the Taylor micro-scale. The first Lagrangian measurements for single particles statistics in high Reynolds number regimes ($R_\lambda < 1000$) were obtained by LaPorta et al. [3], using silicon strip detectors (initially developed for high energy particles detection) at a frame rate up to 70kHz. They observed a strong Lagrangian intermittency characterized by strong acceleration events with non Gaussian fluctuations. Recently, Bourgoin et al. [4] have developed a high resolution 3D-PTV facility using ultrafast cameras at a repetition rate of 27kHz, which allows the tracking of several hundred of particles in high Reynolds number regimes ($R_\lambda < 1000$). They studied relative dispersion and Lagrangian structure functions, which also exhibit strong intermittency [5].

All the techniques mentioned so far are optical and give the particles position. Particles velocity and acceleration are obtained by differentiating numerically once and twice the position. This is a very noise sensitive procedure which requires an important oversampling in order to get a good signal to noise ratio for the particles velocity and ultrafast optical systems need to be used to study highly turbulent flows. An alternative option is given by ultrasonic techniques. Mordant et al. [6] used Doppler frequency shift of acoustic waves scattered by tracer particles to obtain Lagrangian *velocity* in a closed von Kármán flow of water. The frequency shift is directly proportional to the particle velocity, no differentiation is required to get the instantaneous velocity, and a single differentiation gives the acceleration of the particles. Even though the tracers they used were bigger than the Kolmogorov length of their flow (acoustic scattering impose a minimal tracer size) their results clearly show strong Lagrangian intermittency : the probability density functions of velocity increments are Gaussian for large time separation and strongly non Gaussian for small time separation.

In neither of the previous studies a clean comparison between Lagrangian and Eulerian statistics in turbulent flows were made, either because Eulerian measurements were not available [3, 6], or because Reynolds number was too low, so that turbulence could not be considered as fully developed [1, 2, 7]. In the present article, we describe an acoustical Lagrangian setup well adapted to measurements in turbulent open flows of gas where simultaneous Eulerian measurements can be easily done (with classic hot wire anemometry for instance). We report the first measurements, carried out in an open air jet at $R_\lambda \simeq 320$. As tracers, we use small helium-filled neutral soap bubbles seeded from a position upstream of the jet nozzle. The three components of the Lagrangian velocity are obtained by acoustical Doppler effect. First, the detection setup, as well as the tracers are described. Then, we present the signal processing techniques used to detect the passage of bubbles and to extract their velocities from the frequency shift. finally we report some results on the main statistical properties of the velocity signals.

2 Experimental setup

The measurements have been conducted in a round air jet, emerging from a 2.25 cm conic nozzle and expanding freely in the room. The flow becomes self-similar at a distance of approximately 40 times the nozzle diameter [8]. Measurement has been done at distances up to 120 diameters. In all experiments the Reynolds number based on the Taylor micro-scale is about 320. Jet characteristics are based on hot-wire measurements and classical isotropic relations [9, 10].

2.1 Ultrasonic Doppler velocimetry

Figure 1 presents the principle of one-component ultrasonic velocimetry. It is based on the Doppler frequency shift of an acoustic wave scattered by a moving particle. A transducer emits a continuous ultrasonic wave at a given frequency, typically 110 kHz $\leq \nu \leq$ 130 kHz, with a propagating direction $\mathbf{n}_i$ towards the jet. The wave scattered in a specific direction $\mathbf{n}_d$ ($\theta = (\mathbf{n}_i; \mathbf{n}_d)$ is the scattering angle) by particles transported in the flow, is recorded by a receiver. Because of the particles motion, the scattered wave is Doppler shifted an its frequency ν' differs from ν:

$$\frac{\nu' - \nu}{\nu} = \frac{\mathbf{V} \cdot (\mathbf{n_i} - \mathbf{n_d})}{c} = -2\frac{V_{//}}{c} \sin(\theta/2), \tag{1}$$

where c is the speed of sound in the experimental conditions. For a given incoming frequency ν and a given scattering angle θ, the instantaneous frequency shift $\nu' - \nu$ gives a direct measurement of the projection, $V_{//}$, of the tracer velocity along $\mathbf{n}_i - \mathbf{n}_d$. Note that this is an algebraic measurement : the sign of $V_{//}$ is given by the sign of the frequency shift.

The electric signal originating from the receiver is digitized by a HP E1430A card at 65536 Hz after digital heterodyne demodulation. Series of 1,048,576 samples are recorded. Within each series, several isolated particles are successively detected. Transducers are capacitive electro-acoustical circular pistons of Sell-type, with a diameter of 24 cm. They are reciprocal, highly directive and linear. Thanks to the transducers high directivity, tracers can only be detected when they are located in the volume defined by the intersection of the incoming and the detection transducer beams, which will be called in the following the "measurement volume". Its shape is sketched on figure 1. Dimensions are $L_s \simeq 50$ cm along the jet axis and 25 cm across the stream.

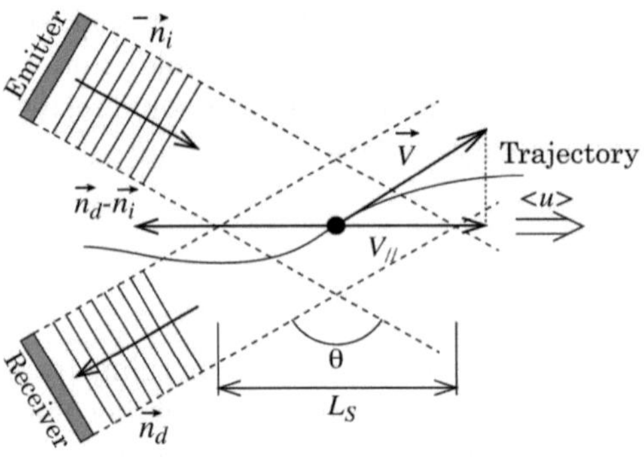

Fig. 1. Principle of velocity measurements. Particles can be detected in the intersection of the emitter and receiver acoustic beams (dashed lines). In the configuration we used : $L_s \simeq 50$ cm.

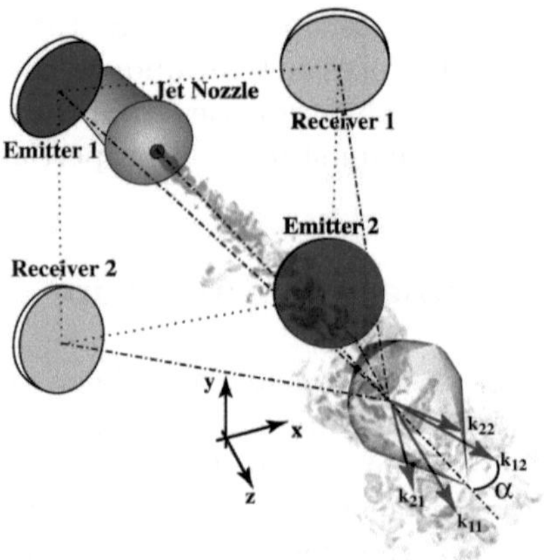

Fig. 2. Transducers arrangement for the three-component measurement. $D =$ 24 cm. Emitters are referred to as E1 and E2, receivers as R1 and R2. $\alpha = 26^\circ$, yielding a scattering angle $\theta = 128^\circ$.

One challenge that arises in Lagrangian measurement is to follow the particle over a long enough trajectory in order to describe its dynamics up to scales of order the integral scale of the flow. Thus, the length L_s of the measurement volume is chosen so that the time of flight of a particle traveling along is larger than the Lagrangian integral timescale of the flow T_L.

2.2 Three-component measurement

It is possible to measure the complete three velocity components by combining several one-component measurements from multiple pairs of acoustic transducers. The optimal configuration uses four transducers (two emitters and two receivers) placed at the vertices of a square, tilted so that their axes cross at the same point on the jet axis, in a square-based pyramid configuration (figure 2). Distances from transducers to the intersection with the jet axis are all identical, ensuring that wave propagation times are all identical. Transducers are reverse-facing the nozzle, to avoid bubble impacts on their active surfaces and they are sufficiently apart from each other not to perturb the flow. The pyramid is elongated along the axis, with an angle α at its top of 26°

The first emitter E1 generates a continuous wave at frequency $\nu_1 =$ 110 kHz, scattered by each bubble, and then recorded in two different directions by the two receivers (resp. R1 and R2). This gives access to two projections of the tracer velocity, respectively along vectors $\mathbf{k}_{11}$ and $\mathbf{k}_{12}$. Similarly, the wave emitted by E2, at frequency $\nu_2 = 122$ kHz (different from ν_1) is also scattered and then recorded by the two receivers, giving access to two more projections along vectors $\mathbf{k}_{21}$ and $\mathbf{k}_{22}$. Provided the bubble lies in the intersection of the four acoustic beams, its velocity can be obtained through four non-orthogonal projections, which we will denote by $(v_{11}, v_{12}, v_{21}, v_{22})$. Components along and perpendicular to the jet axis (u,v,w) can then be computed by a simple matrix transformation.

$$u = \frac{v_{11} + v_{12} + v_{21} + v_{22}}{4 \cos \alpha} \qquad v = \frac{v_{12} - v_{21}}{2 \sin \alpha} \qquad w = \frac{v_{22} - v_{11}}{2 \sin \alpha} \tag{2}$$

2.3 Tracers

We have first used the acoustic velocimetry technique to study Lagrangian statistics of single phase turbulence. This requires to have Lagrangian tracers which must match the carrier fluid density in order to cancel buoyancy forces. Whereas solid particles can be used for liquid, particles filled with light gases must be used in air flows. We use helium-inflated soap bubbles. Once the soap film has been made thin enough, the overall bubble density can match the one of air. Helium-inflated bubbles have also the additional benefit of offering a high contrast of acoustic impedance that increases the scattered amplitude.

Due to the evaporation of the liquid film, bubbles life time is around one or two minutes, which is much larger than the time needed to travel across the whole measurement volume (less than .1 s). Bubbles are produced by a dedicated machine to the desired density, and injected just upstream of the nozzle, so as not to disturb the flow. Statistics for bubble diameters have been obtained with the help of a video camera. It has been found that bubbles diameter has a very monodisperse distribution (2 mm $\pm$ 6 %). This ensures that limitations due to bubble size will be at a constant scale. Indeed, the finite size of the bubbles sets the small-scale limitation of the measurement. With our present apparatus bubbles can't be made smaller than about 2 mm. This is below the Taylor micro-scale (from 4.4 mm to 7.3 mm), but still above the Kolmogorov scale η (from 0.12 mm to 0.20 mm). Thus, we expect the bubble dynamics to reflect a substantial part of the inertial range, up to the integral scale, but very small-scale dynamics might be filtered due to the particles size (previous work suggests that tracers should be smaller than about 5η to behave as perfect fluid particles [11]).

Production frequency has to be low enough to ensure that most of the time a single bubble is detected in the measurement volume. Injection rates of about 5 bubbles per second gave good results.

The acoustic technique can also be used to track non Lagrangian particles, in order for instance to study inertial effects. This work is in progress at present. The first measurements aim to study finite size effects of neutrally buoyant particles as well as effects of fluid to particle density ratio when heavy bubbles are used (obtained by thickening the soap film and/or by filling the bubbles with a heavier gas, such carbon dioxide).

3 Signal processing

3.1 Channels separation

Figure 3a shows the power spectrum of a typical acoustic signal recorded on one of the receivers over several second (during the period of the record, of order 10 particles have traveled successively in the measurement volume). One observes two sharp high-amplitude spikes exactly at the emitting frequencies. They correspond to direct propagation between emitters and receiver, as well as echoes. Beside each spike, at lower frequencies, stands a broader-band bump which corresponds to Doppler frequencies of the acoustic wave scattered by the particles. The maximum of the band reflects the average velocity of the particles, while its width reflects the velocity fluctuations. All time information is lost in the Fourier space representation of the acoustic signal. The main point of the processing of the acoustic signals will be to obtain simultaneously time and frequency information in order to extract the instantaneous velocity of single particles. This requires first to filter out the spikes (which correspond to direct acoustic waves propagation and do not carry any information on the particles velocity) and to separate the channels for each emitter-receiver pair. Filtering out the spikes removes a small part of the Doppler frequencies because of overlapping. This enforces a limitation on the smallest velocity that can be extracted. Two notch filters are applied to remove the spikes, then a demodulation followed by a low-pass filtering allow the separation of the two scattered signals. We emphasize the fact that demodulation is performed by multiplication with a complex exponential (not a real cosine), yielding an analytic signal. The same operation is applied to the two signals recorded on the two receivers, leading to four different frequency-modulated signals, which will be denoted hereafter by s $(s_i \dots s_l)$.

3.2 Detection of tracers traveling into the measurement volume

Each acquisition of the acoustic signal is typically 20 seconds long. During each acquisition, about 100 bubbles travel successively in the measurement volume. Only the portions of signal corresponding to the presence of a particle in the measurement volume are analyzed for extraction of Lagrangian velocity. Figure 3b shows the real part of a typical s signal (only a fraction of order 0.8 second is represented). The presence of a bubble in the measurement volume corresponds to high amplitude events easily identifiable on the figure. When no bubble is present, amplitude is not strictly zero. This remaining amplitude is mostly due to sound scattering by the vorticity field (see [12]) and will be considered as noise here. For the velocity vector to be computed, the bubble should be detected at least on three signals simultaneously. Another selection procedure is thus applied to keep only intervals that have a common part on the four channels.

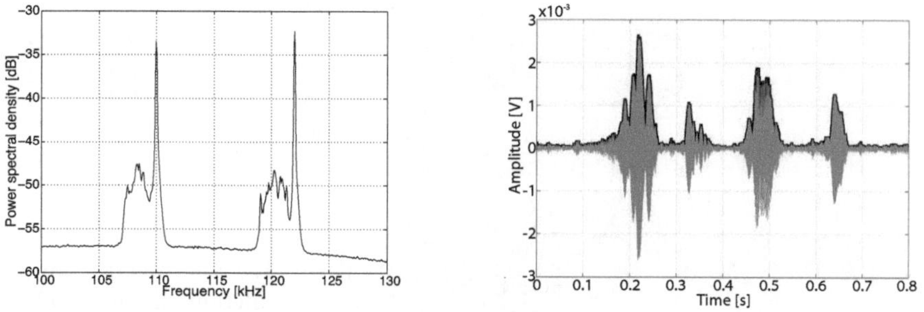

Fig. 3. (a) Power spectral density of the signal recorder by one of the receivers. The two spectral lines are the emitter direct frequencies (110 kHz and 122 kHz), and correspond to acoustic waves propagating straight between the emitters and the receiver. (b) Real part (grey line) and amplitude (black line) of separated acoustic signal (s)

3.3 Extracting velocity from acoustic signal

In order to extract the instantaneous velocity of the individual tracers, we have to combine the spectral and the time representations of the acoustic signal. Indeed, the spectral representation loses all time information and can only give time-averaged information such as the mean velocity of the bubbles (which corresponds to the peak of the Doppler shift) and the time representation is well suited to detect in time individual tracers traveling in the measurement volume but it does not give any information on their velocity. The determination of the instantaneous velocity of the tracers relies on an estimation of the instantaneous frequency of the acoustic signal. Cohen class energetic estimators are classical tools for this purpose. We chose the Choï-Williams distribution, for its moment-preserving property ([13]). If x is a complex-valued signal, its Choï-Williams distribution C_x is defined by:

$$\forall t, f \quad C_x(t,f) \mathrel{\widehat{=}} \int_{-\infty}^{\infty}\int_{-\infty}^{\infty} \frac{1}{|\tau|} \mathrm{e}^{-2(s-t)^2/\tau^2} x\left(s+\frac{\tau}{2}\right) x^*\left(s-\frac{\tau}{2}\right) \mathrm{e}^{-2\mathrm{i}\pi f\tau}\,\mathrm{d}s\,\mathrm{d}\tau \tag{3}$$

A sample result is shown on figure 4. Frequency is along the vertical axis, time along the horizontal. Gray-scale levels quantifies the energy level $C_x(t,f)$. Two crooked lines are visible, they correspond to two distinct bubbles. The lines where the signal energy is concentrated reflect the time evolution of the instantaneous Doppler frequency shift. The time-frequency transformation (3) gives a 2D representation from which the instantaneous frequency shift $\nu'(t)$ is extracted as the frequency average weighted by the energy distribution C_x and equation (1) directly gives the corresponding velocity component.

This velocity extraction is applied to each s signal, on the selected intervals. Coordinate transformation yields lots of small (2000 to 8000 points) velocity signals which will be called "velocity segments" in the following. No time

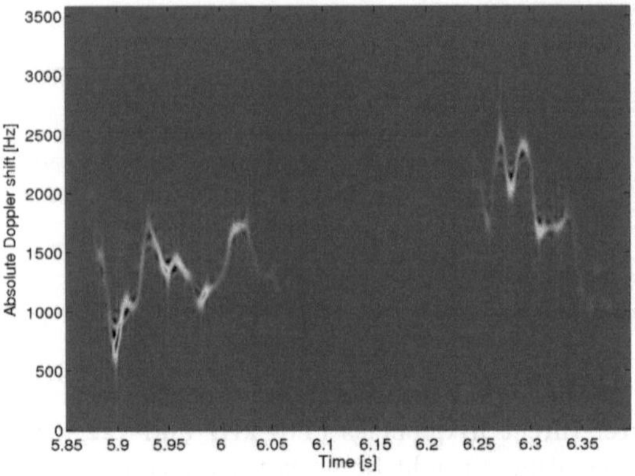

Fig. 4. Time-frequency representation. Two bubble signals are visible.

D	N	$\frac{L_s}{L_E}$	$\frac{\langle T_s \rangle}{T_L}$
58	3552	8.1	1.48
80	9584	5.7	1.22
93	7444	5.0	1.20
111	1358	4.2	1.02

Table 1. Experimental parameters at different distances D from the nozzle and the center of the measurement volume. D is measured in multiples of the nozzle diameter. N is the number of velocity segments.

continuity exists between velocity segments, they all correspond to different bubbles. Such a procedure leads to a large set of independent realizations of Lagrangian velocities. The j-th point (time) of the i-th segment (realization) will be denoted by $\mathbf{v}_i(j)$.

4 Results

4.1 Data set

As already discussed, in order to resolve not only the small-scale dynamics but also the large-scale dynamics of the particles we need the measurement volume dimensions to be comparable to the integral scale of the flow. Therefore, we carried the experiments in an air jet with a small nozzle (2.25 cm in diameter) compared to the transducers diameter. As a consequence, only moderate Reynolds number (up to $R_\lambda = 320$) were achievable. Series of recordings were made at four distances (D) from the nozzle. Every measurement corresponds to the same Reynolds number, as it is constant along the jet, but to different integral length scales [9, 8, 10]. The measurement volume was centered on the jet axis, to preserve cylindrical symmetry as much as possible.

Table 1 lists the main parameters of the different measurements. The number of velocity segments exceeds 1000 for all measurements, ensuring good statistical convergence. Measurement volume length is always several times larger than the Eulerian integral length scale (L_s/L_E), and the ratio of the average time-of-flight to the Lagrangian integral time scale ($\langle T_s \rangle / T_L$) is everywhere above one.

4.2 Velocity probability density function

The normalized velocity probability density functions (PDF) for the longitudinal velocity component measured at different distance from the nozzle are represented on figure 5. No significant change in shape can be seen between the four curves, indicating that the variation of L_s/L_E does not break self-similarity. The same remark is true for transverse components (figure 6). All curves are Gaussian, but small departures exist. For the longitudinal component (figure 5), PDF edges are largely sub-Gaussian due to limitations of the velocity extraction algorithm and has no physical meaning. For transverse components (figure 6), edges are over-Gaussian because of noise introduced by the velocity extraction algorithm.

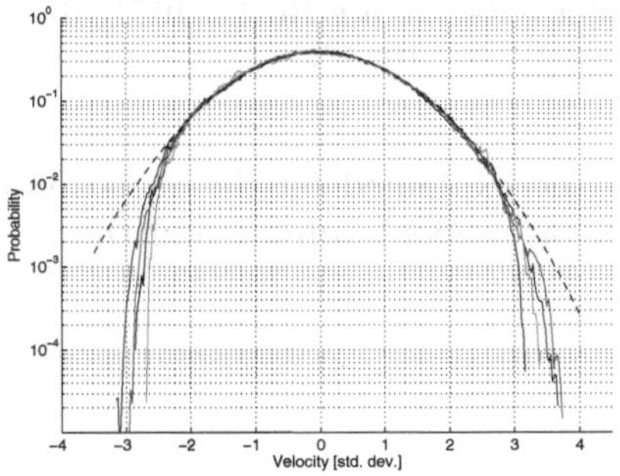

Fig. 5. Longitudinal velocity PDF with zero mean and unity variance. Corresponding Gaussian curve is plotted in dashed line.

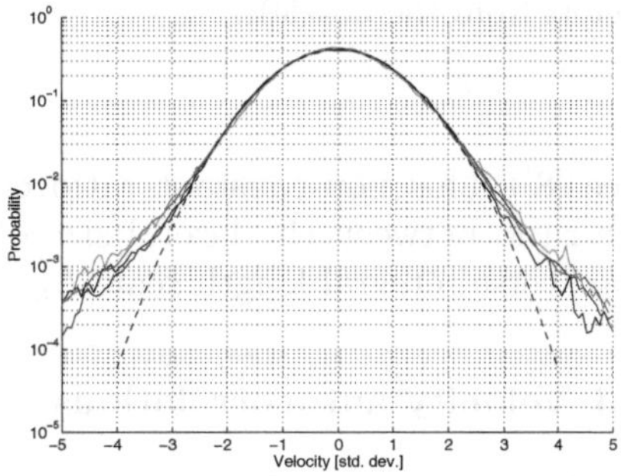

Fig. 6. Transverse velocity PDF with zero mean and unity variance. Corresponding Gaussian curve is plotted in dashed line.

Figure 7 shows the comparison of the Lagrangian PDF ($P(u)$) with the corresponding Eulerian one (the hot-wire was located near the center of the Lagrangian measurement zone). A reasonable agreement is found. A slightly higher mean velocity is found in the Eulerian case (5 % higher), and the standard deviation is higher for the Lagrangian velocity. These effects result from the inhomogeneity of the flow inside the acoustic measurement volume, which tend to under estimate the Lagrangian mean velocity on the axis and over estimate its fluctuations but is not visible on the Eulerian measurement which is carried out at a fixed point.

Figure 8 shows isocontours of the joint PDF $P(u,v)$ of longitudinal u and transverse v Lagrangian velocity. A slightly elliptical shape is visible, indicating that no large-scale isotropy exists (horizontal and vertical coordinates are identical). Standard deviation of the longitudinal component is higher than the corresponding one for the transverse component, by a factor ranging from 1.1 to 1.25, depending on the position along the jet (resp. farthest and nearest from the nozzle). A similar behavior exists for Eulerian velocity components

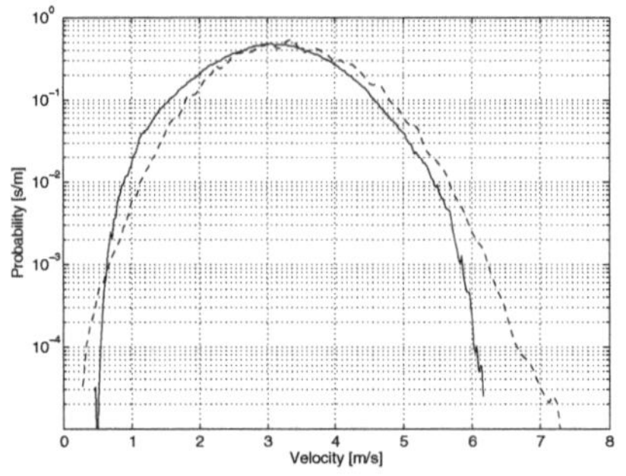

Fig. 7. Lagrangian (solid line) and Eulerian (dashed line) velocity PDF (80 diameters from the nozzle).

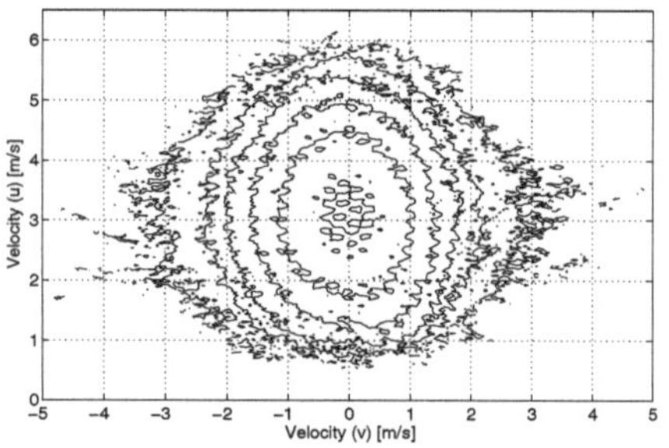

Fig. 8. Isocontours of joint Lagrangian velocity PDF. Contour values, from the center outward, are $10^{-3.5}$, 10^{-4}, $10^{-4.5}$, 10^{-5}, $10^{-5.5}$, $10^{-6.5}$.

(see [8]) This non-constant ratio can also be explained by the variation of the ratio between the lateral size of the measurement volume and the local transverse integral length scale L_E.

4.3 Velocity autocorrelations

We have measured the velocity autocorrelation function for the Lagrangian and the Eulerian signals. The Lagrangian velocity correlation time T_L plays an important role in modeling turbulent diffusion of passive tracers[14]. Moreover accurate measurements of the ratio T_E/T_L are of particular interest in the frame of numerical models such as RANS calculations [15] where this ratio is a parameter to be calibrated.

The statistical estimation of the Lagrangian autocorrelation function has been obtained with an unbiased estimator, which also compensates to second order the axial inhomogeneity of the flow, inherent to open flows situation [16]. For the Eulerian autocorrelation, the integral scale L_E is estimated from the hot wire measurement using a Taylor hypothesis based on the local mean velocity and the integral time T_E is then defined as $T_E \widehat{=} L_E/\sigma_E$, where σ_E is the Eulerian velocity standard deviation.

Figure 9 shows the autocorrelation function of the Lagrangian velocity components and the Eulerian longitudinal velocity for a measurement performed at 80 diameters from the nozzle. The two curves for Lagrangian transverse components are almost identical, in accordance with the cylindrical symmetry of the flow. The longitudinal component exhibits a slightly longer time scale.

We denote in the following the longitudinal and transverse Lagrangian integral time scales by T_L^l and T_L^t respectively. These values are computed by fitting an exponential curve on the autocorrelation. Corresponding values for Eulerian components are denoted by T_E^l and T_E^t. Only T_E^l can be readily obtained from measurements, because of the necessity of a Taylor hypothesis. As no measurement of transverse Eulerian velocity has been performed, T_E^t is estimated from the longitudinal value, assuming that $L_E^l/L_E^t \simeq 2.3$ and $\sigma_E^l/\sigma_E^t \simeq 1.2$ (as found in [8]).

All these values are listed in Table 2 for the different positions of the measurements. We note that the transverse integral time scales are smaller than the longitudinal. In the Eulerian case, the ratio T_E^l/T_E^t is constant as a consequence of the previous hypotheses. On the contrary, we observe that ratio T_L^l/T_L^t tends to increase with the distance D from the nozzle. Several reasons may be responsible for that. On the one hand, the jet self-similarity can be broken. Wygnansky and Fiedler [8] have shown that self-similarity can be violated for distances as large as 100 nozzle diameters, depending on the quantity considered, in which case actual measurements of Eulerian time scales would lead to similar results. On the other hand the velocity profile varies linearly with the distance to the nozzle, while the measurement volume size is constant, so that the flow homogeneity in the measurement volume depends on the position in the jet. As T_L^l/T_L^t increases when D increases, this indicates that large-scale isotropy either does not exist whatever the distance, or is recovered very slowly. Lagrangian times T_L can be considered as a rough measure of eddy life-time, whereas T_E is related to the eddy turnover time. These results show that whatever the component considered, both times are very close, the turnover time being slightly longer. Obtained ratios are compatible with the predicted value of $1/0.78 \simeq 1.28$ [17]. A simple phenomenological analysis leads to $T_L \simeq T_E$ [18]. A larger Eulerian time scale can be explained by sweeping effects. The advection of the internal scales by the energy-containing scale leads to broadening of the Eulerian autocorrelation in comparison with the Lagrangian one [19].

D	T_L^l	T_L^t	$\frac{T_L^l}{T_L^t}$	T_E^l	T_E^t	$\frac{T_E^l}{T_E^t}$	$\frac{T_E^l}{T_L^l}$	$\frac{T_E^t}{T_L^t}$
58	35	26	1.35	48	25	1.92	1.37	0.98
80	62	49	1.27	98	51	1.92	1.58	1.04
93	79	59	1.34	129	69	1.92	1.63	1.16
111	117	75	1.56	205	98	1.92	1.75	1.30

Table 2. Eulerian and Lagrangian time scales in milliseconds. T_L^l, T_L^t and T_E^l measured. T_E^t computed from T_E^l (see text).

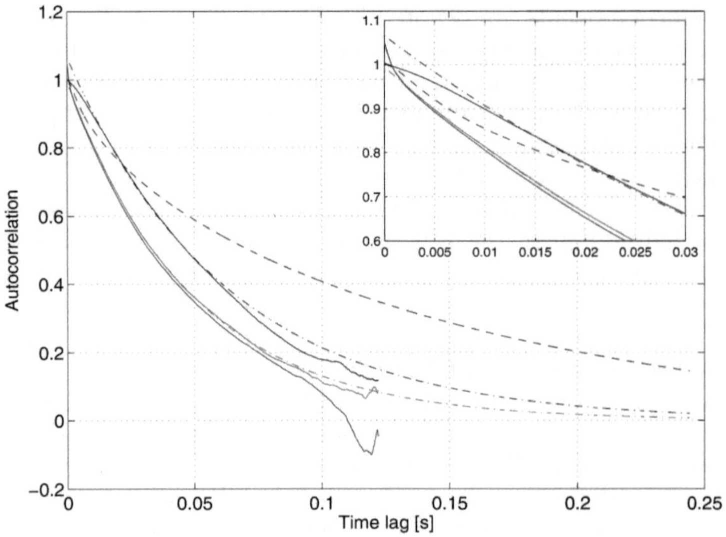

Fig. 9. Lagrangian velocity autocorrelation (solid line) for longitudinal and transverse components. Eulerian velocity autocorrelation (dashed). An exponential fit has been superimposed to Lagrangian correlations (dot-dashed).

5 Conclusion

Lagrangian measurements in a free turbulent air jet were performed using acoustical Doppler effect. This method is adapted to collecting large data sets without tremendous memory requirement, contrary to visualization methods. A single tracer at a time can be detected, with the time- and space- dynamics of the measurements comprising a large part of the inertial scales, comparable to previously-obtained results [6]. Simultaneous Eulerian measurements were performed.

We show that the Eulerian integral time is larger than the Lagrangian one. This might be a consequence of the Eulerian statistics sensitivity to sweeping effects, which instead do not affect Lagrangian statistics. This result holds for distances in the jet ranging from 60 nozzle diameters up to 110 nozzle diameters. The ratio T_E^l/T_L^l is found of order 1.4, with a slight dependence on the distance from the jet nozzle.

The acoustic technique is now being adapted to study two phase flows laden with inertial particles. The first experiments aim to explore Stokes number dependence of individual particles dynamics, with a particular focus on the effect of particles finite size and of the particle to fluid density ratio.

References

[1] Virant, M., and Dracos, T., 1997. 3D PTV and its application on Lagrangian motion. Measurement science and technology, 8, pp. 1539-1552

[2] Ott, S., and Mann, J., 2000. An experimental investigation of the relative diffusion of particle pairs in three-dimensional turbulent flow. Journal of Fluid Mechanics, 422, pp. 207-223

[3] LaPorta, A., Voth, G. A., Crawford, A. M., Alexander, J., and Bodenschatz, E., 2002. Fluid particle accelerations in fuly developped turbulence. Nature, 409, February, p. 1017

[4] Bourgoin, M., Ouellette, N. T., Xu, H., Berg, J., and Bodenschatz, E., 2006. The role of pair dispersion in turbulent flow. Science, 311, February, p. 835

[5] Xu, H., Bourgoin, M., Ouellette, N. T., and Bodenschatz, E., 2006. High order Lagrangian velocity statistics in turbulence. Physical Review Letters, 96, January, p. 024503

[6] Mordant, N.,Metz, P.,Michel, O., and Pinton, J.-F., 2001. Measurement of Lagrangian velocity in fully developed turbulence. Physical Review Letters, 87(21), p. 214501

[7] Sato, Y., and Yamamoto, K., 1987. Lagrangian measurement of fluid-particle motion in an isotropic turbulent field. Journal of Fluid Mechanics, 175, pp. 183-199

[8] Wygnanski, I., and Fiedler, H., 1969. Some measurements in the self-preserving jet. Journal of Fluid Mechanics, 38(3), pp. 577-612

[9] Pope, S. B., 2000. Turbulent flows. Cambridge University Press

[10] Tennekes, H., and Lumley, J. L., 1992. A first course in turbulence. MIT press

[11] Voth, G. A., la Porta, A., Crawford, A. M., Alexander, J., and Bodenschatz, E., 2002. Measurement of particle accelerations in fully developed turbulence. Journal of Fluid Mechanics, 469, pp. 121-160

[12] Poulain, C., Mazellier, N., Gervais, P., Gagne, Y., and Baudet, C., 2004. Spectral vorticity and Lagrangian velocity measurements in turbulent jets. Flow, Turbulence and Combustion, 72, pp. 245-271

[13] Flandrin, P., 1993. Temps-frequence. Hermes

[14] Taylor, 1921. Diffusion by continuous movements. Proc. London Math. Soc., 20, p. 196

[15] Lipari, G., Apsley, D. D., and Stansby, P. K., 2006. Numerical particle tracking studies in a turbulent jet. in the present Proceedings of Euromech Colloquim 477

[16] Batchelor, 1957. Diffusion in free turbulent shear flows. Journal of Fluid Mechanics, 3, pp. 67-80

[17] Yeung, P. K., 2002. Lagrangian investigations of turbulence. Annual Review of Fluid Mechanics, 34, pp. 115-142

[18] Corrsin, S., 1963. Estimates of the relations between Eulerian and Lagrangian scales in large Reynolds number turbulence. Journal of the Atmospheric Sciences, 20(2), pp. 115-119

[19] Kraichnan, R. H., 1964. Relation between Lagrangian and Eulerian correlation times of a turbulent velocity field. Physics of Fluids, 7(1), pp. 142-143

Lagrangian multi-particle statistics

Beat Lüthi, Jacob Berg, Søren Ott and Jakob Mann

Risø National Laboratory, Wind Energy Department, P.O. Box 49,
Frederiksborgvej 399, DK-4000 Roskilde, Denmark beat.luthi@risoe.dk

Summary. Combined measurements of the Lagrangian evolution of particle constellations and the coarse grained velocity derivative tensor $\partial\widetilde{u}_i/\partial x_j$ are presented. The data is obtained from three dimensional particle tracking measurements in a quasi isotropic turbulent flow at intermediate Reynolds number. Particle constellations are followed for as long as one integral time and for several Batchelor times. We suggest a method to obtain quantitatively accurate $\partial\widetilde{u}_i/\partial x_j$ from velocity measurements at discrete points. We obtain good scaling with $t_* = \sqrt{2r^2/15S_r(r)}$ for filtered strain and vorticity and present filtered R-Q invariant maps with the typical 'tear drop' shape that is known from velocity gradients at viscous scales. Lagrangian result are given for the growth of particle pairs, triangles and tetrahedra. We find that their principal axes are preferentially oriented with the eigenframe of coarse grained strain, just like constellations with infinitesimal separations are known to do. The compensated separation rate is found to be close to its viscous counterpart as $1/2\left(dr^2/dt\right)/r^2 \cdot t_*/\sqrt{2} \approx 0.11 - 0.14$. It appears that the contribution from the coarse grained strain field, $r_i r_j \widetilde{s}_{ij}$ filtered at scale $\Delta = r$, is responsible only for roughly 50% of the separation rate. The rest stems from contributions with scales $\Delta < r$.

1 Introduction

An important consequence of turbulence is effective mixing and dispersion of advected Lagrangian particles [1]. Recent work on two particle dispersion [2, 3] raised the question to what degree two particle separation in the inertial range is governed by the coarse grained velocity derivative field $\widetilde{A}_{ij} = \partial\widetilde{u}_i/\partial x_j$. Moreover, it has been recognized for a few years now that constellations with more than two particles have a rich structure at scales smaller than the integral scale $\mathcal{L}$ [4, 5, 6, 7, 8]. Work that started with [5] and currently is being further developed by [9] is relating the dynamics of $\widetilde{A}_{ij}$ to the evolution of tetrahedra and a stochastical model has been developed for its simulation. Experimental and numerical studies have investigated some of the properties of $\widetilde{A}_{ij}$ [10, 11]. The most important finding is that coarse grained velocity derivatives exhibit

Bernard J. Geurts et al. (eds), Particle Laden Flow: From Geophysical to Kolmogorov Scales, 257–269.

roughly the same properties like their small scale counterparts. Probably the most important property is that $\left\langle \widetilde{\Lambda}_2 \right\rangle > 0$, where $\widetilde{\Lambda}_i$ are the eigenvalues of the rate of strain tensor $\widetilde{s}_{ij} = 1/2\left(\partial \widetilde{u}_i/\partial x_j + \partial \widetilde{u}_j/\partial x_i\right)$. It means that also for inertial range scales the field of velocity derivatives experiences self-amplification.

In this contribution, we present for the first time experimental results that attempt to combine measurements of $\widetilde{A}_{ij}$ with measurements of the evolution of particle pairs, triangles, and tetrahedra. The filter scale covers a good part of the inertial range and the particle constellations are followed as long as the integral time, $\mathcal{T}$, and for several Batchelor times, $\tau_B = R_0^{2/3}/\varepsilon^{1/3}$, where R_0 is the scale of the constellation at $t = 0$. Since Batchelor [12] it is known that for the case of two particle separation at τ_B the relative separation regime changes from $\langle r^2 - r_0^2 \rangle \propto t^2$, known as the ballistic regime, to $\langle r^2(t) \rangle = g\varepsilon t^3$, which is known as the Richardson law. The importance of having observation times $t > \tau_B$ can also be expressed in terms of kinetic energy of relative motion in a particle swarm of size R with N points, $E = 1/2 \left\langle \left(\mathbf{u}^N - \mathbf{U}\right)^2 \right\rangle_R$: Only if tracking times are long enough a transition from the regime where $dE/dt < 0$ to a regime with $dE/dt > 0$ can be observed [13]. The former regime is essentially governed by Eulerian dynamics while the latter is governed by the Lagrangian evolution of particle constellations.

One can define the tensor $\widetilde{A}_{ij}$ coarse grained at scale Δ as

$$\widetilde{A}_{ij} = \frac{1}{V_\Delta} \int_V \frac{\partial u_i}{\partial x_j} d^3x, \tag{1}$$

where $V_\Delta \approx \Delta^3$. If we provide an at least one time differentiable approximation to the velocity field as

$$\widetilde{\mathbf{u}}(\mathbf{x}) \approx \frac{1}{V_\Delta} \int_V \mathbf{u}(\mathbf{x} + \mathbf{x}')\, d^3x' \tag{2}$$

we overcome the difficulty of having to measure $\frac{\partial u_i}{\partial x_j}$ directly but can instead differentiate the filtered velocity field to obtain

$$\widetilde{A}_{ij} = \partial \widetilde{u}_i/\partial x_j. \tag{3}$$

The left hand side of eqn. 2 can be approximated by least square fitting linear polynomials to discrete velocities of at least $n = 4$ points. For $n \to \infty$ this operation becomes equivalent to top-hat filtering the spatial velocity derivative field. Different to [14] here spherical polynomials that by definition are incompressible and orthogonal are used. Since for $\Delta > \eta$ the velocity field is not smooth $n > 4$ is necessary to obtain convergence. As we will demonstrate below in the result section we have found that $n > 12$ is sufficiently high.

2 3D-PTV Experiment

In our attempt to simultaneously measure $\widetilde{A}_{ij}$ and the evolution of particle constellations we have performed a Particle Tracking Velocimetry (PTV) experiment in an intermediate Reynolds number turbulent flow. PTV is by now a well established non-intrusive flow measuring technique [15, 16, 17, 18, 19, 14, 20, 2] which naturally allows to probe a flow's Lagrangian properties. To meet the competing goals of high tracer seeding density to allow for coarse graining, and high trackability of particle constellations to reach $t > \tau_B$ some trade off's in the experimental design had to be made: Typically 900 particles are tracked in an observation volume of $15 \times 15 \times 15\text{cm}^3$. This results in an average particle distance of $d_p \approx 50\eta$ and tracking lengths longer than integral scales $t_T > \mathcal{T}$ and $t_T > 10\tau_B$. For the sake of 'good' statistics the total recording time is $t_R \approx 500\mathcal{T}$.

The flow is forced with eight rotating propellers placed in the corner of a water tank of $32 \times 32 \times 50\text{cm}^3$ and neutrally buoyant tracer particles are recorded with four synchronized, 50Hz CCD cameras. To suppress the development of a mean flow the propellers change their rotational direction after 0.5s of stirring and after an additional 0.5s of pausing. A typical propeller tip velocity is 50cm/s. Further details of the experiment are described in [2]. The characteristic flow properties are summarized in table 1. A recent modification of tracking 3d particle positions through consecutive time frames allows to connect tracked particle trajectories that are only interrupted by one 'missing' point. The main impact of this feature is a drastic increase of the number of long trajectories. The number of tracks with length $t_T > \mathcal{T}$ has more than doubled while the number of tracks with $t_T > 2\mathcal{T}$ is one order of magnitude larger.

Table 1. Flow properties of the turbulent flow as already reported in [2].

η	$\mathcal{L}$	τ_η	$\mathcal{T}$	ε	σ_u	$\mathcal{L}/\eta$	Re_λ
0.25mm	48mm	0.07s	2.45s	$168\text{mm}^2/\text{s}^3$	23mm/s	190	172

3 Properties of $\partial\widetilde{u_i}/\partial x_j$

In this section we present Eulerian results for $\widetilde{A}_{ij}$ for $100 < \Delta/\eta < 300$, where $\eta = \left(\nu^3/\varepsilon\right)^{1/4}$ is the Kolmogorov constant. The lower bound of Δ is defined by our experimental tracer seeding density. Only for volumes larger than $(100\eta)^3$ the number of particles is $n > 12$. In fig. 1(a) we plot the averages of $\widetilde{s}^2$ and $\widetilde{\omega}^2$ as a function of filtering scale Δ/η. The comparison with the straight dashed line that is proportional to $r^{-4/3}$ shows that there is no clear K41 scaling, which is not surprising at the relatively low scale separation of $\mathcal{L}/\eta = 190$. Based on the longitudinal second order structure function $S_2(r)$ we construct a time, $t_*(r)$ with which a better compensation of $\widetilde{A}_{ij}$ can be obtained. We define

$$t_* = \sqrt{\frac{2r^2}{15 S_2(r)}}, \tag{4}$$

which for $r \ll \eta$ gives $t_*(r) = \sqrt{2}\tau_\eta$ in order that $\langle s^2 \rangle \cdot t_*^2 = 1$ and in homogeneous flows $\langle \omega^2 \rangle \cdot t_*^2 = 2$. For $\eta \ll r \ll \mathcal{L}$ eqn. 4 is proportional to $r^{2/3}$. With the parameterized form of $S_2(r)$ employed by [21]

$$\widetilde{S}_2(r) = 2\left(1 - \exp\left(-\frac{r}{(15C_k)^{3/4}\eta}\right)\right)^{4/3} \cdot (\varepsilon\mathcal{L})^{2/3}\left(\frac{r^4}{\frac{64\mathcal{L}^4}{C_k^6} + r^4}\right)^{1/6} \tag{5}$$

the second order structure function is expressed as a function of separation r, the viscosity ν, the flow properties $\mathcal{L}$ and ε and the Kolmogorov constant C_k. Again in fig. 1(a) we show $\langle \widetilde{s}^2 \rangle \cdot t_*^2$ and $\langle \widetilde{\omega}^2 \rangle \cdot t_*^2$ as a function of Δ/η. Clearly the more general scaling with eqn. 4 holds over our entire range of $100 < \Delta/\eta < 300$ as $\widetilde{s}^2 \cdot t_*^2 \approx 1$. To see how far off the approximation of $\widetilde{A}_{ij}$ is with a too low number of points we show in fig. 1(b) the same quantities for $n = 4$ points. Up to $\Delta/\eta \approx 160$ there is a seemingly nice K41 scaling. However, at $\Delta/\eta = 100$ we have $\langle \widetilde{s}^2 \rangle \approx 50$, which is much too high if we keep in mind that for $\varepsilon = 168\text{mm}^2/\text{s}^3$ at the smallest scale $\langle s^2 \rangle \approx 80$. Furthermore, the compensated strain is much too high with $s^2 \cdot t_*^2 \approx 8$.

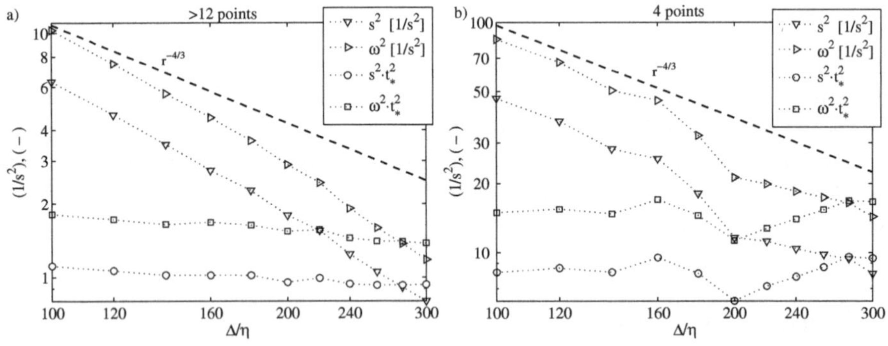

Fig. 1. Coarse grained mean strain $\langle \widetilde{s}^2 \rangle$ and enstrophy $\langle \widetilde{\omega}^2 \rangle$ are plotted versus filtering scale Δ/η along with the compensated values $\langle \widetilde{s}^2 \rangle \cdot t_*^2$ and $\langle \widetilde{\omega}^2 \rangle \cdot t_*^2$. In (a) results are obtained from $n > 12$ points per least square fit to linear polynomials and in (b) only $n = 4$ points are used.

The qualitative difference between $\widetilde{A}_{ij}$ as obtained from $n = 4$ or $n > 12$ becomes even clearer if we look at the PDFs of the strain 'shape' Λ_2/Λ_1. It is known that in the viscous range $\langle \Lambda_2 \rangle / \langle \Lambda_1 \rangle \approx 0.15$ over a wide range of Reynolds numbers [22, 23, 14] and also for inertial scales it is reported that $\left\langle \widetilde{\Lambda}_2 \right\rangle > 0$ [10, 11]. This means that also larger scale $\widetilde{A}_{ij}$ have a self-amplifying nature. In fig. 2(a) we see that the PDFs for the shapes obtained for $100 < \Delta/\eta < 300$ almost collapse and yield a mean value of $\left\langle \widetilde{\Lambda}_2 / \widetilde{\Lambda}_1 \right\rangle \approx 0.15$. Contrary, in fig. 2(b) the same PDFs but obtained from only $n = 4$ points peak at zero and $\left\langle \widetilde{\Lambda}_2 / \widetilde{\Lambda}_1 \right\rangle \approx 0$, i.e. one of the most important turbulent properties is lost completely.

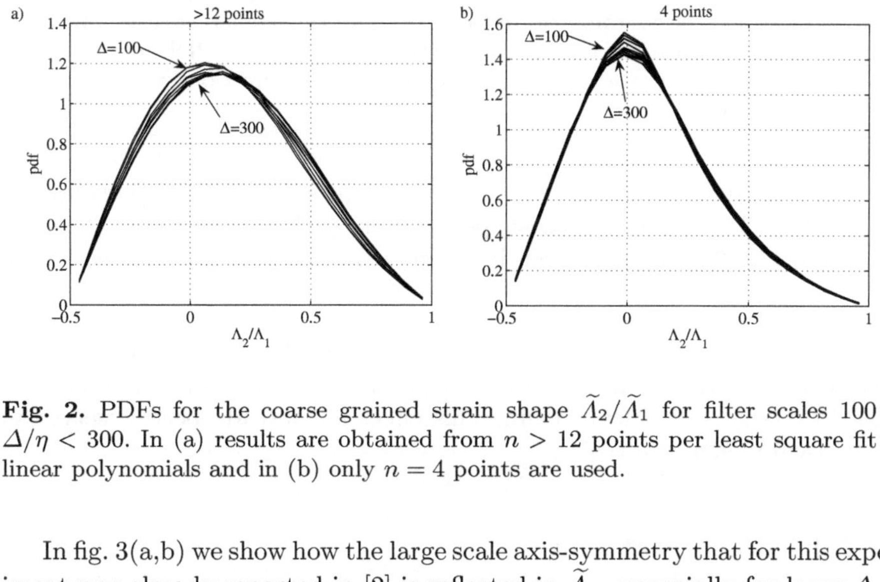

Fig. 2. PDFs for the coarse grained strain shape $\widetilde{\Lambda}_2/\widetilde{\Lambda}_1$ for filter scales $100 < \Delta/\eta < 300$. In (a) results are obtained from $n > 12$ points per least square fit to linear polynomials and in (b) only $n = 4$ points are used.

In fig. 3(a,b) we show how the large scale axis-symmetry that for this experiment was already reported in [2] is reflected in $\widetilde{A}_{ij}$, especially for large Δ/η. In fig. 3(a) we see the preferential orientation with $\mathbf{x}_3$ of the most stretching principal strain axis $\widetilde{\lambda}_1$ of $\widetilde{s}_{ij}$. $\mathbf{x}_3$ is the vertical tank axis, which with $32 \times 32 \times 50\text{cm}^3$ is higher than wide. Towards smaller scales a slow relaxation of this anisotropy can be observed, similar to reports of [9]. For $\widetilde{\omega}$ the situation is slightly different as is shown in fig. 3(b). Consistent with the tank dimensions it seems that large scale vorticity is preferentially aligned with the longer vertical tank axis but equally distributed over both directions of rotation. Towards larger scales the symmetry is broken slightly as the horizontal component of vorticity starts to align with $+\mathbf{x}_1$ and $-\mathbf{x}_2$.

Finally we show in fig. 4 the topological property of measured $\widetilde{A}_{ij}$ by means of the two invariants R and Q [24]. The normalized invariants are defined as

$$Q = -\frac{1}{2} Tr\left(\widetilde{A}_{ij}^2\right) / \langle \widetilde{s}^2 \rangle, \; R = -\frac{1}{3} Tr\left(\widetilde{A}_{ij}^3\right) / \langle \widetilde{s}^2 \rangle^{3/2}. \qquad (6)$$

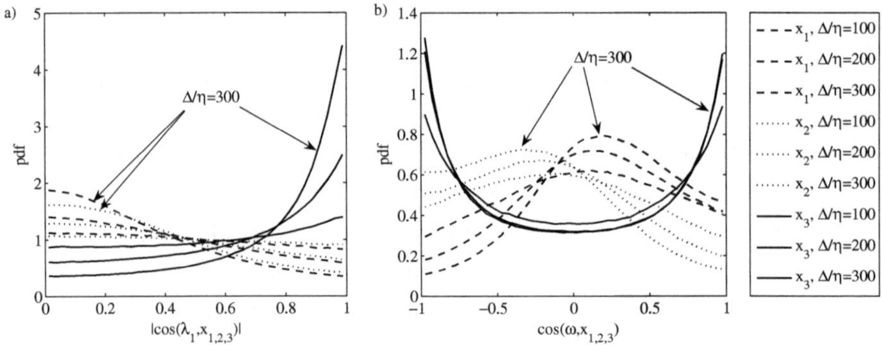

Fig. 3. (a) PDFs of the cosines between the most stretching principal coarse grained eigenvector $\widetilde{\lambda}_1$ and the coordinate directions $\mathbf{x}_{1,2,3}$ for three different scales Δ/η. $\mathbf{x}_{1,2}$ are the horizontal directions and $\mathbf{x}_3$ is the vertical direction. (b) PDFs of the cosines between coarse grained vorticity $\widetilde{\omega}$ and the coordinate directions $\mathbf{x}_{1,2,3}$ for three different scales Δ/η.

For plots obtained from only $n = 4$ points we see that essentially for all scales $100 < \Delta/\eta < 300$ the RQ-shapes look like such obtained from Gaussian velocity fields [22]. Very differently for $n > 12$ the well known 'tear drop' shapes are recovered for all scales. This is at first surprising since we would expect Gaussian RQ-shapes for $\Delta > \mathcal{L}$. Indeed, for $\Delta/\eta = 200$ we observe the most symmetric RQ-shape for the $n > 12$ figures. The only explanation we have is that the observed 'tear drop' shapes at larger than integral scales are caused by large scale mean strain. This effect has already been observed in stochastical model results [9].

4 Multi point statistics

In the previous section we established that the measured $\widetilde{A}_{ij}$ is approximating well the actual coarse grained velocity derivative tensor. In this section we show how particle pairs, triangles and tetrahedra grow in time and how their principal axes are oriented with respect to $\widetilde{A}_{ij}$. In addition, we check to what degree the kinematic relation for the growth of pairs r

$$\frac{1}{2}\frac{dr^2}{dt} = r_i r_j s_{ij}, \tag{7}$$

which in the viscous range is exactly true, also holds in the inertial range with $\widetilde{s}_{ij}$ instead of s_{ij}.

We start with the growth of particle constellations. In fig. 5(a,b) we show how normalized separations, areas and volumes grow in time. Times are normalized with τ_B that correspond to initial scales R_0, which for pairs are r_0.

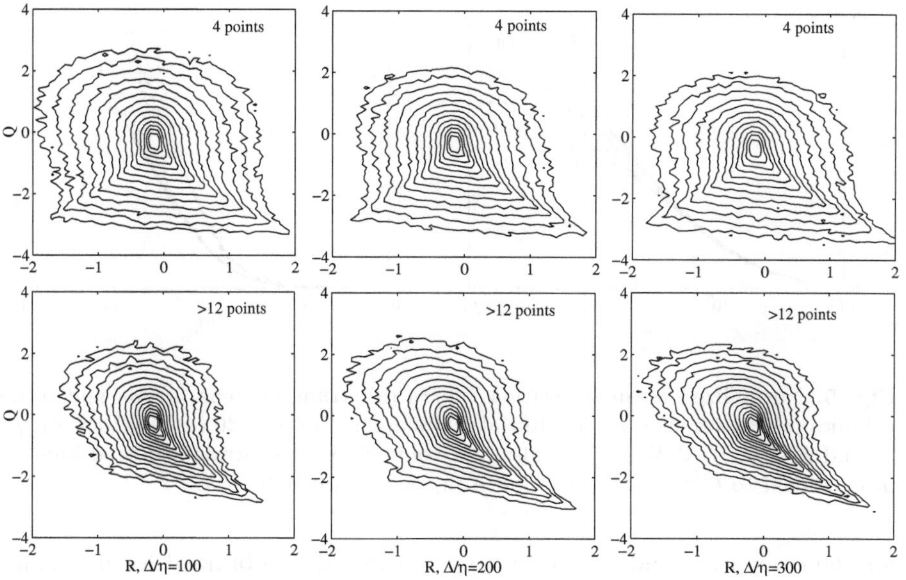

Fig. 4. Joint PDFs of the invariants R and Q as defined in eqn. 6. Shown are results from $n > 12$ points per least square fit to linear polynomials (bottom row) and $n = 4$ points (top row) for three different filtering scales $\Delta/\eta = 100, 200, 300$. The isoprobability contours are logarithmically spaced.

For evolving triangles or tetrahedra we use $R_0 = \sqrt{g_1}$, where g_1 is the largest eigenvalue of the moment of inertia tensor

$$
\begin{aligned}
g^{ab} &= \sum \rho_i^a \rho_i^b \qquad (8)\\
\rho_1 &= (\mathbf{x}_1 - \mathbf{x}_2)/\sqrt{2}\\
\rho_2 &= (2\mathbf{x}_3 - \mathbf{x}_1 - \mathbf{x}_2)/\sqrt{6}\\
\rho_3 &= (3\mathbf{x}_4 - \mathbf{x}_1 - \mathbf{x}_2 - \mathbf{x}_3)/\sqrt{12}
\end{aligned}
$$

that was introduced in [5]. Growth is scaling nicely with the Batchelor time τ_B. However, neither pairs, nor triangles nor volumes reach a convincing Richardson scaling of $\propto \varepsilon^{1/2}t^{3/2}$, εt^3, or $\varepsilon^{3/2}t^{9/2}$, denoted by the straight dotted lines. We attribute this to the too low scale separation of our experiment. In fig. 5(a) it can be seen how shortly before $r_0 = \mathcal{L}$ (denoted by circles) the growth rate starts weakening. In the case of triangles and volumes fig. 5(b) the statistics become too sparse even before integral scale is reached.

Fig. 6 shows the temporal evolution of the eigenvalues of the tensor g^{ab} defined in eqn. 9 and the evolution of the mean shape factors $\langle w \rangle$ and $\langle I_2 \rangle$ for triangles and tetrahedra respectively. These shape factors are a measure

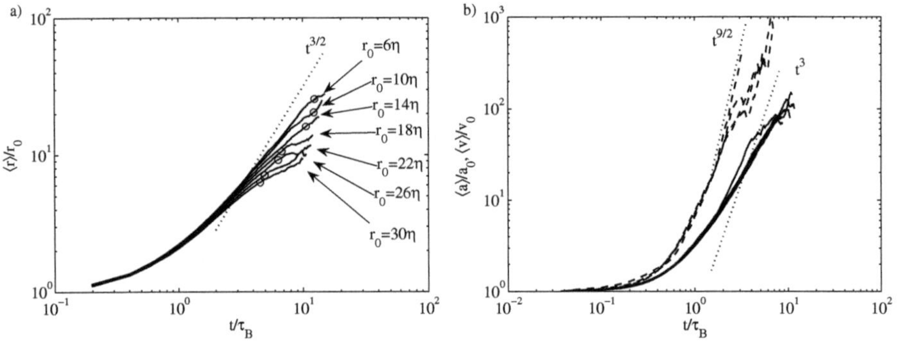

Fig. 5. Normalized mean separation $\langle r \rangle / r_0$ (a), triangle areas $\langle a \rangle / a_0$ and tetrad volumes $\langle v \rangle / v_0$ versus t/τ_B (b). Initial scales are $6 < r_0/\eta < 30$ (a) and $14 < r_0/\eta < 30$ and $22 < r_0/\eta < 30$ for triangles and volumes(b). The straight dotted lines are proportional to $t^{3/2}$, t^3, $t^{9/2}$. Circles in (a) denote when $\langle r \rangle = \mathcal{L}$.

for the elongation and $w = 0$ and $I_2 = 0$ would be obtained for needle-like objects. Following [6] they are defined as

$$I_2 = g_2/R^2 \tag{9}$$

$$w = 2\sqrt{I_2\left(1 - I_2\right)} \tag{10}$$

where $R^2 = \rho_1^2 + \rho_2^2 + \rho_3^2$ is the radius of gyration. Again, the eigenvalues do not reach Richardson scaling that is denoted by straight dotted t^3 lines in fig. 6(a,b). For the tetrahedra it can be observed how at early times, where due to small scales the velocity field is still quite smooth, the volumes are almost conserved. This is reflected in initially decreasing mean values of g_3, which are the most compressed directions of the tetrahedra. Both shape factors reach a stable plateau after a short transient time in which the initially regular shapes assume their intermediate state. It is difficult to decide if these intermediate states reflect self-similarity or just 'Gaussianity'. The noise level is relatively high and our inertial range is very small. In addition, as can be seen by the straight dotted lines of fig. 6(c), the values for self-similar and Gaussian shapes are fairly close together. From slightly higher Reynolds number DNS simulations [8] deduced $I_2 \approx 0.16$ for the inertial range while the Gaussian value is $I_2 \approx 0.22$ [6]. Our data lies in-between for tetrahedra and, since three points are 'easier' to follow, probably also for triangles for which we don't know the corresponding values.

We now look at how evolving particle constellations are oriented with respect to the strain eigenframe spanned by the coarse grained eigenvectors $\widetilde{\lambda}_i$. For the following Lagrangian results we use as an evolving filtering scale $R_0(t) = r(t)$ or $R_0(t) = \sqrt{g_1(t)}$. For infinitesimal separations and also for infinitesimal areas it is well known that after a transient time of $t/\tau_\eta > 1$ separation vectors are predominately aligned with the most stretching axis $\widetilde{\lambda}_1$

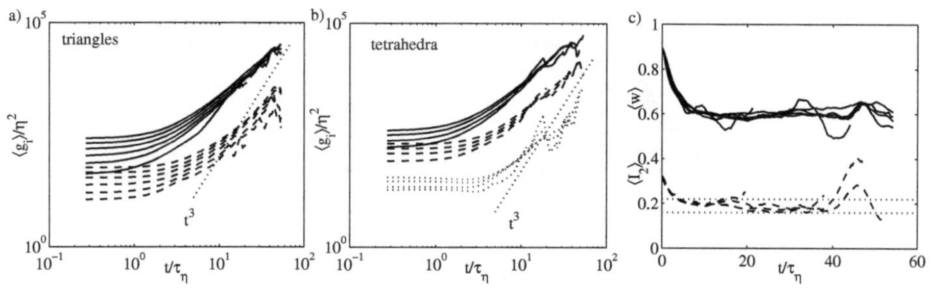

Fig. 6. Temporal evolution of mean eigenvalues of the tensor g^{ab}, solid lines for g_1, dashed lines for g_2 and dotted lines for g_3(a,b). Initial separations range from $14 < r_0/\eta < 30$ for triangles and from $22 < r_0/\eta < 30$ for tetrahedra with 4η bins. (c) Mean shape factors $\langle w \rangle$ and $\langle I_2 \rangle$ for triangles and tetrahedra. The straight dotted lines denote inertial range value $I_2 \approx 0.16$ and the Gaussian value $I_2 \approx 0.22$.

and that surface normals are predominately aligned with the most compressing axis $\widetilde{\lambda}_3$ [25, 22, 14, 26]. It is natural to expect the larger scale counterparts r_0, g_1, and g_3 of pairs, areas and volumes to behave similarly. In fig. 7 and 8 we present experimental evidence for this. PDFs for all cases are shown for $t > \tau_B$, i.e. τ_B is replacing τ_η as the relevant time scale. Initial scales range from $6 < \eta < 30$, $14 < \eta < 30$, and $22 < \eta < 30$, for pairs, triangles and tetrahedra respectively. This is reflecting that it is more difficult to find e.g. four points close by and to be able to track them for a long time than it is to find and track 'just' a pair. In all cases we observe a clear alignment of r and g_1 with $\widetilde{\lambda}_1$ (fig.7), and moreover, the PDFs for pairs, triangles and tetrahedra are almost identical also on a quantitative level. For the surface normals of triangles and the smallest eigen-direction of tetrahedra, g_3, we see a clear preferential alignment with the compressing principal axis $\widetilde{\lambda}_3$ (fig. 8). These alignments are one way to explain why in the inertial range flat and elongated structures can be observed as it is reported in [6, 8].

Such alignments to principal axes only affect separations if the corresponding strain field $\widetilde{s}_{ij}$ is strong enough, e.g. as one would expect from K41 type arguments. As we have seen above in fig. 1 this seems to be the case. For particle pairs we now directly check how much the coarse grained counterpart of eqn. 7

$$\frac{\frac{1}{2}\frac{dr^2}{dt}}{r^2} \cong \frac{r_i r_j \widetilde{s}_{ij}}{r^2} \tag{11}$$

is balanced. In other words, we check to what degree particle separation $dr(t)/dt$ is governed by the strain field filtered at the local scale $\Delta = r(t)$ as it is assumed in [3]. Fig. 9(a) shows the temporal evolution of averages of the l.h.s. and r.h.s. of eqn. 11 for initial separations $6 < r_0/\eta < 30$. For the r.h.s. values are only given if separations are large enough to find $n > 12$ particles inside volumes of $v = r^3(t)$. We see how at early times the terms

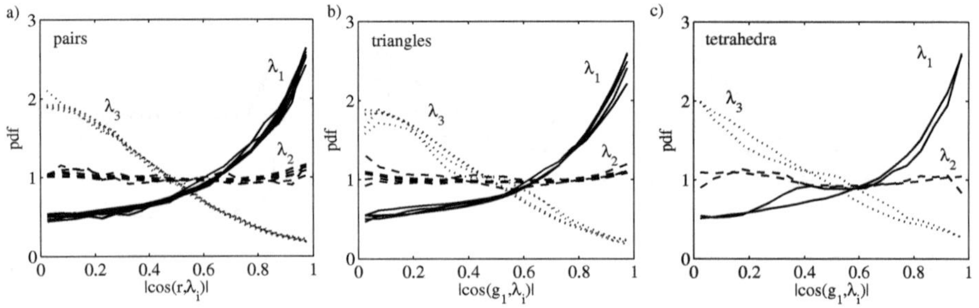

Fig. 7. PDFs of the cosines between the most elongated axis of particle constellations and the eigenframe of the filtered strain tensor $\widetilde{s}_{ij}$ at $t > \tau_B$. (a) Particle pairs with initial separations r_0 of $6 < r_0/\eta < 30$. (b) Particle triangles with initial $R_0 = \sqrt{g_1}$ of $14 < R_0/\eta < 30$. (c) Particle tetrahedra with initial $R_0 = \sqrt{g_1}$ of $22 < R_0/\eta < 30$.

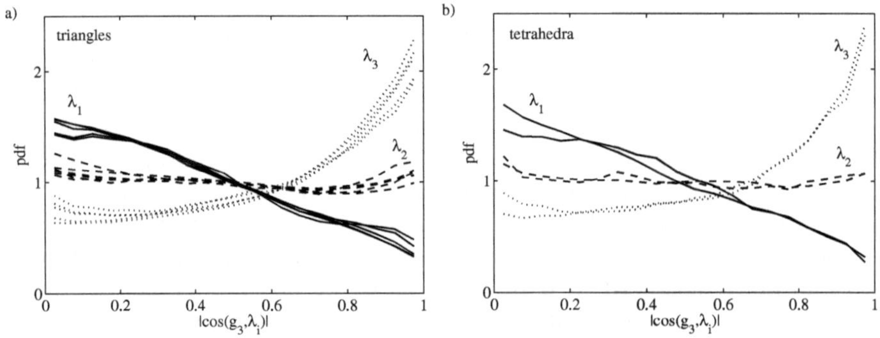

Fig. 8. PDFs of the cosines between the shortest axis of particle constellations and the eigenframe of the filtered strain tensor $\widetilde{s}_{ij}$ at $t > \tau_B$. (a) Triangles with initial $R_0 = \sqrt{g_1}$ of $14 < R_0/\eta < 30$. (c) Tetrahedra with initial $R_0 = \sqrt{g_1}$ of $22 < R_0/\eta < 30$.

are roughly balanced but at later times it is clear that $1/2dr^2/dt > r_i r_j \widetilde{s}_{ij}$. This means that a significant part of separation must be caused by scales $\Delta < r(t)$. In fig. 9(b) we show the same quantities but compensated with $t_*/\sqrt{2}$ that was introduced above with eqn. 4. We observe a good scaling for $1/2\left(dr^2/dt\right)/r^2$ while the compensated $r_i r_j \widetilde{s}_{ij}/r^2$ continues to decrease. The two straight dotted lines at 0.11 and 0.14 denote the interval of the seemingly universal stretching rate for the viscous scales [25, 22, 14, 26, 2]. Our data for the compensated large scale separation rates fall into this range. It thus appears, that the total separation rate indeed does behave like its viscous counterpart, also on a quantitative level, as it was assumed in [3]. However, we infer that the total separation must be the sum of contributions that stem

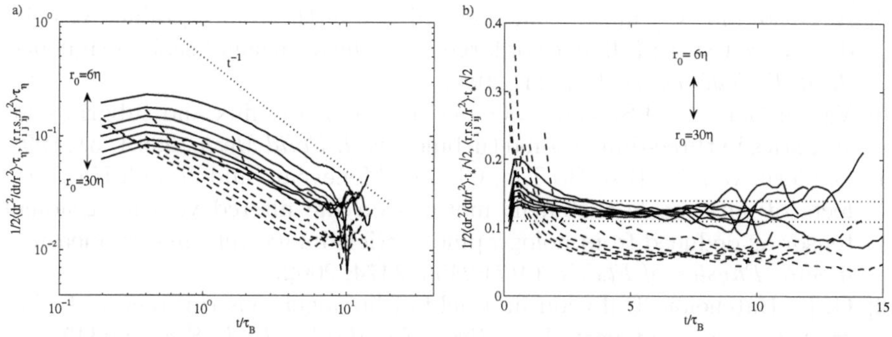

Fig. 9. (a) Temporal evolution of the means of l.h.s (solid lines) and r.h.s. (dashed lines) of eqn. 11. Initial separations are $6 < r_0/\eta < 30$ with 4η spacing. (b) Same data as in (a) but all quantities are compensated with $t_*/\sqrt{2}$ of eqn. 4. The two straight dotted lines denote the interval between 0.11 and 0.14.

also from smaller scales $\Delta < r$. The contribution from $r_i r_j \widetilde{s}_{ij}$, filtered at scale $\Delta = r$, is in our case roughly 50%.

References

[1] Brian Sawford. Turbulent relative dispersion. *Ann. Rev. Fluid Mech.*, 33:289–317, 2001.

[2] Jacob Berg, Beat Lüthi, Jakob Mann, and Søren Ott. An experimental investigation: backwards and forwards relative dispersion in turbulent flow. *Phys. Rev. E*, 74(1):016304, 2006.

[3] Beat Lüthi, Jacob Berg, Søren Ott, and Jakob Mann. Self similar two particle separation model. *1st revision, Physics of Fluids*, 2006.

[4] Laurent Mydlarski, Alain Pumir, Boris I. Shraiman, Eric D. Siggia, and Zellman Warhaft. Structures and multipoint correlators for turbulent advection: Predictions and experiments. *Phys. Rev. Lett.*, 81(20):4373–4376, 1998.

[5] Misha Chertkov, Alain Pumir, and Boris I. Shraiman. Lagrangian tetrad dynamics and the phenomenology of turbulence. *Physics of Fluids*, 11(8):2394–2410, August 1999.

[6] Alain Pumir, Boris I. Shraiman, and Misha Chertkov. Geometry of Lagrangian dispersion in turbulence. *Phys. Rev. Lett.*, 85(25):5324–5327, December 2000.

[7] A. Naso and A. Pumir. Scale dependence of the coarse-grained velocity derivative tensor structure in turbulence. *Phys. Rev. E*, 72:056318, 2005.

[8] L. Biferale, G. Boffetta, A. Celani, B. J. Devenish, A. Lanotte, and F. Toschi. Multiparticle dispersion in fully developed turbulence. *Physics of Fluids*, 17(11):111701, 1–4, 2005.

[9] A. Naso and A. Pumir. Scale dependence of the coarse-grained velocity derivative tensor: Influence of large-scale shear on small-scale turbulence. *J. of Turbulence*, 7(41):1–11, 2006.

[10] Vadim Borue and Steven A. Orzag. Local energy flux and subgrid-scale statistics in three-dimensional turbulence. *J. Fluid Mech.*, 366:1–31, 1998.

[11] Fedderik van der Bos, Bo Tao, Charles Meneveau, and Joseph Katz. Effects of small-scale turbulent motions on the filtered velocity gradient tensor as deduced from holographic particle image velocimetry measurements. *Physics of Fluids*, 14(7):2456–2474, 2002.

[12] G. K. Batchelor. Diffusion in a field of homogeneous turbulence. II The relative motion of particles. *Proc. Cambridge Phil. Soc.*, 48:345–363, 1952.

[13] Alain Pumir, Boris I. Shraiman, and Misha Chertkov. The Lagrangian view of energy transfer in turbulent flow. *EuroPhys. Lett.*, 56(3):379–385, 2001.

[14] Beat Lüthi, Arkady Tsinober, and Wolfgang Kinzelbach. Lagrangian measurement of vorticity dynamics in turbulent flow. *J. Fluid Mech.*, 528:87–118, 2005.

[15] T. Chang and G. Taterson. Application of image processing to the analysis of three-dimensional flow fields. *Opt. Engng*, 23:283–287, 1983.

[16] R. Racca and J. Dewey. A method for automatic particle tracking in a three-dimensional flow field. *Experiments in Fluids*, 6:25–32, 1988.

[17] Marko Virant and Themistocles Dracos. 3D PTV and its application on Lagrangian motion. *Meas. Sci. Technol.*, 8:1529–1552, 1997.

[18] Søren Ott and Jakob Mann. An experimental investigation of the relative diffusion of particle pairs in three dimensional turbulent flow. *J. Fluid Mech.*, 422:207–223, 2000.

[19] A. La Porta, Greg A. Voth, Alice M. Crawford, Jim Alexander, and Eberhard Bodenschatz. Fluid particle accelerations in fully developed turbulence. *Nature*, 409:1016–1017, Feb. 2001.

[20] Mickaël Bourgoin, Nicholas T. Ouellette, Haitao Xu, Jacob Berg, and Eberhard Bodenschatz. Pair Dispersion in Turbulence. *Science*, 311:835–838, 2006.

[21] Michael S. Borgas and P.K. Yeung. Relative dispersion in isotropic turbulence. Part 2. A new stochastic model with Reynolds-number dependence. *J. Fluid Mech.*, 503:125–160, 2004.

[22] A. Tsinober. *An Informal Introduction to Turbulence*. Kluwer Academic Publishers, 2001.

[23] M. Kholmyansky, A. Tsinober, and S. Yorish. Velocity derivatives in the atmospheric surface layer at Re-lambda=10^4. *Physics of Fluids*, 13(1):311–314, 2001.

[24] B.J. Cantwell. Exact solution of a restricted euler equation for the velocity-gradient tensor. *Physics of Fluids*, 4(4):782–793, 1992.

[25] S. S. Girimaji and S. B. Pope. Material-element deformation in isotropic turbulence. *J. Fluid Mech.*, 220:427–458, 1990.

[26] Michele Guala, Beat Lüthi, Alex Liberzon, Arkady Tsinober, and Wolfgang Kinzelbach. On the evolution of material lines and vorticity in homogeneous turbulence. *J. Fluid Mech.*, 533:339–359, 2005.

Simultaneous measurements of the fluid and the solid phases in homogeneous turbulence: preliminary results at $Re_\lambda = 250$

Michele Guala[1], Alexander Liberzon[1,2], Klaus Hoyer[1], Arkady Tsinober[3], and Wolfgang Kinzelbach[1]

[1] Institute of Environmental Engineering, ETH Honggerberg, CH-8093 Zurich, Switzerland guala@ifu.baug.ethz.ch, liberzon@ifu.baug.ethz.ch, hoyer@ifu.baug.ethz.ch, kinzelbach@ifu.baug.ethz.ch

[2] Department of Fluid Mechanics and Heat Transfer,Tel Aviv University, Israel

[3] Department of Aeronautics, Imperial College London Great Britain a.tsinober@imperial.ac.uk

1 Introduction

Two-phase turbulent flows are of great interest for a number of most important and extremely diverse application such as combustion, sedimentation, fluidized beds, settling and resuspension of organic material and contaminants in lakes, ocean, waste water management, evolution of clouds, rain formation and dust storms (see for example Crowe et al., 1998, Gyr and Kinzelbach, 2003, Burton & Eaton 2005 and references therein). In a turbulent flow the interaction between the second phase and the carrier fluid is important since, for example the nonlinear processes that redistribute energy between the scale of the energy containing eddies and the scales associated with the field of velocity derivatives, are two-way coupled with the spatial distribution of the second phase. In the case of dilute suspensions of solid particles, the small scale processes of turbulence were observed to be modified due to the effects such as inertial response of particles to fluid acceleration, gravitational effects and preferential concentration of particles (see e.g. Ferrante & Elghobashi, 2003). In particular, in the flow regions where the dispersed solid phase tends to accumulate (i.e., clusters), the two way coupling effects may be enhanced, leading to strong modification of the local processes of self-amplification of velocity derivatives (see e.g. Tsinober, 2001).

The Stokes number St, defined as the ratio between the particle response time τ_p and a timescale of the flow τ_f is the characteristic dimensionless group for particle laden flows. The relevance of small scales, in this context, is em-

Bernard J. Geurts et al. (eds), Particle Laden Flow: From Geophysical to Kolmogorov Scales, 271–283.

phasized by the occurrence of clusters at the critical Stokes number of the order of 1 when the Kolmogorov time scale τ_k is chosen as the representative flow time scale, i.e $\tau_f = \tau_k$. Indeed for St $\ll 1$ particles tend to follow the flow closely, for St ~ 1 clustering was observed (Squires & Eaton 1991, Ferrante & Elghobashi, 2003, and Fallon and Rogers 2001, both in standard and micro-gravity conditions) and to affect the settling velocity (Aliseda et al. 2002, Bosse and Kleiser, 2006). For St $\gg 1$ particles respond too slowly to be able to feel the spatial and temporal variations of the ambient flow field and therefore tend to settle like in an undisturbed fluid. It is noteworthy that most of the available experimental knowledge on the interaction between a turbulent flow and a suspension of solid particles comes either from measurements of the liquid phase in water channel or pipe flows or from measurements of the distribution of solid particles in a nearly homogeneous turbulent air flows. A few exceptions regard the cases in which solid particles were tracked simultaneously with the fluid phase (PTV combined with PIV measurements in water channel flows, Kiger & Pan, 2002) and the case of the study of particle settling velocity in oscillating grid turbulence in water (e.g. Yang & Shy 2003). In both cases measurements were limited to a two-dimensional velocity field.

The main goals of this experimental study are: i) to observe clustering in a dilute suspension of solid particles in a homogeneous turbulent flow through measurements of local particle concentration in three-dimensions, ii) to quantify the effect of the local particle concentration on the properties of small scale turbulence. It is thus essential to study the interactions between the two phases by assessing simultaneously both the location and distribution of the solid particles, and the fields of velocity and velocity derivatives of the flow. In order to obtain such measurements in three dimensions, we adopted the 3D-PTV technique. We choose water, instead of air, to be the carrier fluid and we employ neutrally buoyant particles as the fluid tracers and silica-gel particles as the dispersed phase (ρ_s=1400 kg m^{-3} and diameter D=900 μm).

In the case of water as the carrier fluid and large solid particles, i.e. mean diameter D which is larger than the Kolmogorov length scale η, as the dispersed second phase, the definition of the particle response time is less straightforward than in the typical case of small solid particles in air. In this case one of the key assumptions of Maxey & Riley (1983) is violated, implying that the particle motion cannot be correctly estimated by using the first principle. In such case a fully resolved simulation of the particle boundary layer has to be implemented as described by Burton & Eaton (2005). According to the authors, the Stokes drag is the dominant term for a fixed particle in homogeneous turbulent flow with diameter twice as large as the Kolmogorov length scale, while the history term and the combination of added mass, fluid pressure gradient and viscous stress were observed to be small. In a turbulent channel flow, for $\rho_p \simeq \rho_f$, Armenio & Fiorotto (2001) observed that, as compared to the Stokes drag, the added mass force is negligible, the Basset force is appreciable and the pressure drag is relevant when the densities of the carrier ρ_f and dispersed ρ_p phases are comparable. Therefore, in the present contribution we assume that i) the equation of motion as described by Maxey & Riley (1983) is still valid in our case, ii) we neglect the Basset force and the Faxen force, iii) we neglect the acceleration of the fluid in the particle location. The particle response time can be thus defined as:

$$\tau_p = \frac{2(\rho_p + \rho_f)/\rho_f D^2}{36\nu_f}, \tag{1}$$

note that for $\rho_p >> \rho_f$ we obtain the definition $\tau_p = \frac{\rho_p D^2}{18\mu_f}$ commonly used for the cases of gas-solid, gas-liquid two-phase flows.

Regarding the steady Stokes drag term $F_D = 1/2\rho_f C_D|u - v|(u - v)$, we must note that the drag coefficient C_D is a function of the Reynolds number based on the slip velocity, i.e. the difference between the particle velocity v and the fluid velocity u, $Re = D|u - v|/nu$. For $Re > 1$ Oseen correction has to be considered. The following definition of the particle response time is given for larger Re, up to 200 (see e.g. Crowe, 1998).

$$\tau_p = \frac{(2\rho_p + \rho_f)/\rho_f D^2}{36\nu_f(1 + 0.15 * Re^{0.687})} \tag{2}$$

We can estimate the particle Reynolds number to be approximately 50, if we assume the slip velocity to be equal to the settling velocity W_s of 5 cm s^{-1} calculated following Clift et al (1978), which is approximately equal to the measured r.m.s. velocity of the ambient fluid. In this case $\tau_p = 30$ ms, while if we neglect the Oseen correction, $\tau_p = 80$ ms. The Stokes number based on the kolmogorov scale $\tau_\eta = 20$ ms, is thus in the range of 1.5 - 4. It is noteworthy that the time scale of the flow which participates to the definition of the Stokes number can also vary (see e.g. Eaton & Fessler, 1994). Due to the fact that D_p is larger than η we consider other time scales, e.g. the one based on the r.m.s velocity, and Taylor microscale $\lambda/u_{rms} \simeq 100ms$. In this case the Stokes number is in the range of 0.3 - 0.8. In either case we may expect to observe clustering since the discussed range of values are close to the critical value of 1.

2 Experimental setup

The experiment has been carried out in a glass aquarium of 120 x 120 x 140 mm^3 in which the flow is forced mechanically from two sides by two sets of four rotating disks with artificial roughness elements. The observation volume of approximately 30 x 30 x 30 mm^3 was centered with respect to the forced flow domain, mid-way between the disks. A description of the forcing device, depicted in fig 1 can be found in Liberzon et al. 2005.

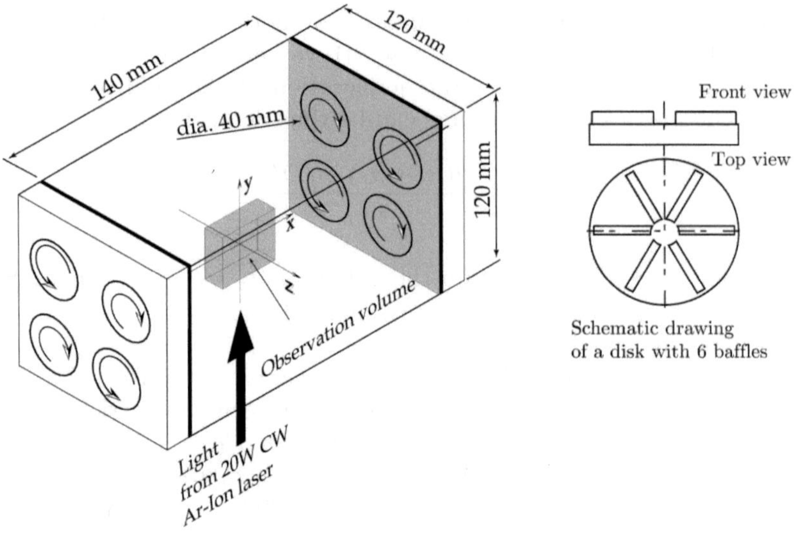

Fig. 1. Sketch of the forcing device.

2.1 Recording system

The experimental approach is based on the three-dimensional Particle Tracking Velocimetry technique (3D-PTV, Luthi et al. 2005). Such technique is of special importance since the second phase (i.e. the solid particles) is an essentially Lagrangian object, so one can track its evolution, distribution (cluster) and dynamics, along with its interaction with the carrier fluid. A sketch of the experimental apparatus is shown in figure 2. For the present experiments two 3D-PTV systems have been synchronized at a frame rate of 500 fps. A Photron-Ultima high-speed camera with a 4 mirror image splitter and a resolution of 1024×1024 pixels was set to record images of the solid phase only, i.e. the Rhodamine labeled silica gel particles. A dichroic red filter and a low aperture were employed to filter out the light scatter by the $40 \div 60$ μm neutrally buoyant tracer particles. Another system, consisting of 4 CMOS cameras (Mikrotron GmbH, 1280×1024 pixels) linked to a real-time digital video recording system (IO Industries) was used for the fluid phase. These cameras were equipped by dichroic green filters. Typical images of the tracers and the solid particles are shown in figure 3, where we can see that there is no contamination of the silica particle images in the system devoted to the fluid tracers recording and vice versa. The observation volume of $30 \times 30 \times 30$ mm^3 was illuminated by a continuous 20 W Ar-Ion laser.

2.2 Mutual calibration of the two 3D-PTV systems

Both 3D-PTV systems are calibrated on the same measurement volume. In order to achieve this we used a planar (x,y) calibration target and we acquired

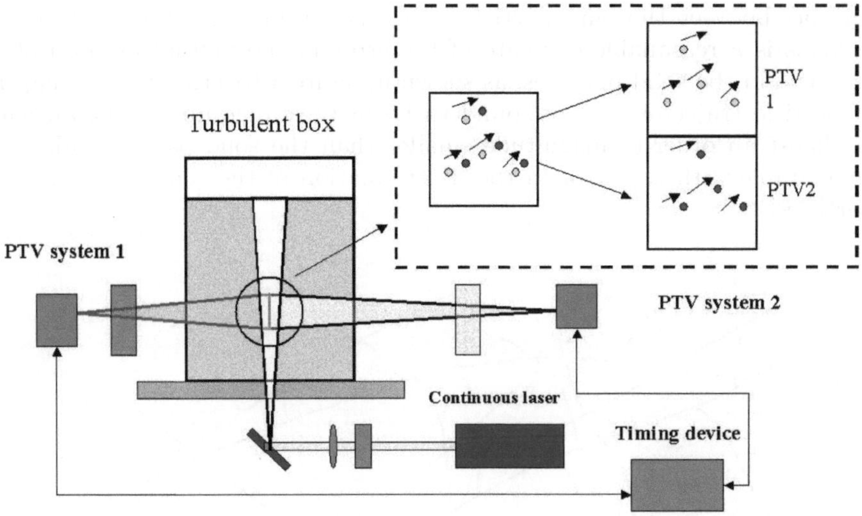

Fig. 2. Sketch of the experimental setup.

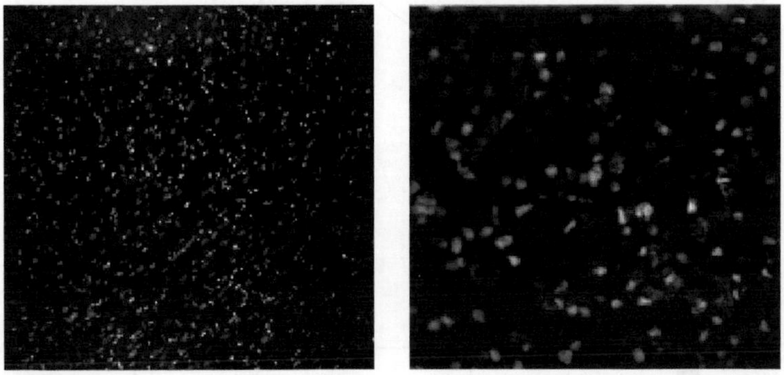

Fig. 3. Sample images of tracers particles (1000 pixels, left) and silica particles (512 pixels, right)

calibration images at different z-locations in depth (multi-plane calibration). The calibration target consists of two symmetric aluminum black painted plates with conically shaped holes forming a grid of 5 mm mesh size. A thin aluminum foil ($< 10\mu m$) was positioned in between the two plates in order to scatter the light in correspondence of the grid points in both directions of the two acquisition systems. In such way the x, y, z location of each point used for the calibration is the same for the two systems besides the thickness of the aluminum foil in the z direction. The calibration procedure was tested in a single phase experiments where the seeding was kept low. Both systems recorded the same tracers particle within the same observation volume. The distance between the same particles at the same time as obtained by the two systems is a reasonable estimate of the error in the mutual location of the tracers and the solid particles: as shown in figure 4 together with a snapshot of particle trajectories, it is found to be in the range up to $100\mu m$, which is almost an order of magnitude smaller than the solid particle radius and comparable with the error in the determination of the centroid of the solid particles.

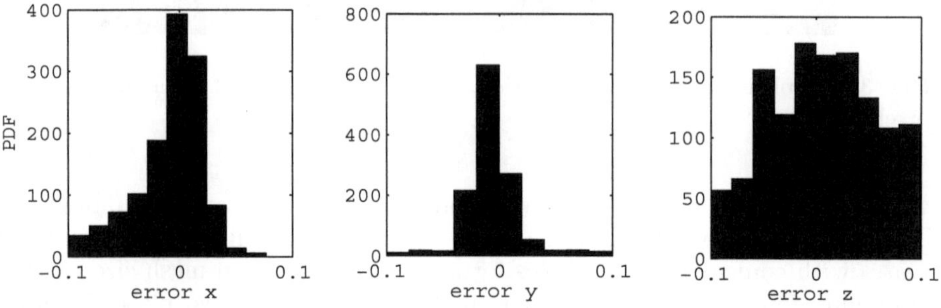

Fig. 4. A snapshot from simultaneous measurements of tracer particles (top). PDF of the displacement components between the location of the same tracer particle as reconstructed by the two 3D-PTV systems.

3 Results

3.1 Solid phase: estimate of local concentration

The three dimensional location of solid particles is the first output of 3D-PTV. In the following we present only statistics related to the position and concentration of the second phase. We can thus use a larger dataset of matched particles (reconstruction of the 3D location in space from the 2-D location in the image plane of at least three cameras) as compared to the (slightly) smaller dataset of tracked particles (reconstruction of 3D particle trajectory in time). The particle image of the dispersed solid phase is rather large, with respect to the observation volume. Thus at this stage only with a limited number of particles (order of 100 per image) it was possible to obtain simultaneous measurement of the fluid phase. This is due to the fact that solid particles, though transparent to laser light, are not refractive-index matched to the fluid. This implies that the tracers particles located behind them in the optical path from the camera sensor, cannot be reconstructed. Therefore increasing the solid particle concentration has the side effect of reducing the spatial resolution of the flow. The relatively low concentration is the major limitation of the current setup. Nevertheless, an estimate of the local concentration as a function of space x, y, z and time t is possible, by counting the number of solid particles N_p which can be found in a sphere of a known radius centered on each solid particle. The first evidence of non Gaussian distribution of particles, though weak, is given in figure 5 in which the distribution of $N_p$is provided for different sphere radii.

3.2 The fluid phase: turbulent quantities and scales

From the trajectories of the fluid tracers it is possible to extract the three dimensional velocity and velocity derivatives fields, together with the Lagrangian acceleration, as a function of space and time. With respect to the previous results (Luthi et al. 2005, Liberzon et al. 2005) the estimate of the velocity derivatives is not as accurate. This is because in the latter experiments the inter-particle distance was of the order of the Kolmogorov length scale, while in this case, is of the order of 10 Kolmogorov length scale. Since the silica gel particles are as well larger than the Kolmogorov length scale, such estimate of ω_i and s_{ij} is not considered as a limiting factor for the present investigation. We present the point-wise checks on the Lagrangian, Eulerian and convective acceleration ($Du_i/Dt = \partial u_i/\partial t + u_j \partial u_i/\partial x_j$) in figure 6. the shape the Joint PDF could serve as a qualitative check on the estimation of temporal and spatial velocity derivatives, according to Luethi et al 2005.

Despite of the aforementioned technical limitation, the qualitative behavior of the small scales of turbulence is still captured as we can see from figures 7, 8. In figure 7 we observe the typical shape of the PDFs of strain, enstrophy and their respective production terms in the left panel and Joint PDFs of

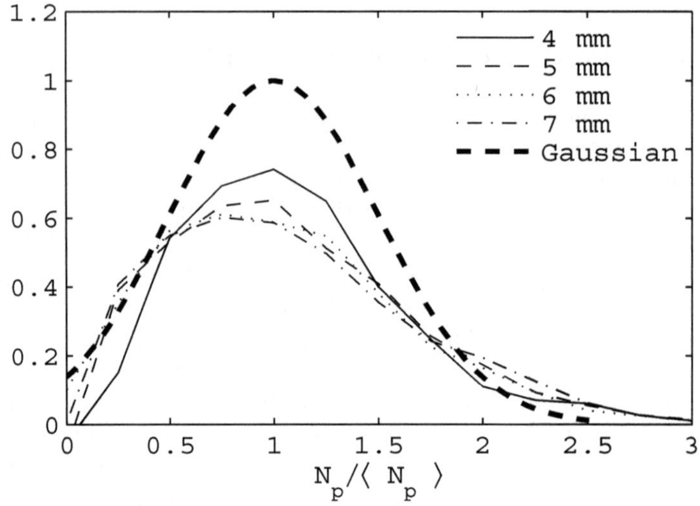

Fig. 5. PDF of $N_p/\langle N_p \rangle$ for different sphere radii as compared to the Gaussian distribution.

these quantities on the right side. In figure 8 the important features of small scale turbulence, as the alignment between vorticity ω and the eigenframe λ_i of the rate of strain tensor and the vortex stretching vector W_i are shown. The distribution of these quantities is in qualitative agreement with previous results (see for example Tsinober, 2001).

The scales of the flow are reported in table 1 and the comparison with some of the relevant quantities from the single phase experiments of Liberzon et al. (2005) is provided in table 2. The Kolmogorov scales are based on the estimate of the dissipation rate ϵ. The latter was slightly overestimated by $\epsilon = 2\nu s_{ij} s_{ij}$, compared to the definition based on the r.m.s. velocity u_{rms} and integral length-scale u_{rms}^3/L. Using the estimated standard deviation of one component of the fluid acceleration $a = 0.8$ ms^{-2}, we obtain the dimensionless parameter $a_0 = var(a)/(\epsilon^{3/2}\nu^{-1/2}) = 4.77$ which is, at this value of $Re_\lambda = 250$, in agreement with results of Voth et al., 2002, among others. The Taylor microscale was estimated as $\lambda = 15\nu u_{rms}/\epsilon$ to be equal to 5 mm.

L[cm]	u_{rms} [cms^{-1}]	ϵ [m^2s^{-3}]	$\langle a_i^2 \rangle^{0.5}$[$ms^{-2}$]	η [μm]	τ_η [ms]
4	5	2.5 10^{-3}	0.8	195	20

Table 1. Some relevant scales of the flow derived from the estimates of the velocity, strain and acceleration fields; Taylor microscale $\lambda = 5$ mm.

Re_λ	$\langle\Lambda_1\rangle$	$\langle\Lambda_1\rangle$	$\langle\Lambda_1\rangle$	$\langle s^2\rangle$	$\langle\omega^2\rangle$
250	14.9 $[s^{-1}]$	1.4 $[s^{-1}]$	-16.4 $[s^{-1}]$	510 $[s^{-2}]$	890 $[s^{-2}]$
250	0.30	0.03	-0.34	0.2	0.36
50	1.3 $[s^{-1}]$	0.26 $[s^{-1}]$	-1.52 $[s^{-1}]$	4.9 $[s^{-2}]$	11.8 $[s^{-2}]$
50	0.30	0.06	-0.35	0.26	0.62

Table 2. Dimensional quantities are reported in the first row, while dimensionless quantity normalized by Kolmogorov time scale $\tau_\eta = 0.02s$ are reported in the second row. In the following two rows are reported the same dimensional and dimensionless mean values from Liberzon et al. (2005) with $\tau_\eta = 0.23s$.

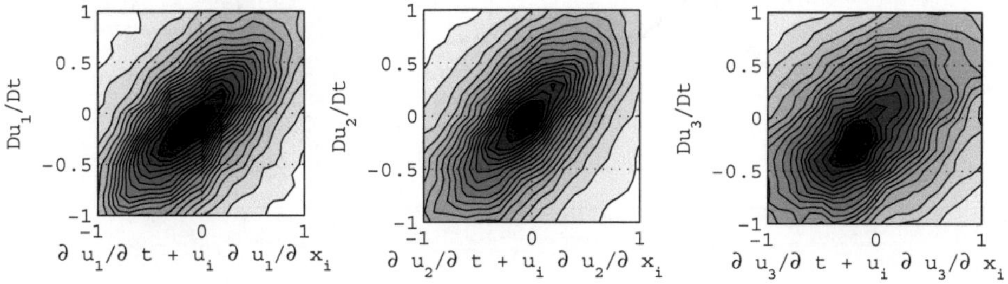

Fig. 6. Joint PDF between the components of Lagrangian acceleration and the sum of the Eulerian and convective components of acceleration

3.3 Joint statistics: local particle concentration in the field of velocity derivatives

In order to relate the local concentration to the flow topology, we condition the statistics of N_p on the value of $Q = \omega^2 - 2s^2$, averaged on a sphere of radius R_Q. The results of joint statistics is given in figures 9. The conditional average of $(N_p - \langle N_p\rangle)/\langle N_p\rangle$ is shown as a function of $Q/\langle\omega^2\rangle$ for different R_Q. There is an indication that particles tend to accumulate in high strain regions ($Q/\langle\omega^2\rangle < 0$) rather than in high enstrophy regions ($Q/\langle\omega^2\rangle > 0$). This preliminary result suggests that large solid particles exhibit local preferential concentration, though weak, in nearly homogeneous turbulent flow in water.

4 Summary and discussion

Simultaneous measurements of a dilute suspensions of large solid particles in liquid turbulent flow at $Re_\lambda \simeq 250$ in nearly homogeneous conditions are

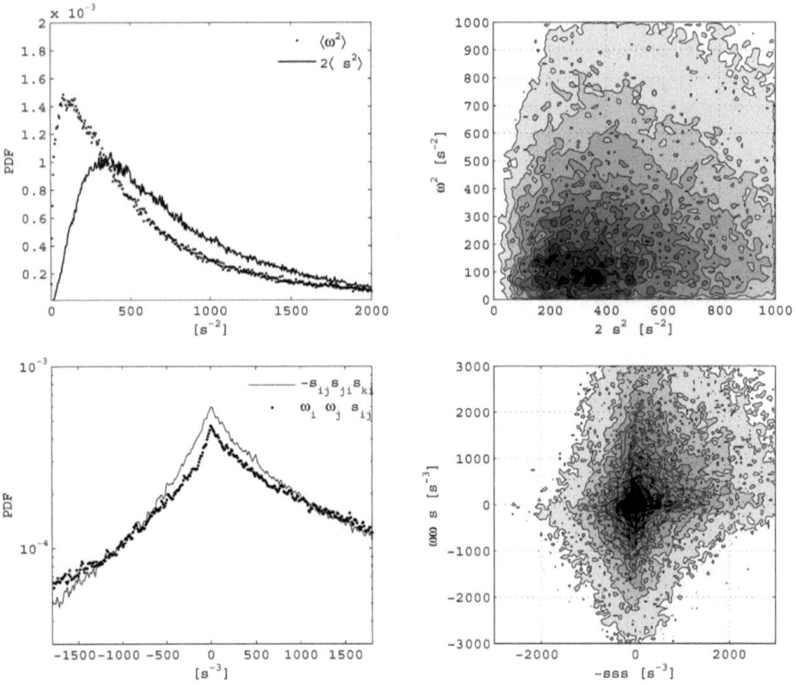

Fig. 7. PDF and Joint PDF of strain, enstrophy and their production terms

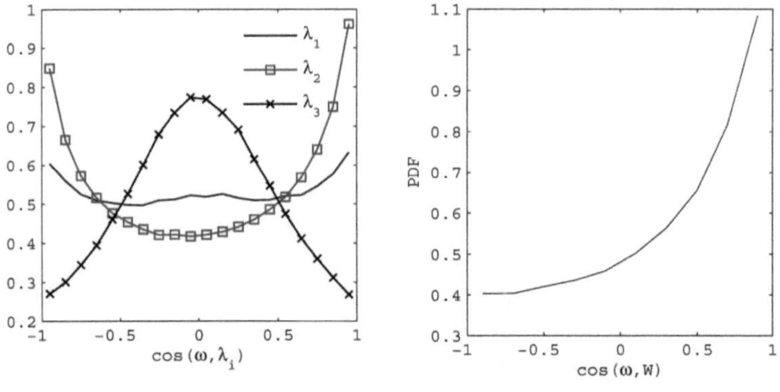

Fig. 8. PDF of the alignments between ω and the eigenvalues λ_i of the rate of strain tensor(left) and the vortex stretching vector $W_i = \omega_j s_{ij}$ (right).

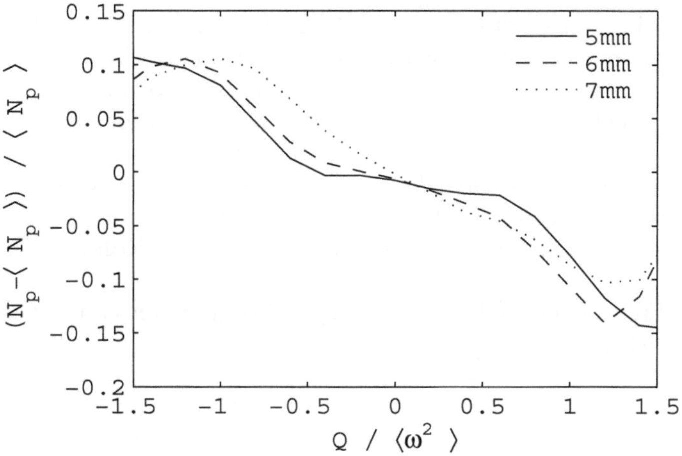

Fig. 9. Normalized average of N_p as computed on a sphere of a fixed 5mm radius, conditioned on the value of $Q = \omega^2 - 2s^2$ in the surrounding of the particle. Different curves correspond to different radii of the sphere in which ω^2, s^2, are locally averaged.

provided. The spatial distribution of solid particles, together with the fields of velocity and velocity derivatives, are obtained by means of two synchronized 3D-PTV systems. Preliminary tests on the mutual calibration of the two systems, on the optical filtering and particles labeling and on the performance of 3D-PTV in high Reynolds number flows, were performed. The scales of the flow were consistently estimated and a series of checks on the quality of the measurements of the liquid phase were made and compared to previous results at $Re_\lambda \simeq 50$ (see Luthi et al., 2005 and Liberzon et al., 2005). Results on the strain and vorticity fields show that the key features of turbulent flows are captured, despite of the coarse spatial resolution. Regarding the solid phase, the local concentration is estimated by counting the number of particles N_p in the close neighborhood of each particle, within a sphere of radius comparable to the Taylor microscale. This approach is adopted due to the limited number of tracked solid particles per frame. Joint statistics between the local particle concentration and the strain and vorticity fields show that solid particles have a weak tendency to concentrate in strain dominated regions rather than in vorticity dominated regions. This is a preliminary result which suggests that large solid particles ($D > \eta$) in water (ρ_p comparable with ρ_f) tend to form clusters in strain dominated regions. The clustering effect observed in this study is significantly weaker as compared to the numerical results of Squires and Yamazaki (1995). We understand that the main difference is due to the settling velocity which is in our case almost 10 times larger than the Kolmogorov velocity scale. We may expect to observe stronger preferential concentration for particle of similar size but lower density. One has to be aware of the limitations of this experiment. A limited number of solid particles in a small observation volume were tracked. Therefore the occurrence of clusters cannot be visualized in snapshots. Moreover N_p provides only a rough estimate of particles distribution and it depends on the radius of the sphere in which the particles are counted. In the further experiments it is necessary to implement a method of refractive index matching allowing for a significant increase of solid particle concentration. In addition, the occurrence and the intensity of clustering of large solid particles in liquids, remains to be investigated for different Stokes numbers and settling to Kolmogorov velocity ratios.

We gratefully acknowledge funding by ETH research commission under grant TH 15/04-2.

References

[1] Aliseda A., Cartellier A., Hainaux F and Lasheras J.C. (2002), Effect of preferential concentration on the settling velocity of heavy particles in homogeneous isotropic turbulence J. Fluid Mech., 468, 77-105.

[2] Bosse T. and Kleiser L. (2006), Small particles in homogeneous turbulence: Settling velocity enhancement in two way coupling, Phys. Fluids, 18, 027102.

[3] Armenio and Fiorotto (2001), The importance of the forces acting on particles in turbulent flows Phys. Fluids, 13(8), 2437-2440.

[4] Burton T.M. and Eaton J.K. (2005), Fully resolved simulations of particle-turbulence interaction J. Fluid Mech. Vol. 545, pp. 67-111.

[5] Crowe C., Sommerfeld and Tsuji Y. (1998) Multiphase flow with droplets and particles, CRC Press, Boca Raton, Boston.

[6] Eaton J.K. and Fessler J.R. (1994), preferential concentration of particles by turbulence. Int. J. Multiphase Flow (20), pp 169-209

[7] Fallon T. and Rogers C.B., (2002), Turbulence induced preferential concentration of solid particles in microgravity conditions Exp. Fluids, 33, 233-241.

[8] Ferrante A. and Elgobashi S. (2003), On the physical mechanisms of two way coupling in particle laden isotropic turbulence, Phys. Fluids, 15(2), 315-329.

[9] Gyr A. and Kinzelbach W. (2003) Sedimentation and sediment transport: at the crossroad of physics and engineering, Kluwer, Dordrecht.

[10] Kiger K.T. Pan C. (2002), Suspension and turbulence modification effects of solid particulates on a horizontal turbulent channel flow. Journal of Turbulence (3) 019

[11] Liberzon A., Guala M., Lüthi B., Kinzelbach W. and Tsinober A. (2005). Dilute polymers in turbulence. Phys. Fluids, **17**, 031707.

[12] Lüthi B., Tsinober A. and Kinzelbach W., 2005, Lagrangian measurement of vorticity dynamics in turbulent flow. J. Fluid Mech. Vol. 528, pp. 87 - 118.
[13] Maxey M.R. and Riley J.J. (1983), Equation of motion for a small rigid sphere in a non uniform flow. Phys. Fluids, 26(4), 883-888.
[14] Squires K.D. and Eaton J.K., (1991), Phys. Fluids, Preferential concentration of particles by turbulence. A 3(5), 1169-1178.
[15] Squires K.D. and Yamazaki H., (1995), Preferential concentration of marine particles in isotropic turbulence. Deep-Sea Research 42 (11/12), 1989-2004.
[16] Tsinober A. (2001) An informal introduction to turbulence, Kluwer, Dordrecht.
[17] Voth G.A. La Porta A., Crawford A.M. Alexander J. and Bodenshatz E.(2002), Fully resolved simulations of particle-turbulence interaction J. Fluid Mech. Vol. 469, pp.121-160.
[18] Yang T.S. and Shy S.S. (2005), The settling velocity of heavy particles in an aqueous near-isotropic turbulence Phys. Fluids, 15(4), 860-880.

Laboratory model of two-dimensional polar beta-plane turbulence

G.F. Carnevale[1], and A. Cenedese[2], S. Espa[2], M. Mariani[2]

[1] Scripps Institution of Oceanography, University California San Diego, La Jolla, California, USA gcarnevale@ucsd.edu

[2] Dipartimento di Idraulica, Trasporti e Strade, Univrsita di Roma 'La Sapienza', Rome, Italy stefania.espa@uniroma1.it

Summary. The evolution of a two-dimensional turbulent decaying flow is experimentally analyzed in a rotating system considering the effect of the change of the Coriolis force with latitude. The flow is generated using an electro-magnetic (EM) cell, i.e., by electro-magnetically forcing a thin layer of a saline solution, in a rotating reference frame. A Feature Tracking (FT) technique is used to measure the flow field allowing the reconstruction of high resolution velocity and vorticity fields. In agreement with theoretical prediction and previous experiments, results corresponding to high values of the beta parameter show a preferential transfer of energy towards zonal modes and the consequent organization of a weak anticyclonic circulation in the polar zone. Moreover, the analysis of the one-dimensional energy spectra shows a scaling steeper than Kolmogorov's law and a peak near the Rhines scale indicating a soft barrier of the energy transfer towards low wave-numbers.

1 Introduction

Large scale flows in the ocean and in the atmosphere of the Earth and other planets are significantly influenced by rotation and stratification. Both of these effects cause the inhibition of the vertical velocity; as a consequence, motion on these scales can be represented using a quasi-two-dimensional approximation. Also, turbulent flows subjected to rotation develop spectral anisotropy [1]. Kolmogorov's $k^{-5/3}$ scaling, which characterizes the inverse energy cascade developing towards large scales [2, 3] in the absence of rotation, under strong rotation occurs in the wave-number domain only away from the zonal axis [4]. The dynamics of 2D turbulence in the presence of strong rotation differs from 2D classical turbulence in at least two main respects: the inverse energy cascade is arrested at a characteristic wavelength [5], known as Rhines scale, above which the flow is predicted to show high anisotropy and opposite sign vortices organize along the meridional axis forming zonal jet like structures [6] observed in several geophysical systems (i.e., in the atmosphere of Giant Planets [7]).

Bernard J. Geurts et al. (eds), Particle Laden Flow: From Geophysical to Kolmogorov Scales, 285–297.

Rhines' argument to explain the energy transfer towards zonal modes in beta-plane turbulence [5] is based on a competition between nonlinear and beta terms in the quasi-geostrophic vorticity equation [8]:

$$\frac{Dq}{Dt} = \partial_t \zeta + J(\Psi, \zeta) + \beta \partial_x \Psi = -\lambda_E \nabla_H^2 \Psi + \nu \nabla_H^4 \Psi + F \tag{1}$$

where $q = (\zeta + f)/h$ is the potential vorticity (PV), ζ the relative vorticity, h the height of the fluid column, $J = \Psi_x \zeta_y - \Psi_y \zeta_x$ the Jacobian operator , Ψ is the stream function, β the gradient of $f = 2\Omega \sin(\phi)$ (i.e., the Coriolis parameter), $\lambda_E = f E^{1/2}/2$ the bottom friction (Ekman friction, E is the Ekman number) coefficient, ∇_H the horizontal Laplacian operator, ν the kinematic viscosity and F represents the forcing. From the dispersion relation for Rossby waves and estimates of time scales for isotropic turbulence, one can predict that the beta effect term will dominate only at large scales. The anisotropy of the dispersion relation results in energy being transferred to large-scale zonal motions more easily than to other large-scale motions. Thus the energy distribution becomes strongly anisotropic at large scale producing a tendency toward the formation of zonal jets. The associated characteristic scale separating the regions of wave-vector space where either beta or nonlinear effects dominate respectively is also anisotropic; however, an isotropic average of this dividing scale, expressed in terms of rms velocity and beta coefficient i.e., the Rhines scale, can be written as [5]:

$$k_{Rh} = \sqrt{\frac{\beta}{2V_{rms}}} \tag{2}$$

Where linear wave effects dominate, energy transfer rates are diminished with respect to turbulent transfer rates. Thus k_{Rh} can be said to represent the arrest scale of the inverse cascade, a soft barrier to energy transfer towards smaller wave numbers. The presence of this barrier leads to a steeper power law spectrum for $k > k_{Rh}$:

$$E(k) = C\beta^2 k^{-5} \tag{3}$$

where the value of the constant $C \approx 0.5$ has been estimated in numerical beta plane turbulence simulations [4, 7]. Subsequent studies [4, 9] clarified Rhines theory relating the inhibition of non-zonal modes to triad interactions. The scaling (3) has been recovered in numerical simulations of the flow on the Cartesian beta plane [9, 10] as well as in spherical geometry [4]. Also, a region characterized by a homogeneous distribution of PV surrounded by large PV gradients has been seen in the polar zone in simulations of the flow on a rotating sphere [11]. Chekhlov et al. [1] show that the scaling (3) develops in a very narrow sector of the wave numbers space close to the zonal axis while a -5/3 slope, indicative of an isotropic-turbulence-like behavior, prevails in almost all other regions. The role of bottom friction has been discussed by Danilov and Gurarie [12] who introduce a measure of the zonal strength of the flow in terms of the ratio $\gamma = k_{Rh}/k_{fr}$ where k_{fr} represents the arrest scale of the cascade in absence of rotation due to dissipative effects. Anisotropy arises if $\gamma \gg 1$.

Laboratory experiments give an insight into beta plane turbulence dynamics, showing, for instance, the organization of jets and mixing of PV [13]. Starting from experimental data, a detailed investigation of the spectral flow characteristics is still difficult to perform because of two limiting factors: the resolution of the reconstructed velocity field (which depends on the acquisition system and on the adopted measuring technique) and the rapid decay rate due to bottom friction that restricts the time scale of the experiments.

In the recent work Afanasyev and Wells [14] (AW herein), have demonstrated a new experimental methodology that allowed them to study beta-plane turbulence both qualitatively and quantitatively. We intend to use this methodology in conjunction with numerical simulations to further explore the nature of two-dimensional turbulence. We hope, in particular, to understand how both the beta effect and a strong bottom drag interact to determine the width of the observed jets. The preliminary work presented here demonstrates how our laboratory apparatus can be used in this investigation.

In this context, we perform a series of experiments, with various configurations of the velocity field in two different regimes, corresponding to two different rotation rates. Following AW and previous works on 2D non rotating turbulence [15, 16], the flow is generated using an EM cell where the initial condition of vorticity distribution is generated by the superposition of electric and magnetic field on a thin layer of saline solution The effects of the variation of the Coriolis force with the latitude is modeled by placing the test section on a rotating table in order to make the fluid surface assume a parabolic profile. Flow is measured using a particle imaging based technique; in particular, a feature tracking (FT) algorithm [17, 18] allows the reconstruction of high resolution instantaneous velocity and vorticity fields.

FT allows a Lagrangian description of the velocity field, in particular it reconstructs the displacement field by selecting image features (i.e., salient point characterized by large luminosity intensity gradients) and tracking these from frame to frame. Lagrangian data are then used to reconstruct instantaneous Eulerian velocity and vorticity fields through a re-sampling procedure.

Results here do reproduce the important aspects of the phenomena as previously found by AW: the organization of a weak anticyclonic circulation as well of a zonal jet in the high rotation rate experiments. Also, in the case of high rotation rate, we see the formation of the expected soft spectral energy barrier in the wave-numbers space.

2 The model

In order to reproduce the planetary beta effect in a laboratory, a sloping bottom is generally used [19]; such a configuration, known as topographic beta-plane, is dynamically equivalent to a fluid layer with flat bottom with a Coriolis parameter that is linear in y, which measures the distance in the meridional direction from some reference latitude. If in the polar region an approximation for the variation of the Coriolis parameter would be quadratic in y, this would imply assuming the pole as the reference point (the so-called gamma-plane). In order to reproduce in laboratory a similar effect of the quadratic variation of f with y, we can use the parabolic height profile assumed by the free surface of a rotating fluid (figure 1), considering that topographic variations can be seen in the quasi-geostrophic approximation to be equivalent to variations in f when appropriately scaled.

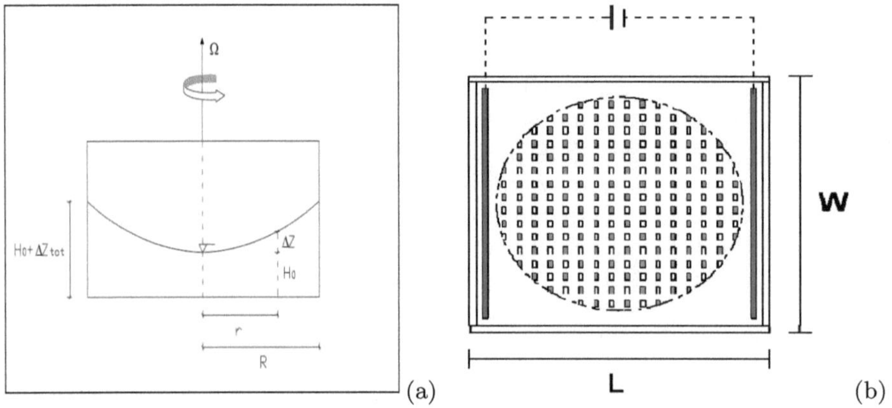

Fig. 1. (a) Side view and (b) plan view of plexiglas tank.

In stationary rotating flow, the centrifugal force and the hydrostatic pressure are balanced when:

$$\Delta z = \frac{\Omega^2}{2g} r^2 = sr^2 \tag{4}$$

where $s = \Omega^2/(2g)$, Ω is the rotation rate, g the gravitational acceleration, and r indicates the north-south distance (note that here r is used instead of y and that with the origin at the pole it assumes only positive values). In this model, the point of maximum depression of the fluid surface represents the pole while the peripheral areas corresponds to the lower latitudes.

Considering now the PV expression:

$$q = \frac{\zeta + f}{H_0 + sr^2} \tag{5}$$

where H_0 is the height of the fluid at $r = 0$, assuming the constant $r_0 = R/2$ (see figure 1) as a reference radius in order to express in terms of y the small shifts of r from r_0 ($r = r_0 + y$; $y \ll r_0$), an expression for the laboratory beta-plane is obtained with the local value of beta given by:

$$\beta = \frac{2sr_0f_0}{H_0 + sr_0^2} \tag{6}$$

2.1 Experimental set-up

The experimental set-up has then been built from the superposition of two standard mechanisms commonly used for generating 2D turbulence in the laboratory [20]: a rotating table and an EM cell. The rotating table has an adjustable angular speed. The sense of rotation used is counter-clockwise in order to simulate flows in the northern hemisphere.

The EM cell (figure 1b) consists of a plexiglas tank whose dimensions are $L = 33\ cm$, $W = 35\ cm$, it is filled with an electrolyte solution of water and NaCl. The cell is placed flat on the rotating table. The center of the tank is lined up with the rotation axis of the table. The flow is generated by a superposition of electric and magnetic fields, i.e., using the effects of Lorentz force. The magnetic field is generated by 97 Neodymium permanents magnets covering the area of a circle on a metallic plate just below the bottom surface of the tank. The tank is filled by injecting the fluid from the lateral reservoirs surrounding the test section. These reservoirs have been designed to house the copper electrodes providing the forcing current, a voltage signal of fixed amplitude generated by a computer controlled power supply. In the experiments, the fluid thickness when non-rotating is 10 mm.

The combined effect of electric and magnetic forcing induces the continuous formation of opposite signed vortical structures whose vorticity is positive or negative according to the phase of the resulting Lorentz force, and whose characteristic length scales are related to the distance between magnets, (2 cm here).

2.2 Experimental procedure

All the experiments were performed as follows:

- the free fluid surface is seeded using styrene particles ($d_m \sim 50\mu m$);
- the electromagnetic forcing is switched on. The early flow field is a regular array of closely packed dipolar vortices, each corresponding to one magnet. As this flow becomes stronger, it proves unstable and begins to deform under the effect of strong nonlinear interactions between the vortices;
- the tank is brought up to the desired rotation rate by imposing a constant angular acceleration ($\sim\ 1rpm$ every 3 s). An approximately parabolic profile is assumed by the fluid free surface in a time interval $5\ min < T < 10\ min$ after the target rotation rate is reached;
- the forcing is switched off. The subsequent evolution of the flow field depends on the value of the beta parameter, i.e., on the imposed rotation rate. In both high and low beta experiments dissipation effects start dominating quite quickly after the forcing has stopped;

- the fluid flow is recorded using a standard speed video camera (acquisition frequency=25 Hz, resolution 720 × 570 pixels) placed orthogonally to the tank surface and co-rotating with the system. Acquired images are then digitized and post-processed using an FT image-processing technique to reconstruct the velocity field evolution.

A series of experiments characterized by various rotation rates ($0s^{-1} \leq \Omega \leq 3.1s^{-1}$), and, therefore, by different values of β and λ_E, has been performed. Results corresponding to the parameters in Table 2.2 will be commented in the further sections.

#Run	Ω (s^{-1})	s (cm^{-1})	H_0 (cm)	D (cm)	β ($cm^{-1}s^{-1}$)	$2\lambda_E$ (s^{-1})
8	2.5	0.0032	0.37	0.66	0.444	0.51
15	1.6	0.0013	0.75	0.86	0.075	0.31

Table 1: Experimental parameters corresponding to the performed experiments. $D = (1/R)\int_0^R H(r)dr$ is the average depth of the layer.

2.3 Measuring technique

The Feature Tracking method is based on the Lukas-Kanade algorithm [21] and on a subsequent version given by Shi and Tomasi [22] and can be defined as a tracking technique based on correlation windows [17]. The method defines the matching measure between fixed-sized feature windows (i.e., interrogation window surrounding a feature) in two consecutive frames and the window displacement is then evaluated by considering the best correspondence between subsequent images as in classical PIV. In FT this measure is given by minimizing the sum of squared differences (SSD) between the image intensity in two subsequent images - instead of by maximizing the inner product (i.e., the correlation) between intensities - and the problem is reformulated as a minimization in a least square sense providing a system with two equations and two unknowns (i.e., the velocity components $u(x, y)$ and $v(x, y)$). To this regard, it has shown [18] that FT compared to PIV yields higher spatial resolution and detects flow characteristics more exhaustively.

The FT routine can be subdivided in two steps: feature extraction, i.e., of (x, y) coordinate pairs that are good to track, and feature tracking. The feature extraction algorithm is then defined in such a way that the optimal solution of the tracking algorithm is achieved: rather than defining a subjective, a priori, notion of a good feature, this definition is based on the method used for tracking itself: 'a good feature is one that can be tracked well'. With this approach, one overcomes a subjective definition of the object which has to be tracked (such as particle centroids in classical Particle Tracking Velocimetry-PTV) and of its corresponding definition parameters. In particular, it can be proved that the solvability of the SSD minimization problem is guaranteed if the eigenvalues of the correlation matrix of the gradients of the intensity values corresponding to a small window around the feature are both real and positive and if the minimum eigenvalue of this matrix is greater than a minimum prefixed threshold [17].

It follows that the solution of this problem has to be searched where image intensity gradients are not null both along x and y direction. As a consequence, contrary to classical PTV, the seeding density can be very high and trajectories can be followed for large time intervals. After the tracking procedure, Lagrangian data are obtained by evaluating the feature position in different time instants and by dividing the displacement by the time interval between frames. By a re-sampling procedure of sparse velocity vectors over a regular grid, the Eulerian flow picture in terms of instantaneous velocity and vorticity fields is obtained as well. In both sets of experiments, the FT algorithm has allowed the reconstruction of almost 10000 trajectories per processed frame. Consequently, the interpolation procedure for obtaining the Eulerian velocity fields is very accurate and maximizes the information content of the raw data. Both of these aspects are crucial when dealing with experiments [15]. In particular, concerning the Eulerian view-point and related analysis (i.e., spectra evaluation) it provides the possibility of characterizing the flow on a wide range of spatial scales. Concerning the Lagrangian view point and related analysis (i.e., statistics on particle displacement and diffusion coefficients evaluation), it should be noted that FT provides trajectories sufficiently long for observing the asymptotic behavior. In contrast, when classical techniques are used, long trajectories are difficult to obtain and they are often evaluated synthetically by integrating the measured velocity field.

3 Results

In this section we will discuss in detail two representative cases, with relatively high and low values of beta (Table 1: experiments numbered 8 and 15 respectively).

In figures 2-4, we show snapshots of the flow evolution for both of the selected experiments in terms of reconstructed particles' trajectories and velocity and PV fields. Eulerian instantaneous flow fields are obtained by re-sampling sparse velocity vectors over a regular 128x128 grid. As already discussed, the initial condition (figure 2) is similar for both flows while differences clearly arise as time evolves.

In particular, in figures 3a-b results corresponding to the intense beta effect (experiment 8), shortly ($\sim$ 2 s) after the electromagnetic forcing is stopped, are shown. The pictures show the flow pattern - a weak anticyclonic circulation forms at the center of the domain while a cyclonic jet-like zonal current forms around the edge of the central anticyclonic region. Moreover,

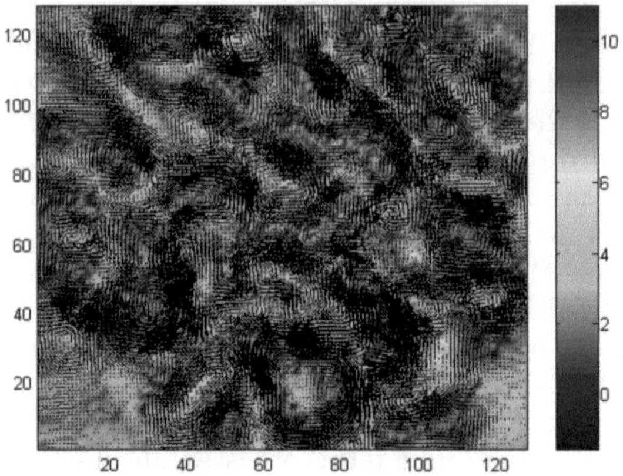

Fig. 2. Velocity and potential vorticity fields for experiment # 8 shortly ($\approx$ a few frames) after stopping of the forcing.

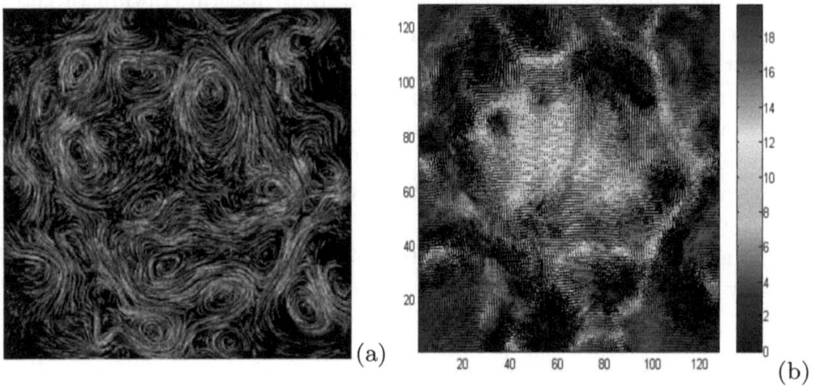

Fig. 3. Snapshot of the flow evolution for experiment #8 at $\approx$ 2 s after stopping of the forcing. Reconstructed trajectories (a), velocity potential vorticity fields (b). PV scale is $cm^{-1}s^{-1}$.

the jet region is subjected to a wave-like perturbation, a Rossby wave, which induces the formation of meanders associated with vortices.

In the weak beta case (figures 4a-b), the resulting flow $\sim$ 4 s after the electromagnetic forcing is stopped, consists of a few vortices characterized by a length scale larger than in the strong beta case, while no noticeable zonal jet is observed. We may conclude that in the low beta case, corresponding to a smaller value of k_{Rh}, the inverse cascade has room to proceed to larger scales than in the strong beta case and conjecture that the jet formation cannot occur for the weak beta case because k_{Rh} is so small that the associated jet width would be large compared with the characteristic size of the domain. Moreover, in this case PV mixing is clearly evident.

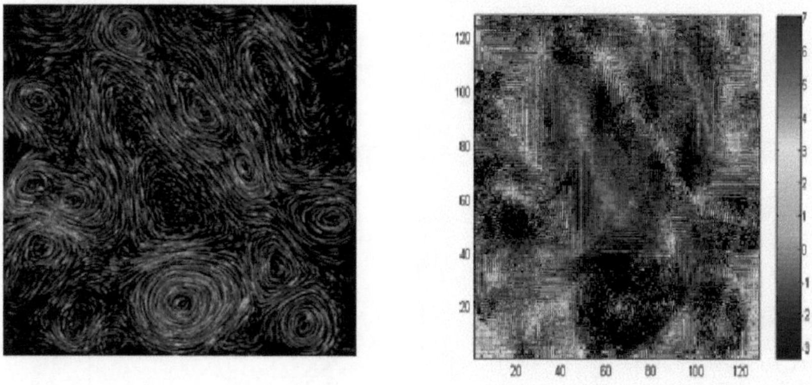

(a) (b)

Fig. 4. Snapshot of the flow evolution for experiment #15 at $\approx$ 4 s after stopping of the forcing. Reconstructed trajectories (a), velocity potential vorticity fields (b). PV scale is $cm^{-1}s^{-1}$.

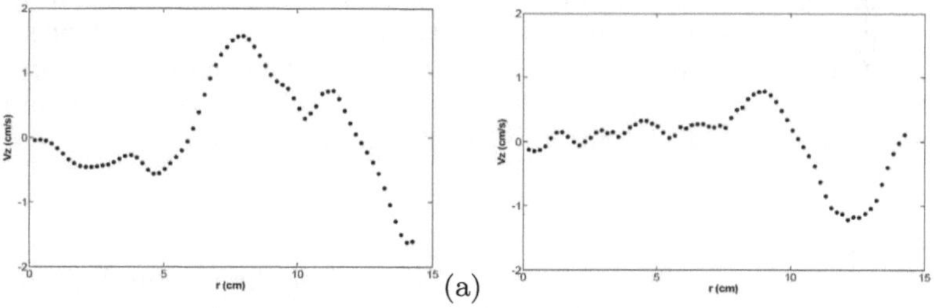

(a) (b)

Fig. 5. Trend of the average zonal velocity V_z (cm/s) with distance r (cm) from the center for the experiment #8 (a) and #15 (b).

The differences between the two regimes can be also highlighted by analyzing the azimuthally averaged zonal velocity versus the distance r evaluated from the center of the domain (pole). In the strong beta case (figure 5a), its distribution is characterized by a narrow peak around $r \approx 8$ cm, corresponding to the jet, and by a relatively weak anticyclonic circulation in the central part of the domain. In the low beta case (figure 5b), the profile is not peaked at large radii, but its maximum seems to be associated with a single strong vortex, not centered on the origin (see figure 4b), forming as a result of the merging process.

Further insight into the flow dynamics can be obtained by spectral analysis; energy spectra evaluation is based on a two-dimensional Fourier transform in Cartesian coordinates. The one-dimensional spectrum with $k = \sqrt{k_x^2 + k_y^2}$ is calculated by averaging both in time, over fixed time intervals, and over direction in wave-number space. In Figure 6 the energy spectra corresponding to the regime of intense beta effect is shown. The spectra are respectively

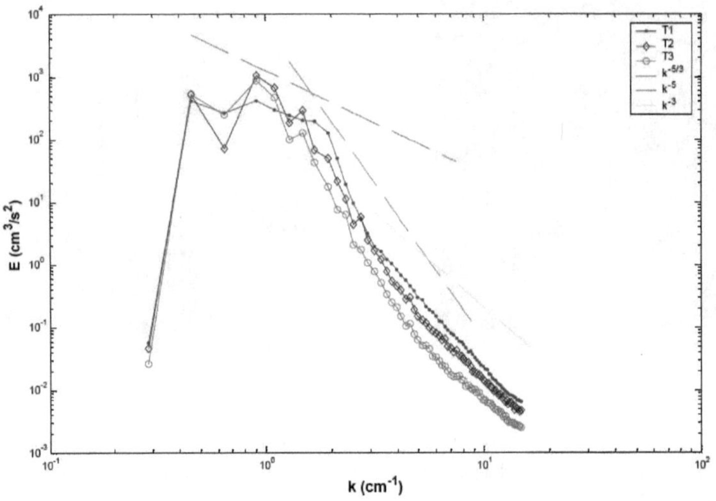

Fig. 6. Energy spectra corresponding to the experiment #8. The spectra are calculated by an average on time intervals concerning various experiment phases (T1 : forced spectrum; T2 : time interval from 10 to 20 frames after the stop of the forcing; T3: time interval from 20 to 45 frames after the stop of the forcing). The characteristic slopes k^{-5}, $k^{-5/3}$ and k^{-3} are shown in the box.

evaluated in three different time ranges: (T1) when the forcing is still active; (T2) in the time interval from 10 to 20 frames after the forcing has stopped; (T3) in the time interval from 20 to 45 frames after the forcing has stopped. Concerning the energy spectra corresponding to the regime of weak beta effect (figure 7), we evaluated them when (T1) the forcing is still activated, (T2) in the time interval from 100 to 150 frames after the forcing has stopped and (T3) in the time interval from 200 to 500 frames after the forcing has stopped.

While spectra corresponding to the forced regime T1 are similar in the two cases, differences arise after the forcing has stopped. In particular, the spectra T2 and T3 corresponding to high beta effects show a peak near $k \sim 1\ cm^{-1}$, close to the theoretical estimate of k_{Rh}, and the slope approximates the k^{-5} scaling.

In the low beta effect experiment, the energy spectra peak seems to be shifted to large scales ($k = 0.3\ cm^{-1}$) indicating that in this case, in analogy with non-rotating case, the cascade process is not arrested. On the other hand, differences between the case with $\beta = 0$ (not shown) can be seen if the slope of the energy spectra is considered: as a matter of fact, a steeper (approximately k^{-4} or k^{-5}) scaling, instead of the classical $k^{-5/3}$ corresponding to the inverse cascade, is recovered.

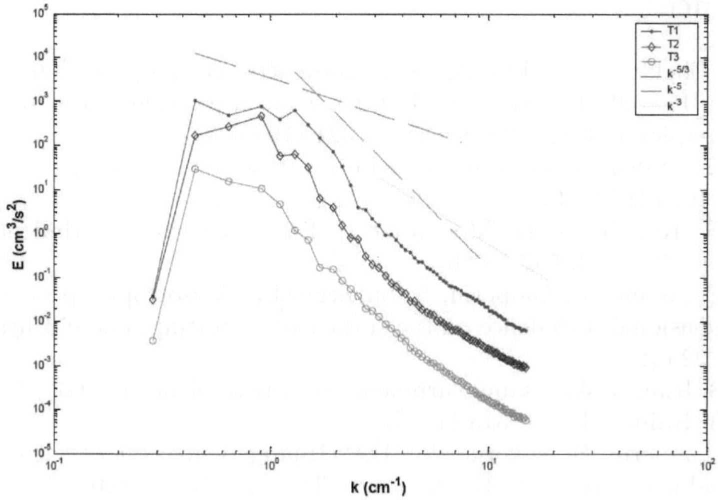

Fig. 7. Energy spectra corresponding to the experiment #15. The spectra are calculated by an average on time intervals concerning various experiment phases (T1 : forced spectrum; T2 : time interval from 100 to 150 frames after stopping the forcing; T3 : time interval from 250 to 500 frames after stopping the forcing). The characteristic slopes k^{-5}, $k^{-5/3}$ and k^{-3} are shown in the box.

4 Conclusions

The experiments on inverse cascade in a rotating system have shown, in case of high beta, the formation of an intense cyclonic zonal jet. The formation of these structures is directly related to the topographic slope of the free surface in the rotating system. In the case of high beta, a barrier to the inverse cascade corresponding to k_{Rh} is evident. The work reported here in large part reproduces the findings of AW, validating the method that we have used. One of the remarkable features of the phenomena investigated here is that the predictions based on stationary turbulence have some predictive ability even though the flow is decaying rapidly (the bottom drag decay rate in on the order of a few seconds in these experiments). Numerical simulations of continually forced as opposed to decaying flows show remarkable differences and pose important questions regarding lack of universality in 2D flows [24]. In future work, we hope to use numerical simulations in conjunction with laboratory experiments to more fully analyze the physical processes that allow jet formation on such a rapid timescale.

Acknowledgments

GFC acknowledges support from: the National Science Foundation (grants OCE 01-29301 and 05-25776); the Ministero dell'Istruzione Universite Ricerca Scientifica-MIUR (D.M. 26.01.01 n. 13).

References

[1] A. Cheklov, S.A. Orszag, S. Sukoriansky, B. Galperin, OI. Staroselsky, The effect of small scale forcing on large-scale structures in two-dimensional flows, Physica D 98, 321 (1995)
[2] R.H. Kraichnan, Inertial ranges in two-dimesional turbulence, Phys. Fluids 10, 1417 (1967)
[3] R.H. Kraichnan, D. Montgomery, Two-dimensional turbulence, Rep. Prog. Phys. 43, 547 (1980)
[4] H.P. Huang, B. Galperin, S. Sukoriansky, Anisotropic spectra in two-dimensional turbulence on the surface of a rotating sphere, Phys. Fluids 13, 225 (2001)
[5] P.B. Rhines, Waves and turbulence on a beta plane, JFM 69, 417 (1975)
[6] P.B. Rhines, Jets 4, 313 (1994)
[7] B. Galperin, S. Sukoriansky, H.P. Huang, Universal n^{-5} spectrum of zonal flows on giant planets, Phys Fluids 13, 1545 (2001)
[8] J. Pedlosky, Geophysical Fluid Dynamics, Springer, (1979)
[9] G.K. Vallis, M.E. Maltrud, Generation of mean flows and jets on a beta plane and over topography, J. Phys. Oceanogr. 23, 1346 (1993)
[10] S. Yoden, Yamada M., A numerical experiment of decaying turbulence on a rotating sphere, J. Atmos. Sci, 50, 631 (1993)
[11] J. Y.-K. Cho, L. M. Polvani, The emergence of jets and vortices in freely evolving, shallow-water turbulence on sphere, Phys. Fl, 8, 1531 (1996)
[12] S. Danilov, and D. Gurarie, Scaling, spectra and zonal jets in beta-plane turbulence, Phys. Fluids 16, 2592 (2004)
[13] J. Aubret, S. Jung, H.L. Swinney, Observation of zonal flows created by potential vorticity mixing in a rotating fluid, Geophys. Research Letters 29, 1876 (2002)
[14] Y. D. Afanasyev, J. Wells, Quasi-2d turbulence on the polar beta-plane: laboratory experiments, Geophys. Astro. Fl. Dyn., 99-1, 1 (2005)
[15] G. Boffetta, A. Cenedese, S. Espa, S. Musacchio, Effects of friction on 2D turbulence: an experimental study, Europhys. Letters 71, 590 (2005)
[16] M.C. Jullien, J. Paret J., P. Tabeling, Richardson pair dispersion in two dimensional turbulence, Phys. Rew. Lett. 82, 2872 (1999)
[17] M. Miozzi, Particle Image Velocimetry using Feature Tracking and Delauny Tessellation, Proceedings of the 12th International Symposium "Application of laser techniques to fluid mechanics", Lisbon, (2004)
[18] A. Cenedese, M. Moroni, Comparison among Feature tracking and more consolidated Velocimetry image analysis techniques in a fully developed turbulent channel flow, Meas. Sci. Technol., 16, 2307 (2005)
[19] G. F. Carnevale, R. C. Kloosterziel, J. G. F. Van Heijst, Propagation of barotropic vortices over topography in rotating tank, J. Fluid Mech., 223, 119 (1991)
[20] P. Tabeling, Two dimensional turbulence: a physicist approach, Phys. Reports 1, 362 (2002)

[21] B.D. Lucas, T. Kanade, An iterative image registration technique with an application to stereo vision, Proceedings of Imaging Understanding Workshop, 121 (1981)
[22] J. Shi, C. Tomasi, Good features to track, In Proceedings of IEEE Conference on Computer Vision and Pattern Recognition, (1994)
[23] A. Cenedese, S. Espa, M. Miozzi, Experimental study of two-dimensional turbulence using Feature Tracking, Proc. 12th International Symposium "Application of laser techniques to fluid mechanics", Lisbon, (2004)
[24] Carnevale, G.F. 2006 Mathematical and Physical Theory of Turbulence John Cannon and Sen Shivamoggi (Eds.) Taylor and Francis

Lagrangian particle tracking in high Reynolds number turbulence

Kelken Chang, Nicholas T. Ouellette, Haitao Xu, and Eberhard Bodenschatz

Max Planck Institute for Dynamics and Self-Organization, 37077 Göttingen, Germany
Laboratory of Atomic and Solid State Physics, Cornell University, Ithaca, NY 14853, USA
The International Collaboration for Turbulence Research (ICTR)

Summary. We describe a Lagrangian particle tracking technique that can be applied to high Reynolds number turbulent flows. This technique produces three-dimensional Lagrangian trajectories of multiple particles, from which both Lagrangian and Eulerian statistics can be obtained. We illustrate the application of this technique with measurements performed in a von Kármán swirling flow generated in a vertical cylindrical tank between two counter-rotating baffled disks. The Taylor microscale Reynolds number investigated runs from 200 to 815. The Kolmogorov time scale of the flow was resolved and both the turbulent velocity and acceleration were obtained and their probability density functions measured. Measurements of the Eulerian and Lagrangian velocity structure functions are presented. The average energy dissipation rates are determined from the Eulerian velocity structure functions.

1 Introduction

Early experimental investigations of turbulence relied on so-called Eulerian measurement techniques, where measurements are made at points fixed with respect to an inertial reference frame. Recent advances in imaging techniques and technology have, however, made Lagrangian measurements of fluid flow, where the trajectories of individual fluid elements are followed, possible. In principle, these trajectories are easily measured by seeding a flow with small tracer particles and following their motion. In practice, this can be a very challenging task. Here, we present a robust optical imaging technique, capable of tracking the motions of multiple particles simultaneously, even in intensely turbulent flow. Intense turbulence is typified by a high Reynolds number. We here report the Taylor microscale Reynolds number, defined as $R_\lambda = \sqrt{15\, u'\, L/\nu}$, where u' is the root-mean-square velocity, L is the correlation length of the velocity field, and ν is the kinematic viscosity of the fluid. The largest length and time scales of the turbulence are L and T_L, where the latter

Bernard J. Geurts et al. (eds), Particle Laden Flow: From Geophysical to Kolmogorov Scales, 299–311.

is the correlation time of the velocity field. According to Kolmogorov [1], the smallest turbulence scales are η and τ_η, defined as $(\nu^3/\varepsilon)^{1/4}$ and $(\nu/\varepsilon)^{1/2}$, respectively, where ε is the mean rate of energy dissipation per unit mass. To investigate the dynamics at the small scales, one must resolve η and τ_η, which, in intense turbulence, can be a demanding task. A full characterization of Lagrangian turbulence also requires following the motion of many Lagrangian particles for long times.

We present here a measurement technique that is capable of tracking the motions of multiple particles simultaneously. We describe briefly the tracking algorithm used to construct the three-dimensional trajectories of tracer particles in Sec. 2. We apply the technique to a von Kármán swirling flow generated in a cylindrical tank between two counter-rotating baffled disks. We validate our technique by measuring the probability density functions (PDFs) of the velocity and acceleration fluctuations and comparing them with known results. Eulerian and Lagrangian measurements of the velocity structure functions are also presented and the energy dissipation rates are measured from the Eulerian structure functions.

2 Particle Tracking

An optical three-dimensional Lagrangian particle tracking algorithm consists of three main steps: first, the particles need to be identified and their positions be determined on the two-dimensional images recorded by the detectors. Next, the three-dimensional coordinates of the particles in real space need to be constructed. Finally, the particles must be tracked in time. Our particle tracking technique is described in detail by Ouellette et al. [2]; the main steps involved are described below.

2.1 Center Finding

The first step in image processing is the determination of the positions of tracer particles on the two-dimensional image plane of the cameras. We identify particles by first assuming that every local maximum in image intensity above some small threshold corresponds to a particle. We then fit two one-dimensional Gaussians to the horizontal and vertical pixel coordinates of each local maximum pixel and its nearest neighbors [2, 3]. An analytical expression for the particle center can be obtained in terms of the coordinates and intensities of the local maximum pixel and its two adjacent pixels. Labeling the horizontal coordinates of these points as x_1, x_2 and x_3, where x_2 is the coordinate of the local maximum, and the corresponding intensities as I_1, I_2 and I_3, we solve the set of equations

$$I_i = \frac{I_0}{\sigma_x \sqrt{2\pi}} \exp\left[-\frac{1}{2}\left(\frac{x_i - x_c}{\sigma_x}\right)^2\right] \tag{1}$$

for $i = 1, 2, 3$ to give the horizontal particle coordinate as

$$x_c = \frac{1}{2} \frac{(x_1^2 - x_2^2) \ln(I_2/I_3) - (x_2^2 - x_3^2) \ln(I_1/I_2)}{(x_1 - x_2) \ln(I_2/I_3) - (x_2 - x_3) \ln(I_1/I_2)} . \tag{2}$$

The vertical position of the particle is defined analogously. We estimate that this algorithm is capable of finding the true particle centers to within 0.1 pixels [2].

2.2 Stereomatching

The second step in the particle tracking technique involves the reconstruction of the three-dimensional coordinates of the tracer particles in the laboratory reference frame from the two-dimensional coordinates of the particles on the camera image planes. For this stereoscopic reconstruction, the characteristics of each camera-lens system and its position in the lab frame must be determined. We discuss this calibration procedure in Sec. 3.3.

Since the particles have no distinguishing features that can be used in the stereoscopic matching, the only information available is the photogrammetric condition. This condition asserts that, for each camera, the camera projective center, the particle image on the camera sensor plane and the particle in the laboratory frame must be collinear and that, therefore, lines of sight from all cameras must intersect at the true location of the particle [4]. The stereomatching algorithm we use is similar to those of Dracos [5] and Mann et al. [6]. We first construct a line of sight from the projective center of one camera through one particle image. This line of sight is then projected onto the image planes of the other cameras and particle images on these image planes that are within some small distance of the projected line are considered to be possible matches for the particle image from the first camera. In this manner, a list of candidate matches for the particle image can be constructed for every other camera. This process is then repeated for every particle image on each camera. Matches in three-dimensional space are then found by performing a consistency check on the lists.

2.3 Tracking

The last step in a Lagrangian particle tracking algorithm is the tracking of particles in time. We have developed a predictive algorithm for this purpose. For each particle in frame n, a velocity is estimated from the camera frame rate and the particle's position in frames $n - 1$ and n. This velocity is used to predict a position for the particle in frame $n + 1$. Particles in frame $n + 1$ that are within some small distance of the predicted position are considered to be possible candidates for the continuation of the track. For each of these candidates, we estimate both a new velocity from the positions in frames n and $n + 1$ and an acceleration from the positions in frames $n - 1$, n and $n + 1$.

This velocity and acceleration are used to predict a position for the particle in frame $n+2$. The particle in frame $n+1$ that gives a predicted position in frame $n+2$ closest to a true particle position is then chosen to continue the track. This process is repeated until a conflict arises or the particle disappears from view. A conflict occurs when a single particle in frame $n+1$ is the best match for multiple particles in frame n. When this occurs, the involved tracks are ended at frame n and a new track is started in frame $n+1$.

We have also developed a way to handle the possible loss of particles for a few frames. Particles might be missing on a frame for a number of reasons, including intensity fluctuations of the illumination, occlusion by other particles or the non-uniform sensitivity of the sensor area within a single pixel. This situation is handled by extrapolating the tracks with estimated positions and looking for a continuation of the track. If no continuation is found within a set number of frames, the track is fully terminated and the estimated positions are dropped.

3 Experimental Details

We have implemented our Lagrangian particle tracking technique in a von Kármán swirling flow confined within a cylindrical tank. Here we briefly describe the details of the experiments.

3.1 Flow Apparatus

Our apparatus has been described in detail previously [7, 8, 9]. A sketch of the experimental setup is shown in Fig. 1. The cylindrical tank has an inner diameter of 48.3 cm, a height of 60.5 cm and contains approximately 120 liters of water. The tank is mounted vertically between two hard-anodized aluminum top and bottom plates. Images are taken through eight round, glass windows, 12.7 cm in diameter and attached symmetrically around the center of the tank, to avoid lensing effects caused by the cylindrical walls of the flow chamber. The top and bottom plates contain channels for cooling water used to control the temperature of the fluid in the apparatus. Turbulence is generated by the counter-rotation of two baffled disks. The two circular disks are 20.3 cm in diameter, 4.3 cm in height and spaced 33 cm apart. Twelve equally spaced vanes are mounted on each disk so that the flow is forced inertially. Each disk is driven by a 1 kW DC motor and its rotation frequency is controlled by a feedback loop. The large-scale flow in the tank is axisymmetric and is composed of a pumping mode and a shearing mode. The measurement volume of approximately $2 \times 2 \times 2\,\mathrm{cm}^3$ is in the center of the tank where the mean flow is negligible. In order to remove dirt, the water in the apparatus is cleaned by pumping it through a filtering loop. Bubbles in the flow are removed by de-gassing the water using a second recirculation loop, with one end open to the atmosphere.

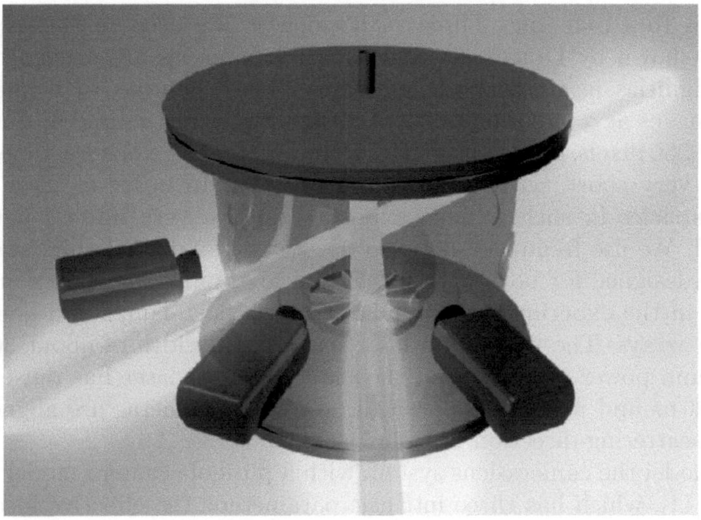

Fig. 1. A sketch of the flow apparatus, cameras and lasers.

3.2 Tracer Particles

To investigate the dynamics of the small scales of turbulence, we must resolve η and τ_η. To resolve η, we use very small tracer particles. The accuracy with which the tracer particles follow the motion of the fluid elements is measured by the Stokes number, defined as

$$\mathrm{St} = \frac{1}{18} \frac{\varrho_p - \varrho_f}{\varrho_f} \left(\frac{d}{\eta}\right)^2 , \tag{3}$$

where ϱ_p and ϱ_f are the densities of the particle and fluid, respectively, and d is the particle diameter. In our experiment, the flow is seeded with polystyrene micro-spheres of diameter 25 µm with a density of $1.06\,\mathrm{g\,cm^{-3}}$, roughly matched to the density of water. The size of these particles is smaller or comparable to the Kolmogorov length scale for all Reynolds numbers investigated and the Stokes number ranges from 5.7×10^{-5} at $R_\lambda = 200$ to 3.9×10^{-3} at $R_\lambda = 815$. Particles with this combination of size and density have been shown to be passive tracers in this flow and thus to approximate fluid elements [10]. We note, however, that our tracking technique is not limited to the tracking of passive tracers. It can also be used to track particles with non-negligible inertia.

3.3 Imaging System and Illumination

To resolve τ_η in our flow, we need an imaging system with very high temporal resolution. We use Phantom v7.1 high-speed CMOS digital cameras developed by Vision Research, Inc. Three such cameras are used in the experiment, since, as shown by Dracos [5], at least three cameras are needed to resolve the ambiguities in stereoscopic matching. The Phantom v7.1 cameras can record images at a maximum rate of 27000 frames per second at a resolution of 256×256 pixels. At such a high frame rate, the exposure time for each frame is very short, with a maximum of 37 µs. In order to illuminate the tracer particles in such a short exposure time, a very intense light source is needed. We use frequency-doubled, Q-switched Nd:YAG solid state lasers, specially designed for both high power and high pulse rates. Two such lasers are used in the experiments, one pumped with flash-lamps and one pumped by diode arrays. The flash-lamp laser has a pulse width of about 300 ns and a maximum power of 60 W, and the diode pumped laser has pulse width of about 120 ns and maximum power of 90 W. The cameras are aligned in the forward scattering direction of both lasers.

We model the camera-lens system with a pin-hole camera model proposed by Tsai [11], which has three intrinsic parameters: the effective focal length, the radial distortion and the aspect ratio of the sensor pixels. Six additional external parameters are needed to determine the three-dimensional position of the camera. By imaging a calibration mask at different positions in the fluid, these nine parameters can be determined.

3.4 Data Acquisition and Post-processing

As noted earlier, the Phantom v7.1 cameras can record images at a rate of up to 27000 frames per second at a resolution of 256×256 pixels. With such a high acquisition speed, the data rate is too large for the cameras to transfer their images to a computer in real time with current technology. Therefore, the cameras store images in an internal buffer and transfer the contents of their buffer over gigabit ethernet. A computer cluster is used for data acquisition. During a data run, all three cameras typically record movies simultaneously for one to two eddy turnover times before transferring their images. The set of three movies is then transferred to one node of the cluster. Once this set of three movies is transferred, the node processes them while the cameras record a new set of movies that will be transferred to a different node. After processing each movie, each node transfers the calculated particle tracks to central data storage and waits for new images. The data acquisition process is entirely automated with no human intervention needed during the data run.

Once the track files are obtained, turbulence statistics involving velocity and acceleration can be calculated by taking time derivatives of the position. Simple finite differences are not adequate for differentiating the tracks; the results obtained from such a method are easily contaminated by errors in the position measurement. Instead, we calculate time derivatives by convolution with a Gaussian smoothing and differentiating kernel [12]. To characterize the mean flow field in our measurement volume, we average the velocity and acceleration measurements for long times. Subsequently, in calculating fluctuating quantities, the mean flow is subtracted.

4 Velocity and Acceleration Statistics

In the measurements presented here, R_λ ranges from 200 to 815. The Kolmogorov length scale η ranges from 192 µm to 23 µm, and the Kolmogorov time scale τ_η ranges from 36.8 ms to 0.54 ms. Figure 2(a) shows the probability density functions (PDFs) of the radial and axial components of the velocity fluctuations at $R_\lambda = 690$. It is well known that, in turbulence, the

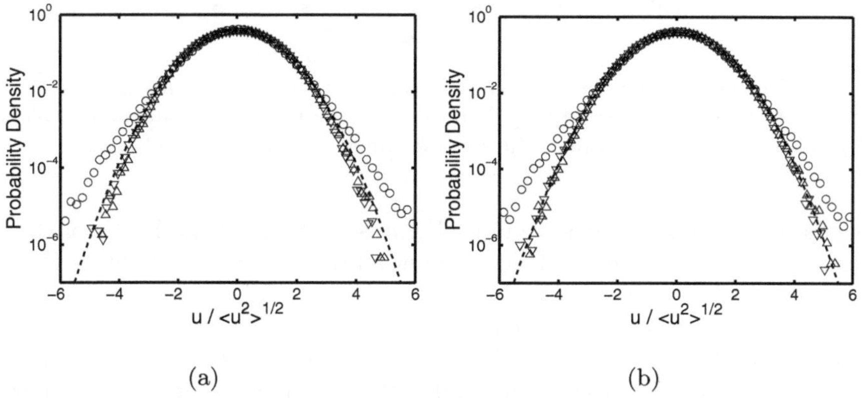

Fig. 2. (a) PDFs of the turbulent velocity fluctuations at $R_\lambda = 690$ for the radial (△ and ▽) and axial (○) components with no residence-time weighting, plotted on a semilogarithmic scale. The velocity has been normalized by its root-mean-square value. The dashed line is a standardized Gaussian. (b) Velocity PDFs for the radial and axial components at $R_\lambda = 690$ with residence-time weighting.

velocity PDF should be Gaussian. Our measured PDFs for both radial components are very close to Gaussian, with a kurtosis of 2.79. Unlike the radial velocity PDFs, the axial velocity PDF deviates from Gaussianity and has a kurtosis of 3.44. In addition, the measured root-mean-square velocity for the radial components differs significantly from that of the axial component, measured to be 0.47 ms^{-1} and 0.31 ms^{-1}, respectively. These results are calculated without considering possible biases that may affect the statistics. Since the measurement volume is finite, fast-moving particles are more likely to enter the measurement volume than slow-moving particles. Slow-moving particles, however, will stay in the measurement volume longer than fast-moving particles. We thus consider the velocity statistics with residence-time weighting, which weights the velocities by the amount of time the particle

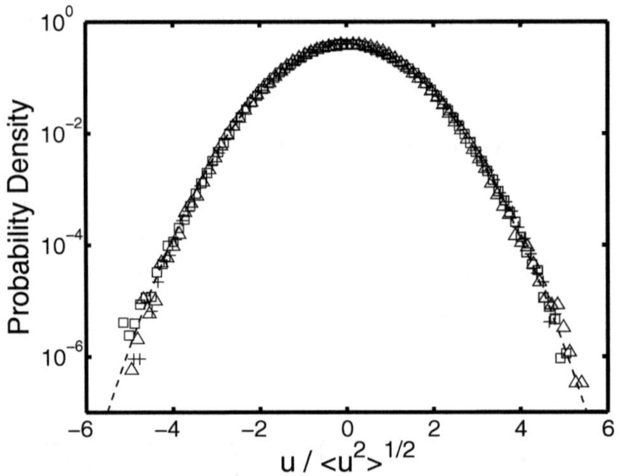

Fig. 3. Velocity PDFs for one radial component of the velocity fluctuations at $R_\lambda = 200$ (+), 690 (△) and 815 (□), normalized by the root-mean-square velocity, with residence-time weighting. The dashed line is a standardized Gaussian.

spends in the measurement volume. Figures 2(b) shows the PDFs of the velocity fluctuations with residence-time weighting. They are slightly closer to Gaussian, with kurtosis of 2.93 and 3.43 for the radial and axial components, respectively. The radial and axial root-mean-square velocities now decrease to $0.44\,\mathrm{ms}^{-1}$ and $0.30\,\mathrm{ms}^{-1}$, respectively. Within experimental uncertainty, the difference between the PDFs with and without residence-time weighting is insignificant, since our measurement volume is large and the finite-volume bias is not strong. The gap between the radial and axial root-mean-square velocities, however, remain significant even after considering residence-time weighting. This difference is most likely due to the effect of the large-scale forcing of the flow. We have also investigated the Reynolds number dependence of the velocity PDF. The PDFs for one radial component of the velocity for $R_\lambda = 200$, 690 and 815 are shown in Fig. 3. The distributions for all three Reynolds numbers show no statistically-significant Reynolds number dependence.

Figure 4(a) shows the standardized PDFs of the acceleration measured in the radial and axial directions without residence-time weighting. The root-mean-square accelerations for the radial and axial components are $105.8\,\mathrm{m\,s}^{-2}$ and $86.2\,\mathrm{m\,s}^{-2}$, respectively. Acceleration PDFs with residence-time weighting are shown in Fig. 4(b). The radial and axial root-mean-square accelerations decrease to $99.3\,\mathrm{m\,s}^{-2}$ and $81\,\mathrm{m\,s}^{-2}$. While strong deviation from Gaussianity is evident, the tails of the PDFs are somewhat depressed compared to previous measurement by Mordant et al. [12] in the same flow at $R_\lambda = 690$ using one-dimensional silicon strip detectors, which can reach a frame rate of 70000 frames per second at a resolution of 512 pixels. The PDFs of one radial component of the acceleration for $R_\lambda = 200$, 690 and 815 are shown in Fig.

5. All three sets of data behave similarly, with narrower tails compared to those measured by Mordant et al. [12]. At $4\,\langle a^2\rangle^{1/2}$, our acceleration PDF is roughly 90% of that of Mordant et al. This difference is most likely due to the poorer temporal and spatial resolution of our cameras. At a resolution of 256×256 pixels, a frame rate of 25 frames per τ_η is not sufficient to resolve very intense accelerations.

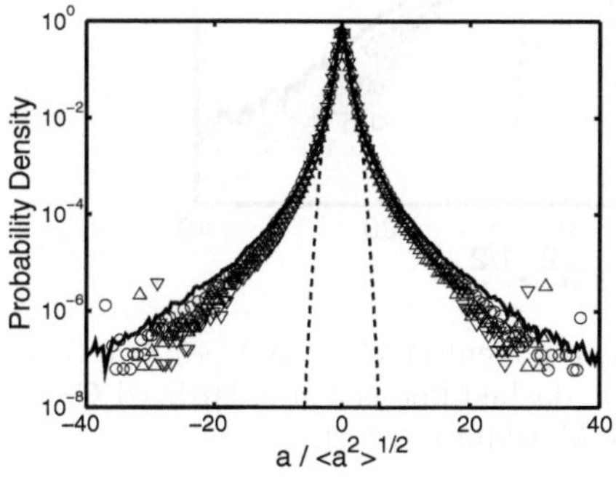

(a)

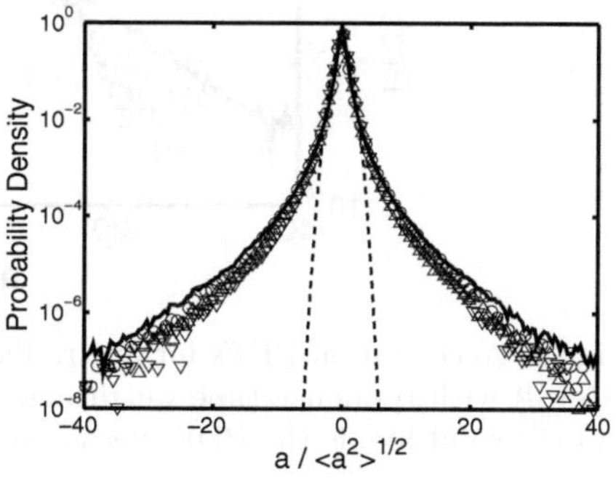

(b)

Fig. 4. (a) Acceleration PDFs at $R_\lambda = 690$ for the radial ($\triangle$ and ∇) and axial ($\circ$) components with no residence-time weighting plotted on a semilogarithmic scale. The acceleration has been normalized by its root-mean-square value. The dashed line is a Gaussian and the solid line is the acceleration PDF measured by Mordant et al. [12]. (b) Acceleration PDFs for the radial and axial components at $R_\lambda = 690$ with residence-time weighting.

5 Structure Functions

Since our particle tracking technique can resolve the trajectories of many particles simultaneously, we can use it to measure multi-point Eulerian statistics as well as Lagrangian statistics. At each instant of time, the three-dimensional coordinates of multiple particles are known, and each frame can be considered to be a collection of Eulerian data.

The Eulerian velocity structure functions, the moments of the spatial velocity differences, have played a fundamental role in describing turbulence. In isotropic turbulence, we can decompose the second-order structure function into two components: a longitudinal component $D_{LL}(r)$ where the velocity is in the direction of the separation vector and a transverse component $D_{NN}(r)$ where the velocity is perpendicular to the separation vector. In the inertial subrange $\eta \ll r \ll L$, Kolmogorov's hypotheses [1] state that turbulence

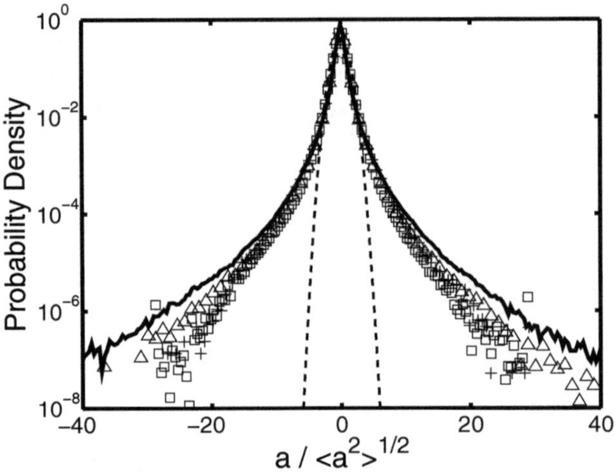

Fig. 5. Acceleration PDFs for one radial component at $R_\lambda = 200$ (+), 690 (△) and 815 (□) with residence-time weighting. The dashed line is a standardized Gaussian and the solid line is the PDF measured by Mordant et al. [12].

statistics have a universal form dictated only by ε. Therefore, the structure functions should scale as

$$D_{LL}(r) = C_2\,(\varepsilon\, r)^{2/3}\ ,\ D_{NN}(r) = \frac{4}{3}\,C_2\,(\varepsilon\, r)^{2/3}\ , \tag{4}$$

in the inertial subrange, where the scaling constant C_2 has a well-established value of 2.13 ± 0.22 [13]. Our measurements of the energy dissipation rate ε are based on these scaling relationships. We determine ε from the plateau value of $(D_{LL})^{3/2}$ and $(D_{NN})^{3/2}$, compensated by $(C_2)^{3/2}\,r$ and $(4\,C_2/3)^{3/2}\,r$ respectively. In addition, we also measured the third-order Eulerian longitudinal structure function $D_{LLL}(r)$ which, in the inertial subrange, scales as [14]

$$D_{LLL}(r) = -\frac{4}{5}\,\varepsilon\, r\ . \tag{5}$$

This result allows an additional estimate of ε from the plateau value of D_{LLL}, compensated by $-\frac{4}{5}\,r$.

Our measurements of the compensated structure functions at $R_\lambda = 690$ are shown in Fig. 6. The average plateau value is $1.22\,\mathrm{m^2\,s^{-3}}$, consistent with a previous measurement in the same flow at the same Reynolds number [10]. While the scaling range for the third-order structure function is short, the value of ε estimated is consistent with those estimated from the second-order structure functions. The short scaling range of D_{LLL} could be due to a number of reasons, including the finite-volume bias or insufficient statistics for the measurement of higher moments.

Using the Lagrangian trajectories recorded by our system, we can also measure Lagrangian statistics. Let us define the Lagrangian velocity increment

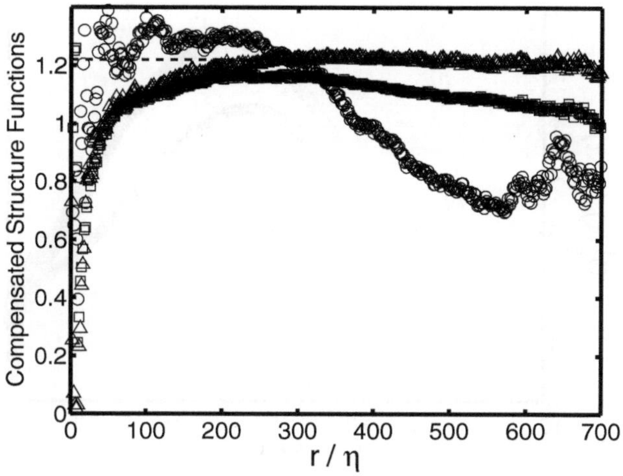

Fig. 6. Compensated second- and third-order Eulerian velocity structure functions at $R_\lambda = 690$. □ and △ are the longitudinal and transverse second-order structure functions respectively and ○ is the third-order longitudinal structure function. The dashed line marks the average plateau value of $1.22\,\mathrm{m}^2\,\mathrm{s}^{-3}$.

as $\delta u_i(\tau) = u_i(t+\tau) - u_i(t)$, taken along the trajectory of a fluid element. The second-order Lagrangian velocity structure functions, defined as $D^L_{ij}(\tau) = \langle \delta u_i(\tau)\, \delta u_j(\tau) \rangle$, scale in the inertial subrange as

$$D^L_{ij}(\tau) = C_0\, \varepsilon\, \tau\, \delta_{ij}\ , \tag{6}$$

where C_0 is assumed to be a universal scaling constant. Figure 7 shows our measurement of the second-order Lagrangian structure functions at $R_\lambda = 690$. It is well known that, at a given Reynolds number, the Lagrangian inertial subrange is narrower than its Eulerian counterpart [15]. Even at $R_\lambda = 690$, we do not see a distinct scaling regime. Nevertheless, we can estimate C_0 by the taking the values of the peaks of the compensated structure functions, obtaining 6.0 and 4.9 for the radial and axial components, respectively. The estimated values are consistent with the asymptotic values 6.2 ± 0.3 and 5.0 ± 0.4 for the radial and axial components measured by Ouellette et al. [16]. The difference in the value of C_0 measured from the radial and axial components is most likely due to the large-scale anisotropy of the flow [16].

6 Conclusions

We have presented a particle tracking technique suitable for both Eulerian and Lagrangian measurements. We validated the technique by measuring the turbulent velocity and acceleration PDFs and comparing them with known results. We found that the PDFs of the two radial components of the velocity

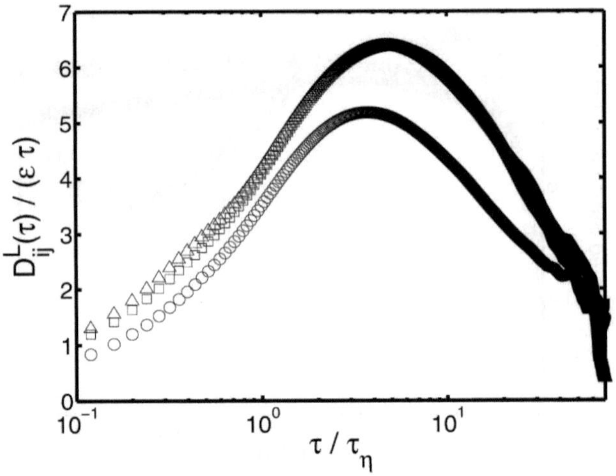

Fig. 7. Compensated second-order Lagrangian velocity structure functions at $R_\lambda = 690$. □ and △ are the radial components and ○ is the axial component. The off-diagonal components (not shown) are indistinguishable from zero.

fluctuations in our flow were very close to Gaussian, but the axial component had tails that deviate from Gaussianity. Extreme deviation from Gaussianity was evident in the acceleration PDFs, though the tails were depressed compared to the measurement of Mordant et al. [12] in the same flow at a higher resolution. We also demonstrated the versatility of our technique with the measurement of the Eulerian and Lagrangian velocity structure functions. Based on the well-established value of the second-order Eulerian velocity structure function scaling constant and a relationship for the third-order velocity structure function, we estimated the average energy dissipation rate in our flow. From the Lagrangian velocity structure functions, we have measured the scaling constant C_0 and found that the large-scale anisotropy of our flow also affects the small-scale statistics of the velocity structure functions.

Acknowledgments

This work was supported by the US National Science Foundation under grants PHY-9988755 and PHY-0216406 and by the Max Planck Society.

References

[1] Kolmogorov AN (1941) Dokl Akad Nauk SSSR 30:299-303
[2] Ouellette NT, Xu H, Bodenschatz E (2006) Exp Fluids 40 (2):301-313
[3] Cowen EA, Monismith SG (1997) Exp Fluids 22:199-211
[4] Maas HG, Gruen A, Papantoniou D (1993) Exp Fluids 15:133-146

[5] Dracos T (1996) In: Three-Dimensional Velocity and Vorticity Measuring and Image Analysis Techniques, 129-152, Kluwer Academic Publisher, Dordrecht, the Netherlands
[6] Mann J, Ott S, Andersen JS (1999) Experimental study of relative, turbulent diffusion, Riso-R-1036(EN), Riso National Laboratory
[7] Voth GA, Satyanarayan K, Bodenschatz E (1998) Phys Fluids 10:2268
[8] Voth GA (2000) Lagrangian acceleration measurements in turbulence at large Reynolds numbers. Ph.D. thesis, Cornell University
[9] Crawford AM (2004) Particle tracking measurements in fully developed turbulence: water and dilute polymer solutions. Ph.D. thesis, Cornell
[10] Voth GA, La Porta A, Crawford AM, Alexander J, Bodenschatz E (2002) J Fluid Mech 469:121-160
[11] Tsai RY (1987) IEEE T Robotic Autom RA-3:323-344
[12] Mordant N, Crawford AM, Bodenschatz E (2004) Physica D 193:245-251
[13] Sreenivasan KR (1995) Phys Fluids 7:2778-2784
[14] Kolmogorov AN (1941) Dokl Akad Nauk SSSR 32:16-18
[15] Yeung PK (2002) Annu Rev Fluid Mech 34:115-142
[16] Ouellette NT, Xu H, Bourgoin M, Bodenschatz E (2006) New J Phys 8:102

Part III

Heavy particles, aggregation and patterns in turbulence

Lagrangian dispersion in coastal applications

Guido Lupieri[1], Stefano Salon[2] and Vincenzo Armenio[1]

[1] DICA-Università di Trieste, Italy glupieri@units.it, armenio@dic.units.it**
[2] OGS, Sgonico, Trieste, Italy ssalon@inogs.it

The role played by the turbulent mixing in the Lagrangian dispersion of tracers in a tidally-driven, mid-latitude, shallow-water environment is here discussed. The Eulerian carrying flow is supplied by a turbulent oscillating boundary layer without and with rotation. In the purely oscillating case, the dispersion of the particulate is stronger in the near-wall region and the diffusivity along the main flow direction is larger than the other ones. For the value of Rossby number herein considered, rotation strongly affects particle dispersion and increases the mixing efficiency of particles within the whole fluid column, up to the near surface region. Diffusion along the spanwise direction is comparable to that in the streamwise one, and consequently the time needed to completely homogenize the physical properties associated to the particulate is reduced.

1 Introduction

Understanding the hydrodynamic characteristics of a dispersed phase in a shallow water basin is an important task especially in practical applications, from pollutant dispersion to biological feeding mechanisms.
In a coastal basin, despite the advective transport is mainly driven by the horizontal components of the velocity field and depends on the characteristics of the coastline, vertical mixing is governed by the three dimensional turbulent regime that develops in the water column. In particular, the following features rule the vertical mixing in coastal applications: 1) the shallowness of the water column; 2) turbulence generated at the bottom boundary layer by a current that drives the flow; 3) turbulent mixing at the free-surface region supplied by wind stress and wave breaking; 4) the presence of thermal and/or haline stratification. Being a geophysical problem, also earth rotation can play an

** corresponding author

Bernard J. Geurts et al. (eds), Particle Laden Flow: From Geophysical to Kolmogorov Scales, 331–343.

important role. In a flow field characterized by such aspects, the dynamics of vertical mixing rules the fate of a dispersed phase.

So far, particle dispersion in a turbulent flow has been mostly investigated in simple cases (i.e. isotropic or homogeneous turbulence, plane channel flow). For an exhaustive review, the reader is referred to [1] and the literature therein reported. Comparatively, only few investigations have been carried out in conditions characterized by either complex geometry or complex physics. Among the others, [2] studied particle segregation in a steady flow over a large amplitude wavy wall, showing that particles tend to be dispersed in different ways depending on the region where they are released in the flow field. Dealing with marine applications, recently [3] studied a problem archetypal of sediment motions over the sea bottom, focusing on the sedimentation mechanism.

The goal of the present study is to get insight into the role of turbulence in the dispersion of particles under the typical conditions of a mid-latitude shallow-water column, focusing in particular on the vertical mixing within the fluid column. In our numerical study, we only consider the contributes to turbulence dynamics given by i) a semi-diurnal tidal current (pure oscillatory flow) that drives the flow, and ii) the Earth rotation. These two features correspond to the minimum level of forcing present in a basin in absence of stratification. The approach to the problem here followed is the Lagrangian-Eulerian one, that considers the dispersed phase as a cloud of Lagrangian particles moving in the Eulerian carrying flow.
This work is part of a research project aimed at improving the knowledge of vertical turbulent mixing in a shallow water basin, subjected to a main tidal current and Earth rotating force. A deep comprehension of the phenomenon can be helpful to understand the role of mixing activity in the marine biological feeding mechanism or in the environmental pollution dispersion. This could be done by considering the micro-organisms or the pollutant agents as small-size particles with inertia comparable with that of the carrying fluid (*i.e.* tracers, as in the present study).

We first study particle dispersion in the case of a purely oscillating flow without rotation. Successively, we consider the rotational effects, comparing the results of the two cases and analyzing the role of rotation in the vertical mixing.

2 The problem formulation

The Stokes-Ekman bottom boundary layer is archetypal of the physical problem of interest in the present research (for details see [4]). This flow is forced by a time-dependent, sinusoidal free-stream velocity and is subjected to the rotation of the frame of reference. A mid-latitude case is here considered, and the horizontal components of the rotation vector are thus included in the governing equations (see also [5]).

The flow is driven by an harmonic pressure gradient aligned with the x-direction:

$$\frac{dP_d}{dx_d}(t_d) = -U_0 \omega \cos(\omega t_d) \quad (1)$$

In Eq. 1, P_d is the dimensional kinematic pressure (pressure divided by the fluid density ρ_0), U_0 is the amplitude of the outer velocity, ω is the angular frequency of the oscillation associated to the tidal flow, t_d is the time and x_d is the direction along which the oscillation develops (the index d refers to dimensional quantities).
The pressure gradient defined above gives the sinusoidal free-stream velocity $U_d(t_d) = U_0 \sin(\omega t_d)$. As the equations are written in a rotational frame of reference, the non-dimensional equations governing the flow read as:

$$\frac{\partial u_i}{\partial x_i} = 0 \quad (2)$$

$$\frac{\partial u_i}{\partial t} + \frac{\partial u_j u_i}{\partial x_j} = -\frac{\partial p}{\partial x_i} + \frac{1}{Re}\frac{\partial}{\partial x_j}\frac{\partial u_i}{\partial x_j} +$$
$$\cos(t)\delta_{i1} + \frac{1}{Ro\ \sin\phi}\epsilon_{ijm} e_j (\sin(t)\delta_{m1} - u_m) \quad (3)$$

In Eq. 2 and 3 the coordinate x_i (hereafter x_1, x_2, x_3 or x, y and z are used interchangeably for the streamwise, spanwise and wall-normal direction) is made non-dimensional with the amplitude of the free stream motion $a = U_0/\omega$, t is the time coordinate made non-dimensional with $1/\omega$, u_i is the i-component of the velocity field (u_1, u_2, u_3 or u, v and w are used for the streamwise, spanwise and wall-normal components) made non-dimensional with U_0 and p is the turbulent pressure (the pressure arising from the turbulent fluctuations) made non-dimensional with $\rho_0 U_0^2$. The third term of the RHS is the non-dimensional harmonic pressure gradient that drives the flow. The fourth term is the non-dimensional Coriolis force associated with the oscillating inertial velocity, with ϵ_{ijm} the Levi-Civita tensor, $e_j = (0, \cos\phi, \sin\phi)$ the unit vector and ϕ the latitude.

The non-dimensional parameters governing the problem are the Reynolds number $Re = aU_0/\nu$ where ν is the kinematic viscosity, and the Rossby number $Ro = \omega/f$ where $f = 2\Omega\sin\phi$ is the Coriolis parameter, and Ω is the Earth angular velocity. Usually, a Reynolds number based on the Stokes laminar thickness $\delta_S = \sqrt{2\nu/\omega}$ is defined as $Re_S = U_0\delta_S/\nu$ (note that $Re = Re_S^2/2$). If we define a Reynolds number $Re_E = U_0\delta_E/\nu$, based on the Ekman penetration length $\delta_E = \sqrt{2\nu/f}$, it can be easily shown that:

$$\frac{Re_E}{Re_S} = \sqrt{Ro}. \quad (4)$$

2.1 The numerical method

The present study is performed by a Large Eddy simulation (LES) technique coupled with the dynamic-mixed model of [6] where the constant is averaged over the planes of homogeneity. The integration of the governing equations is performed using the semi-implicit fractional step method of [7]. Spatial derivatives are discretized by means of second-order accurate centered differences.

The dispersed phase is treated as a swarm of fluid tracers: the position of each tracer is advanced in time following:

$$\frac{dx_{i,p}}{dt} = u_{i,p} \tag{5}$$

where $x_{i,p}$ and $u_{i,p}$ are respectively the i-coordinate of particle position and the i-component of the fluid velocity at particle position. Eq. 5 is integrated by means of an explicit, second-order accurate, Adams-Bashforth scheme.

The interpolation of the Eulerian field onto particle position is performed using a very recent interpolation scheme based on the Taylor series expansion of the Eulerian fluid velocity at the node closest to particle position (see for details [2]). The scheme is suited both for orthogonal and curvilinear grids and its order of accuracy can be easily adapted to that of the Navier-Stokes solver employed for the Eulerian phase.

When the Eulerian field is evaluated by LES, the resolved velocity field misses the energy content associated to the turbulent scales that are parameterized. Armenio et al. [8] showed that when the Eulerian phase is solved by *resolved* LES (a simulation where the wall layer is directly resolved), the contribute of the sub-grid scales (SGS) to particle statistics is small and proportional to the amount of SGS dissipation. In order to minimize such an effect, Kuerten [9] has recently proposed to reconstruct the missed contribute of sub-grid scales performing a suitable deconvolution. Further numerical tests (see [10]) showed that this technique allows to accurately reproduce particle statistics, when compared to the DNS data of the same flow. Here, since the SGS stresses are parametrized by means of a dynamic-mixed model, a first order deconvolution is carried out to de-filter the velocity field. The de-filtered velocity field is thus used to move the particles, following Eq.5.

2.2 Physical and numerical parameters

In the present research we use parameters typical of a real mid-latitude problem (*i.e.* data of the Gulf of Trieste, northern Adriatic Sea, see Fig. 1). Field data and previous investigations suggest that the driving current is dominated by the M2 tidal component ($T_{M2} = 12.42$ hours) and, according to [11], a reference value for the free stream velocity is $U_0 = 0.05$ m/s. The mean water depth is $h = 16$ m and considering the viscosity of the water as $\nu = 1.15 \times 10^{-6}$ m^2/s and a latitude $\phi = 45^o$, we obtain the following parameters:

$$KC = \frac{U_0 T_{M2}}{h} = 140 \qquad Re_S = 5560 \qquad Ro = 1.36.$$

The Keulegan-Carpenter number KC is commonly used to characterize oscillatory flows and represents the ratio between the oscillating period T_{M2} and an inertial time scale h/U_0. These parameters give a value of Re_S laying in the fully developed turbulent regime [12]. Since *resolved* LES at the present value of the Reynolds number would require a number of grid points not affordable with our computational resources, we decided to perform a small-scale experiment, holding the inertial parameters and reducing the Reynolds number of the simulation at an affordable value, nonetheless characterized by the presence of fully developed turbulence along the cycle of oscillation. We chose $Re_S = 1790$ for two reasons: 1) at this value, fully developed turbulence is present in most of the cycle of oscillation [12]; 2) experimental data for the purely oscillatory flow are available for comparison with our results (test n. 8 of [12]).

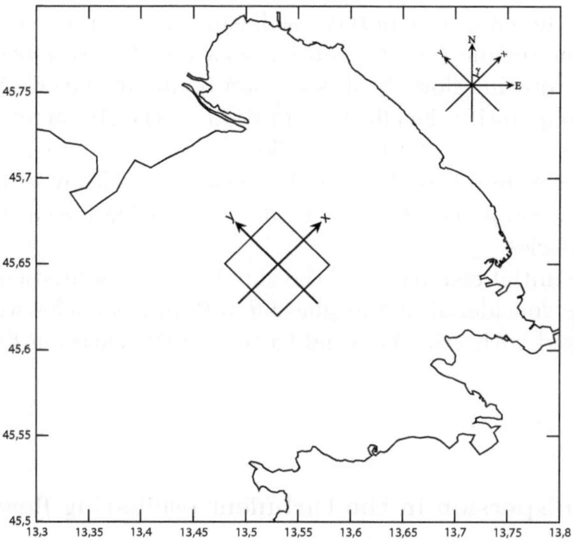

Fig. 1. Sketch of the domain location. The x-direction is aligned with the driving tidal flow, the rotation angle with respect to the North is $\gamma = 45^o$.

The simulations were carried out over a rectangular box, using periodic conditions over the horizontal planes of homogeneity, a stress-free condition at the top boundary and a no-slip condition at the bottom solid wall. The depth of the domain and the amplitude of the free stream velocity were scaled holding the full scale value of KC, whereas the horizontal dimensions of the computational domain were chosen large enough to accurately reproduce the largest scales of the motion. The computational parameters used for the simulations of the Eulerian field are in Table 1.

	$L_x/\delta_S, L_y/\delta_S, L_z/\delta_S$	n_x, n_y, n_z	$\Delta x^+, \Delta y^+, \Delta z^+_{min}, \Delta z^+_{max}$
OF	$50, 25, 40$	$64 \times 64 \times 256$	$62, 31, 2, 22$
OFR	$50, 50, 40$	$64 \times 128 \times 256$	$62, 31, 2, 22$

Table 1. Computational parameters of the simulations at $Re_S = 1790$. The non-dimensional grid spacing is referred to the maximum wall shear stress along the period and consequently to the minimum value of the wall unit $z^* = \nu/u_\tau$, where u_τ is the friction velocity. OF refers to the purely oscillating flow, OFR refers to the oscillating flow in the case with rotation.

The Lagrangian phase consists of tracers, representative of a particulate with inertia comparable with that of the carrying fluid. The simulations were executed in two distinct cases. First we considered the dispersion of a swarm of particles in the case of a purely oscillating flow (Stokes boundary layer) in the turbulent regime (OF in Tab.1), subsequently we moved particles in the oscillatory rotating flow (Stokes-Ekman boundary layer; OFR in Tab.1), with the aim to quantify the effect of rotation on the characteristics of particulate dispersion for a mid-latitude shallow-water environment. In both cases 102400 particles were released in the flow field. Particles were initially placed over 16 longitudinal planes $x_1 - x_3$, respectively using for each plane 80×80 equispaced particles.

As regards the initial conditions of the simulations, a statistically steady Eulerian field was considered at the phase $\theta = 0$ and particles were released in the Eulerian field with velocity equal to that of the carrying fluid.

3 Results

3.1 Particle dispersion in the turbulent oscillating flow field

The Eulerian phase of the turbulent Stokes boundary layer at $Re_S = 1790$ was analyzed and discussed in [13] and in the literature therein reported. A detailed discussion is not repeated here, rather we summarize some results relevant for the present research.

The oscillatory boundary layer is inherently unsteady, and presents a symmetry between the first ($0^o \div 180^o$) and the second ($180^o \div 360^o$) half cycle which, in a statistical sense, repeats identically to the first one, apart that the sign of the streamwise velocity is inverted. The half cycle is characterized by two main phases, namely the phase of acceleration from 0° to 90° and a successive phase of deceleration, up to 180°. The study presented the following scenario: during the late deceleration up to the early acceleration phases, a thin laminar boundary layer develops. At $Re_S = 1790$ this boundary layer undergoes a transition due to shear instability between 30° and 45° accompanied by a sharp increase of the turbulent kinetic energy K, related to the rapid growth of the production rate of K. Fully developed, equilibrium turbulence is observed between 60° and around 150°, when the mean velocity profiles are characterized by the presence of a log-layer and the flow dynamics evolves through a sequence of quasi-equilibrium states. In the early to mid deceleration phases a strong turbulence dissipation is observed related to the explosive, bursting production of small scale vorticity (see also [14]). Around 160° the flow field starts to invert its own direction, first in the near-wall region and subsequently upward in the fluid column. This inversion produces the splitting of the fluid column into two separate regions, a near-wall one where a laminar boundary layer rapidly grows, and an upward one where turbulence has strongly decayed owing to the absence of production rate and, due to a history effect, few large-scale structures coming from the previous phases populate the fluid column. The study has also shown that, at $Re_S = 1790$ significant turbulent activity is detectable in the fluid column up to $25\delta_S$.

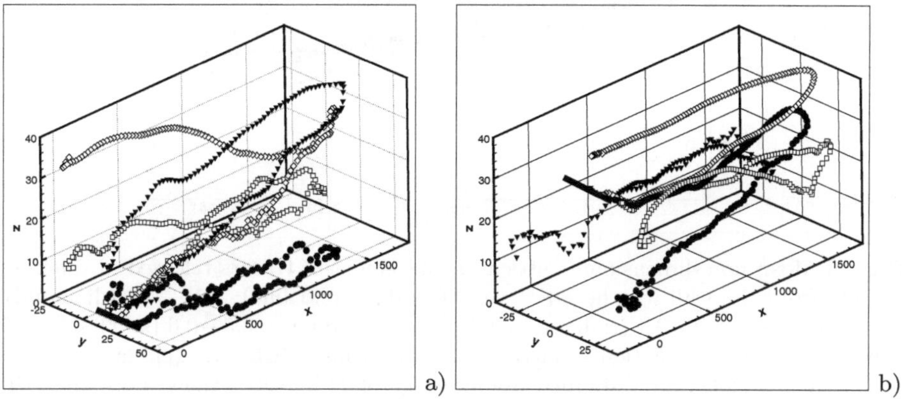

Fig. 2. Trajectories of particles along an oscillatory cycle, released at a fixed horizontal station (plotted as a thick segment) at two different heights in the fluid columns: a) $z = \delta_S$; b) $z = 30\delta_S$. For sake of clarity, the ratio among axis is exaggerated and the ticks are in δ_S units.

Figure 2 shows the 3D trajectories of selected particles released at two different heights in the flow field. The trajectories of particles released in the near-wall region (Fig.2a) show the presence of turbulent mixing during most of the first half cycle of oscillation. The particles undergo a large scale motion during the phases of inversion (around 180°) and again they are subjected to a wide spectrum of turbulent scales during the second half cycle. Conversely, particles released above the turbulent region (Fig.2b) have trajectories that do not exhibit the presence of significant turbulent mixing, except that one marked with circles that penetrates into the turbulent region. In all cases, during the phases when inversion of the flow direction takes place (160°-180°), particles are subjected to a very large scale displacement in the cross-stream direction, which tends to transport them at a fast rate within the fluid column. As clearly shown in [13], this transport cannot be associated to classical turbulent mixing, since at the inversion phases turbulence is nearly suppressed in the near-wall region and only few large-scale structures populate the flow field. However, these weak large-scale structures are able to transport fluid particles over distances comparable with their length scale in the flow field and consequently they still supply a source of mixing for the particulate.

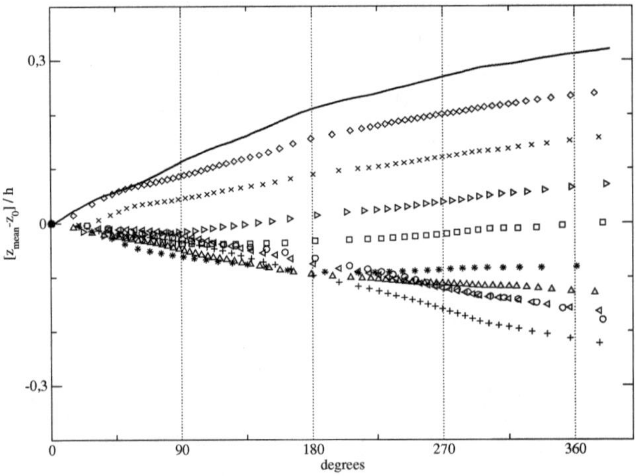

Fig. 3. Mean non dimensional vertical displacement of tracers released at different heights from the wall for the OF case: 0.0 – 0.1 (solid line); 0.1 – 0.2 (diamond); 0.2 – 0.3 (cross); 0.3 – 0.4 (right triangle); 0.4 – 0.5 (square); 0.5 – 0.6 (star); 0.6 – 0.7 (up triangle); 0.7 – 0.8 (left triangle); 0.8 – 0.9 (plus); 0.9 – 1.0 (circle). For sake of clarity we plot values skipped about every 900 time iterations, note that the computational time increment is not constant.

In Fig.3 we observe that the mean vertical displacement of the particulate rapidly increases for particles released in the near-wall region, due to the large mixing. The mean vertical displacement of the particles gradually reduces for particles initially released in the core region of the turbulent layer, associated to nearly symmetric vertical fluctuations, whereas particles released in the free surface region on average tend to be transported toward the wall. This has to be attributed to the low level of turbulence in the top region of the domain; specifically a particle traveling in the top region of the domain (say above $25\delta_S$) tends to remain at the same height or, if trapped in a turbulent structure, it is transported within the near-wall turbulent region. Finally, Fig.3

also shows that the slope of the mean vertical displacement decreases in the second half cycle, as particles spread over a large distance and experience the features of turbulence averaged on a larger slab of fluid.

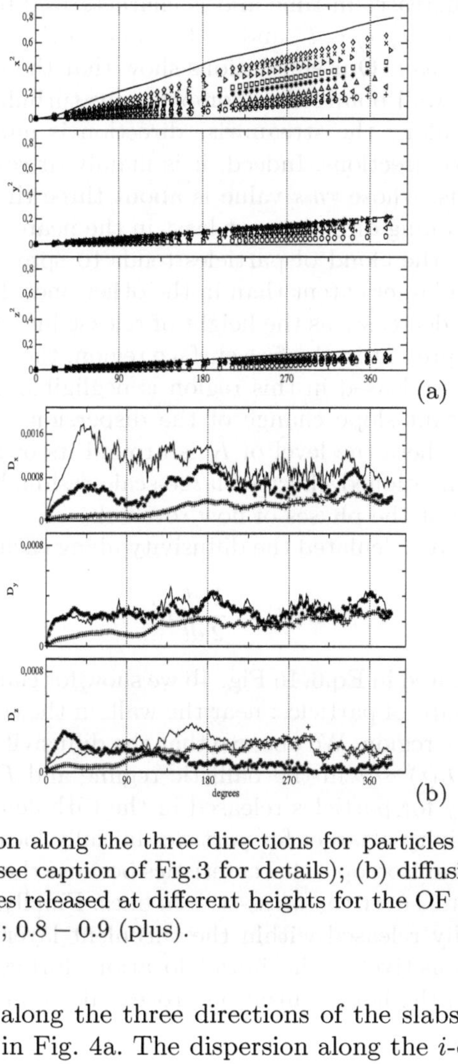

Fig. 4. (a) Dispersion along the three directions for particles released at different heights for the OF (see caption of Fig.3 for details); (b) diffusivity along the three directions for particles released at different heights for the OF case: 0.0 – 0.1 (solid line); 0.5 – 0.6 (star); 0.8 – 0.9 (plus).

The dispersion along the three directions of the slabs of particles above discussed is shown in Fig. 4a. The dispersion along the i-direction is defined as:

$$x_i^2(t) = \frac{1}{N_p} \sum_{j=1}^{N_p} \left[x_{i,j}(t) - \int_0^t \overline{u_i}(t)dt - x_{i,j}(t=0) \right]^2 \qquad (6)$$

where N_p is the number of particles belonging to a slab in the fluid column and $x_{i,j}(t)$ is the i-component of the position of the j-particle of the swarm at time t. The dispersion is related to the fluctuating field only, since the mean displacement given by the averaged velocity field $\overline{u_i}$ ($\int_0^t \overline{u_i}(t)dt$) has been removed. The quantity $\overline{u_i}$ is evaluated at each time step considering the mean velocity of the Lagrangian particles at the position of the j-particle.

Consistently with the classical theory [15], the dispersion in the three directions always increases in time and is characterized by an initial ballistic regime ($t \to 0$) where $x_i^2(t) \sim t^2$ and a Brownian regime ($t \to \infty$) where the dispersion increases as t. Our simulations show that the results of the theory hold also in case of wall bounded, inhomogeneous turbulence.

The dispersion along the streamwise direction is much larger than that along the other two directions. Indeed, it is mainly driven by the streamwise velocity fluctuations, whose *rms* value is about three times larger than that relative to the remaining directions, at least in the near-wall region (see [13]). This indicates that the cloud of particles tends to spread in the streamwise direction at a much larger extent than in the other ones. Figure 4a also shows that the dispersion decreases as the height of release increases. Due to the low level of turbulence present in the free surface region, the dispersion associated to particles initially released in this region is negligible. Interestingly, we do not observe significant slope change of the dispersion during the phases of velocity inversion, where the level of K is around its own minimum. This is due to the above mentioned effect of large-scale residual turbulence present in the fluid column at the phases of flow reversal.

As a final step, we calculated the diffusivity along each direction as in [16]:

$$D_i(t) = \frac{1}{2}\frac{d}{dt}x_i^2(t) \tag{7}$$

where $x_i^2(t)$ is evaluated in Eq.6. In Fig. 4b we show for clarity the diffusivity of three meaningful slabs of particles: near the wall, in the core fluid column and near the free surface region. We observe that the diffusivity obeys the Taylor's two-regime rule ($D_i(t) \sim t$ in the ballistic regime and $D_i(t) \sim$ const in the Brownian one) only for particles released in the turbulent region of the fluid column. The diffusivity is larger for particles initially located in the near-wall region and exhibits anisotropic characteristics because the streamwise velocity component dominates over the cross-stream ones. Finally, Fig.4b shows that, for particles initially released within the turbulent layer, particle diffusivity tends to be less sensitive to the initial location during the second half of oscillation owing to the homogenization process above mentioned.

3.2 Particle dispersion in the rotating oscillating flow field

Although rotation does not vary the amount of energy of a system, it can redistribute energy from mean to turbulent flow and *vice versa*. If we consider the viscous length scales δ_S and δ_E previously defined and respectively associated to the oscillatory and the rotational components of the motion, we obtain $\delta_E/\delta_S = \sqrt{Ro}$, in agreement with Eq.4. Since our simulations are carried out at $Ro = 1.36$, we expect that rotation increases the thickness of the layer in the water column where turbulent activity is present. The results of the simulations (see [4] for details) show that the specific volume-integrated turbulent kinetic energy increases with respect to the OF case through the increase of the vertical and spanwise components (w and v velocity rms) in the fluid column, especially in the half-column near the surface. The presence of the Coriolis force gives rise to a spanwise pressure gradient that oscillates accordingly with the tidal motion. As a result, the mean trajectories of particles follow elliptic paths, and the ratio between the axes varies with the distance from the wall. The horizontal component of rotation gives a background vorticity which combines with the flow mean vorticity thus producing a lack of symmetry in the two half cycles. Specifically, in the second half period, when the mean vorticity associated to the forcing is opposite to the background vorticity, turbulence tends to be enhanced, thickening the turbulent boundary layer and featuring the flow with characteristics very different from those of the non-rotating case. The opposite is true in the first half cycle. Moreover, being all the components of the Reynolds stress tensor not negligible, turbulence assumes a clear three-dimensional character.

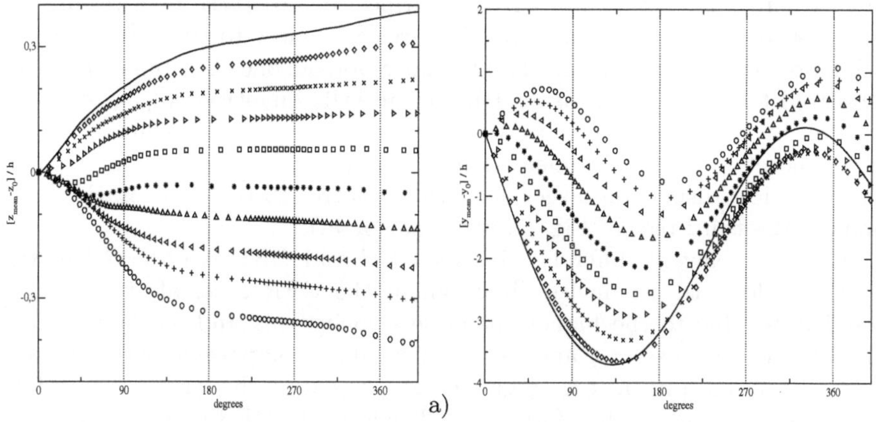

a) b)

Fig. 5. Mean non dimensional vertical (a) and spanwise (b) displacement of tracers released at different heights from the wall for the OFR case (see caption of Fig.3 for details).

Figure 5a shows the mean vertical displacement of the slabs of particles in the OFR case. The rate of increase of the mean vertical displacement is faster than that relative to the OF case in the first quarter of the oscillation cycle, and the largest displacement is observed for the tracers released in the top half column. As shown in [17] and [4], this is related to higher vertical velocity fluctuations between 0° and 90° along the whole water column, and to an increased level of turbulence in the region closer to the free surface. In the OFR case, the vertical mixing is therefore able to move the tracers released close to the free surface region downward to almost half channel, much more than in the OF case. As an example, in the OFR case the tracers released in the slab between 0.9 and 1.0 (Fig.5a) reach, after a complete oscillation cycle, a mean distance that is more than twice that reached in the OF case (Fig.3). Similarly to the OF case, the slopes of the mean vertical displacement decrease in the second half cycle due to the homogenization process above mentioned.

The presence of the cross-stream pressure gradient in the OFR case gives rise to a large spread of tracers in the spanwise direction (Fig.5b). The maximum mean spanwise displacement is observed in the near-wall region, associated to the largest values of the spanwise velocity component. The mean spanwise displacement decreases and eventually changes its sign going up along the fluid column. This is associated to the shape of the mean profile of the spanwise velocity ([4]) that, due to the coupling between oscillating and rotating motions, in the first phases of oscillation has negative values in the bottom region and gets positive in the free surface region.

The dispersion $x_i^2(t)$ along the three directions of the tracers released in the slabs is computed as in Eq.6, and shown in Fig. 6a.
The streamwise velocity rms is slightly affected by rotation in the near-wall region, whereas it increases in the free-surface region ([17]); as a result, the streamwise dispersion related to particles released close to the free surface is much more affected by rotation than the near-bottom ones. A similar behavior is observed for the vertical dispersion, because the quantity w_{rms} is strongly enhanced by rotation in the core region as well as in the free-surface region ([17]). Therefore, particles released in the upper part of the fluid column show a vertical dispersion larger than that relative to the non-rotating case (compare the third plot of Fig. 6a with that of 4a).

The most relevant difference between OF and ORF is observed in the dispersion along the spanwise direction: in the OFR case, after one cycle of oscillation, due to the higher spanwise velocity rms and to the non-zero Reynolds shear stresses τ_{12} and τ_{23}, y^2 has values comparable with x^2, with the tracers released in the core region characterized by the largest dispersion. The turbulent mixing is therefore more efficient in the OFR case: in particular rotation enhances dispersion in the horizontal planes and, at the same time, thickens the fluid layer where a large spreading of the particulate is observed.

Figure 6b shows that the diffusivity along the streamwise direction behaves quite differently in the first and in the second half cycle of oscillation. For particles released in the near-wall region, after a sudden increase similar to the OF case, D_x reaches nearly constant values, comparable with those of the OF case. Conversely, the streamwise diffusivity of tracers released in the core and in particular in the near-surface regions is larger than in the OF case, due to the already discussed increase of turbulence level in the free surface region. As in the previous case, in the second half cycle D_x looses the memory of the tracers' initial location, consistently with the homogenization process. The second half cycle is also characterized by a strong increase in streamwise diffusivity with respect to the OF case, especially for tracers released in the

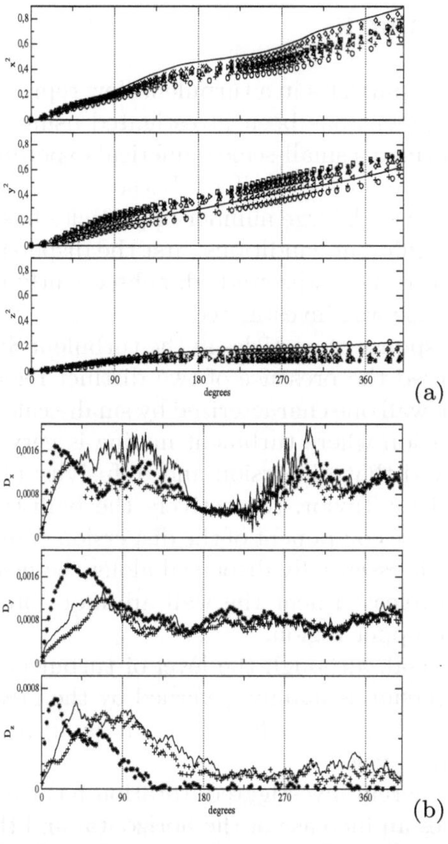

Fig. 6. (a) Dispersion along the three directions for particles released at different heights for the OFR case (see caption of Fig.3 for details); (b) diffusivity along the three directions for particles released at different heights for the OFR case (see caption of Fig.4b for details).

core and the near-surface regions, coherently with the increased slope of the dispersion and with the above mentioned asymmetry in the turbulence dynamics.
In the first half cycle, the vertical diffusivity is larger than that of the OF case, especially for tracers released in the core region. This is due to the augmented level of turbulent mixing in such region, that is basically related to the higher vertical velocity *rms* and to the non-zero correlation between vertical and spanwise velocity fluctuations. In the second half cycle D_z homogenizes and decreases to values comparable to those of the OF case.
As observed for dispersion, the amplitude of the diffusivity in the spanwise component results comparable to the streamwise one. After 90°, D_y appears nearly constant along the whole column, justifying the efficiency of the horizontal mixing.

4 Concluding remarks

The dispersion of fluid particles in a turbulent flow representative of a shallow-water marine environment has been investigated using *resolved* Large Eddy simulation. We performed a small-scale numerical experiment employing physical parameters relative to the Gulf of Trieste, thus considering a tidally-driven, mid-latitude case. A large number of particles has been released in the flow field, under two different conditions: first the dispersion of the particulate in the purely oscillatory flow was studied; subsequently the dispersion in the oscillating-rotating case was investigated.

As regards the response of particles in the turbulent Stokes boundary layer (OF case), we observed the presence of two distinct regions in the water column, namely a near-wall one characterized by small-scale turbulent structures and a free surface region where turbulent mixing is very weak.
Moreover, we observed that dispersion and diffusivity obey the two classical regimes as theorized by Taylor, respectively the ballistic and the Brownian regimes. The streamwise component of the dispersion dominates over the other two, and tracers result essentially dispersed along the mean flow direction. In particular, particles released near the wall are more broadly dispersed than those released in the upper region.
During the flow reversal, although the level of turbulent fluctuations rapidly decays, particle dispersion is mainly governed by the presence of weak, large-scale structures remaining in the fluid column as an history effect from the previous phases of the cycle.

In the OFR case herein investigated, rotation has two main effects on the flow field: 1) it causes an increase of the horizontal and the vertical turbulent mixing; 2) it thickens the turbulent depth in the fluid column.
As a result, both in the core region and in the free surface region the particulate is dispersed at a much larger extent than in the OF case. The presence of a mean spanwise velocity raised by the Coriolis force severely affects the dispersion in the spanwise direction within the whole fluid column. The tracers diffuse mainly on horizontal planes, with intensities that are similar in both the streamwise and the spanwise components.
The results of the present paper do interest a mid-latitude shallow-water problem. Although it is possible to argue that rotation enhances spanwise dispersion of the particulate, more research is needed in order to evaluate particle dynamics dependence on different values of the Rossby number.

Acknowledgments

The present research has been partially supported by the *Istituto Nazionale di Oceanografia e di Geofisica Sperimentale, OGS.*

References

[1] Yeung PK (2002) Annu Rev Fluid Mech 34:115–142
[2] Marchioli C, Armenio V, Salvetti MV, Soldati A (2006) Phys Fluids 18:025102
[3] Chang YS, Scotti A (2006) J Geophys Res 111, C07001, doi:10.1029/2005JC003240
[4] Salon S, Armenio V, Crise A (2005) Large-eddy simulation of an oscillating-rotating turbulent flow. In: Humphrey JAC, Gatski TB, Eaton JK, Friedrich R, Kasagi N, Leschziner MA (eds) Proc. Turbulent and Shear Flow Phenomena 4, Vol.1.
[5] Coleman GN, Ferziger JH, Spalart PR (1990) J Fluid Mech 213:313–348
[6] Armenio V, Piomelli U (2000) Flow Turb and Combustion 65:51–81
[7] Zang Y, Street RL, Koseff JR (1994) J Comput Phys 114:18–33
[8] V. Armenio, U. Piomelli and V. Fiorotto (1999) Phys Fluids 11,10
[9] Kuerten JGM (2006) Phys Fluids 18:025108
[10] Lupieri G, Armenio V (2005) An MPI code for Lagrangian dispersion in turbulent Eulerian field. In: Science and supercomputing at CINECA
[11] Malačič V (1991) Oceanol Acta 14:23–32
[12] Jensen BL, Sumer BM, Fredsøe J (1989) J Fluid Mech 206:265–297
[13] Salon S, Armenio V, Crise A (2006) J Fluid Mech (in press)
[14] Hino M, Kashiwayanagi M, Nakayama A, Hara T (1983) J Fluid Mech 131:363–400
[15] Taylor G (1921) Proc Lond Math Soc 20:196–212
[16] Elghobashi S, Truesdell GC (1992) J Fluid Mech 242:655–700
[17] Salon S (2004) Turbulent mixing in the Gulf of Trieste under critical conditions. PhD Thesis, Universitá di Trieste, Trieste

Influence of Coriolis forces on turbidity currents and sediment deposition.

M.G. Wells

Department of Physical and Environmental Sciences, University of Toronto, 1265 Military Trail, Toronto, Ontario M1C 1A4, Canada

Summary. Using laboratory analogue experiments I show how the Earth's rotation can influence the deposition patterns of large-scale turbidity currents. While it has been previously recognized that the EarthÕs rotation can influence the trajectories of turbidity currents (Middleton 1993; Huppert 1998; Kneller & Buckee, 2000) the experiments discussed in this paper represent the first systematic laboratory study of the Coriolis forces acting upon turbidity currents. The scale at which Coriolis forces become important is best expressed using the Rossby number, defined as $Ro = U/fL$, where U is a depth averaged velocity, L the length scale and the Coriolis frequency, f, is defined by $f = 2\Omega \sin\theta$, where Ω is the Earth's rotation rate and θ is the latitude. Coriolis forces will dominate a current when $Ro < 1$ (Nof 1996). For example a large turbidity current with a velocity of $U = 10\text{m s}^{-1}$ at a latitude of 45° North where $f = 1 \times 10^{-4} s^{-1}$, has $Ro < 1$ for length scales greater than 100 km.

In this paper I discuss two effects of the Coriolis forces upon large-scale turbidity currents. The first series of experiments document how an increase in the Coriolis parameter resulted in a decrease in the rate of turbulent entrainment of overlying sea-water into a density current. The second set of experiments look at the maximum radius of deposition of a turbidity current on a flat plane, and we find that the resulting radius is inversely proportional to the Coriolis parameter. This result implies that there may be a latitudinal dependence upon the radius of turbidite deposition on the flat oceanic abyssal plane. I compare the scaling developed from these idealized laboratory models to field observations of the 300-500 km spatial extent of the turbidites arising during the 1929 Grand Banks earthquake. By making estimates of the velocity we find that this turbidity current had $Ro \sim 1$, so that Coriolis forces may have limited the spatial extent of the resulting turbidite.

1 Introduction

Turbidity currents are underwater flows that are driven by density differences. Such currents can arise by a variety of mechanisms including instances where

Bernard J. Geurts et al. (eds), Particle Laden Flow: From Geophysical to Kolmogorov Scales, 331–343.

sediment laden rivers enter a lake or estuary, or when sloping layers of sediments become unstable due to loading, underground gas release or seismic activity. Turbidity currents can carry large volumes of suspended sediment, and over geological time-scales a series of such turbidity currents can form thick layers of sedimentary rock within deep ocean basins. Turbidite deposits in deep water off the continental shelf can be associated with large petroleum reservoirs, where oil and gas collect within the porous and permeable sands of the turbidites. Such turbidite deposits are now seen as an important area of petroleum exploration, and many of the major marine petroleum reserves in the North Sea and the Gulf of Mexico are within the sands of ancient turbidite deposits. If these turbidite deposits are very large, of the order 10-100 km, then the Coriolis forces might be the limiting factor that determines the spatial extent of large turbidites, and hence the extent of subsurface oil and gas deposits.

The forces that drives turbidity currents are due to the density anomaly from the suspended sediment load, as well temperature and salinity differences. This material is kept in suspension by the turbulence generated at the bottom boundary, and by hindered settling of particles. The entrainment of underlying sediment can make the current denser and move faster, a process called auto-suspension (Parker et al. 1986). The entrainment of water at the upper boundary reduces the density differences and can slow a density current (Turner 1986). Ultimately the turbidity current stops when the turbulence within the current can no longer suspend the sand and mud. If the turbidity current is large then Coriolis forces may begin to be important. While the influence of Coriolis forces upon turbidity currents is acknowledged in several reviews (Middleton 1993; Huppert 1998; Kneller & Buckee 2000) and in theoretical studies (Nof 1996; Emms 1999; Ungarish & Huppert 1999), the present paper is the first experimental investigation to study turbidity currents on a rotating platform. The results in the present paper build upon what has been learned in the oceanographic context, where there have been many previous studies on the influence of Coriolis forces on non-sedimenting density currents (Griffiths 1986; Shapiro & Zatsepia 1997; Jacobs & Ivey 1998; Etling et al. 2000; Hallworth et al. 2001; Cenedese et al. 2004; Davies et al. 2006).

New experimental results for the rates of interfacial entrainment in rotating density currents are presented in §2. The influence of Coriolis force on limiting the spatial extent of sediment deposits is studied experimentally in §3. We finish with a discussion of the observed deposition patterns resulting from the 1929 Grand Banks earthquake in light of our experimental results in §4.

2 Turbulence in rotating density currents

One of the important mechanisms that determines the downward trajectory of a density or turbidity current is the interfacial drag associated with the turbulent mixing of lighter overlying fluid. In this section I will discuss laboratory experiments that quantify the dependence of the mean mixing rates upon the Coriolis parameter f. In the oceanographic literature the mixing of lighter water into a denser current by shear driven turbulence is referred to as "entrainment" (Turner 1986), and should not be confused with the entrainment of underlying sediments that makes a turbidity current denser. In figure 1 is an illustration of the forces that act on a dense current as it flows down a slope. The magnitude and direction of the dense mixture of water and sediments is controlled by rotation, turbulent mixing, bottom drag, slope angle and local stratification resulting in a trajectory at some angle to the maximum slope (Griffiths 1986).

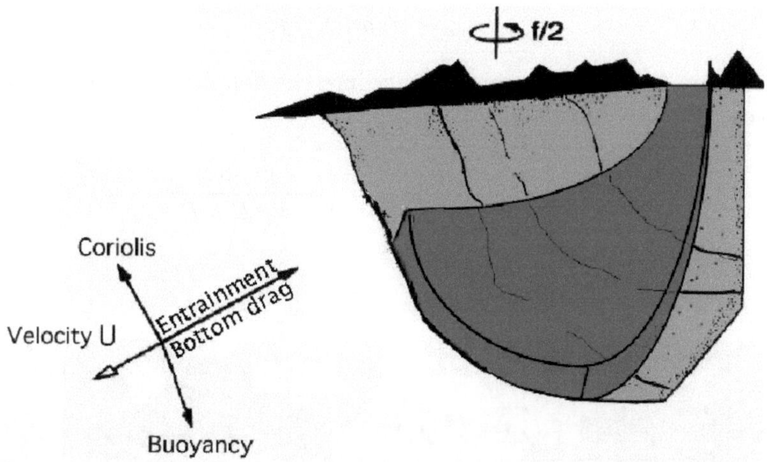

Fig. 1. The depth averaged velocity U and the downwards trajectory of a density or turbidity current in the ocean is set by a balance between buoyancy, Coriolis forces, bottom drag and interfacial entrainment. Figure modified from Baringer & Price (1993).

The turbulence at the upper stratified interface of the density current leads to drag upon the current and dilution of the current with the overlying seawater. The rate at which the current thickens in the downstream direction is known as the "entrainment velocity" and represents the speed at which lighter fluid is mixed into the dense current of thickness h. This entrainment velocity is defined as

$$w_e = dh/dt \tag{1}$$

This entrainment velocity then allows the definition of the dimensionless entrainment ratio (Ellison & Turner 1959) as

$$E = w_e/U. \tag{2}$$

where U is the depth averaged velocity of the current (figure 1).

The entrainment ratio has been found experimentally to depend upon the local Froude number of the flow, which is defined as $Fr = U/\sqrt{g_o'h\cos S}$, where $g' = g\Delta\rho/\rho_o$ is the reduced gravity and S the slope angle (Turner 1986). For a Froude number greater than one, the interfacial entrainment ratio E is proportional to Froude number Fr and E decreases rapidly for $Fr < 1$ (Turner 1986). For low angle slopes the speed is largely independent of slope and the speed scales as $U \sim \sqrt{g'h\cos S}$. If a turbidity current lasts for a period similar to the inertial period $T_{in} = 2\pi/f$, (i.e. at 45°, T_{in}= 19 hours) then rotation becomes important and the gravity current comes into geostrophic balance. In this case the depth averaged velocity U is shown by Nof (1996) to scale with a velocity of

$$U_{geo} \sim g'tan(S)/f. \tag{3}$$

Substitution of this geostrophic velocity into the definition of the Froude number gives

$$Fr_{geo} = \frac{U_{geo}}{\sqrt{g'h\cos S}} = \sqrt{\frac{g'}{h\cos S}}\frac{\tan S}{f}. \tag{4}$$

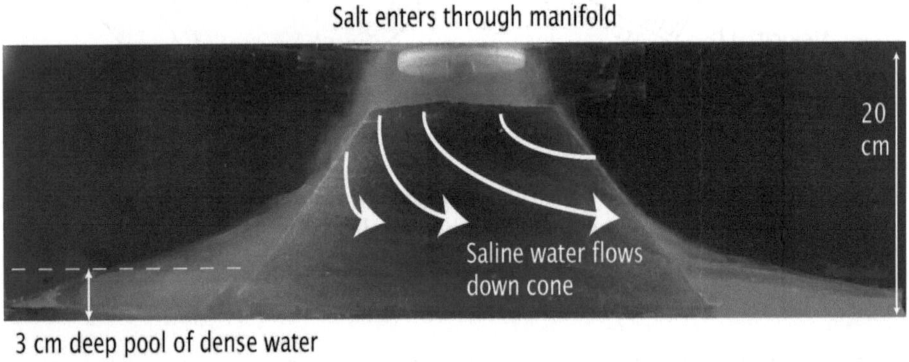

Fig. 2. A side view of the laboratory experiment. Dense saline fluid is pumped through small holes in a circular pipe onto the top of the truncated cone. This dense fluid then flows down the 45° slope and forms a pool of dense water at the base. The depth of the pool of dense water depends upon how much entrainment has occurred on the slope. In this image a thin layer of dense fluid 3 cm deep has formed at the end of the experiment.

Hence if the entrainment ratio is proportional to the Froude number as $E \propto Fr_{geo}$, then the entrainment ratio is inversely proportional to the Coriolis parameter and proportional to the square root of the reduced gravity, for the same slope S and initial thickness h.

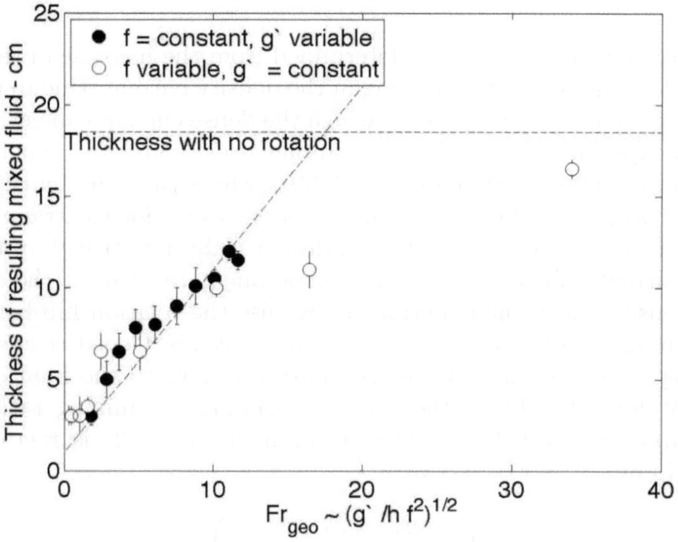

Fig. 3. The depths of the dense pool of fluid were taken from images such as that shown in figure 2. Two sets of experiments are shown, where both the rotation rate and the initial density anomaly were varied. When the resulting depths are plotted against Fr_{geo} (equation 4) the two sets of experiments collapse to the same trend.

To explore the predictions of (4), a series of experiments were conducted on a rotating table using a similar experimental design to that of Shapiro & Zatsepin (1997) and Jacobs & Ivey (1998). The experiments were conducted in a 1 m^2 tank that contained a truncated cone with a flat shelf in the center, as shown in figure 2. Dense saline water is pumped through a manifold located 1.5 cm above the 20 cm diameter shelf. This dense water then flows down a slope of 45°, with a vertical drop of 16 cm. To simulate the variation of Corilios force with latitude the rotation rates were varied between $0.01 < f < 1.5$ rad s^{-1}. The density of the incoming fluid was also varied from $g' = 4.9$ to 196 cm s^{-2}. In each experiment 6 liters of dense saline fluid was pumped through the manifold at a rate of 5 $cm^3 s^{-1}$ for 20 minutes. A mirror located to the side of the tank, at a 45° angle, allows both a top and side view of the experiment to be recorded. The resulting depth of the dense water pool was determined from digital photographs. After the dense current was stopped the depth of the pool of dense water was measured and results are plotted in figure 3. With no mixing, the dense saline water in the gravity current would only form a 1.5 cm deep layer of fluid, but due to the entrainment a layer of between 2-16 cm depth could form. Thus, even when the same buoyancy flux was applied in each case, there could be a 10-fold reduction in mixing for more rapidly rotating experiments. The results of two series of experiments are plotted in figure 3, where both the rotation rate f and the initial density anomaly g' were varied.

The entrainment ratio E was determined from the increase in the volume of dense saline fluid caused by mixing in the density current. Due to the radial symmetry of the cone, the area over which the dense current can entrain is the same in all experiments. This is an advantage over the related experiments of Etling et al. (2000) or Cenedese et al. (2004) where the dense current flowed down a flat slope and hence the surface area over which entrainment into the dense currents can occur increased dramatically for strongly rotationally deflected currents. To determine the entrainment ratio from the measured depth of dense pool in the experiments, we use the relationship between the amount of mixing and entrainment ratio, introduced by Cenedese et al. (2004). The mixing ratio r is the ratio of the initial density to the final density of the density current. This is the same as defining the mixing ratio as $r =$ *input volume/output volume*. The entrainment ratio (2) is related to the mixing ratio by

$$E \equiv \frac{w_e}{U} = \frac{lh}{A}\left(\frac{1}{r} - 1\right) \tag{5}$$

where l the width of the current, and A the area over which the entrainment occurs. The initial current thickness h was observed to stay approximately constant in the experiments where g' varied, and both l and A stay constant because of the radial symmetry of the cone.

In the figure 4 we plot the inferred entrainment ratios against the geostrophic Froude number. For comparison we have also plotted the entrainment ratios against the Froude number from density currents measured in the previous experimental observations of Ellison & Turner (1959), Alavain (1986) and Cenedese et al. (2004) and the field observations of Price & Baringer (1993), Dallimore et al. (2001), Princevac et al. (2005). The new experimental results in figure 4 are within 10% of the entrainment ratio values measured in similar experiments of Ellison & Turner (1959) or Wells & Wettlaufer (2005). Based on figures 3 and 4 we would thus expect turbidity currents to move more slowly at higher latitudes (where f is smaller) and consequently entrain less overlying fluid.

3 Deposition patterns of rotating turbidity currents

To study the influence of Coriolis forces upon the depositional patterns of turbidity currents, a simple series of experiments were conducted where a sediment laden flow was released in the corner of square tank. The density of the sediment laden flow was kept constant, and the dynamics of the turbidity current were observed for increasing background rotation rates. The experimental apparatus is shown in figure (5). The dynamics of the release of a dense turbidity current at the boundary of this tank are analogous to the

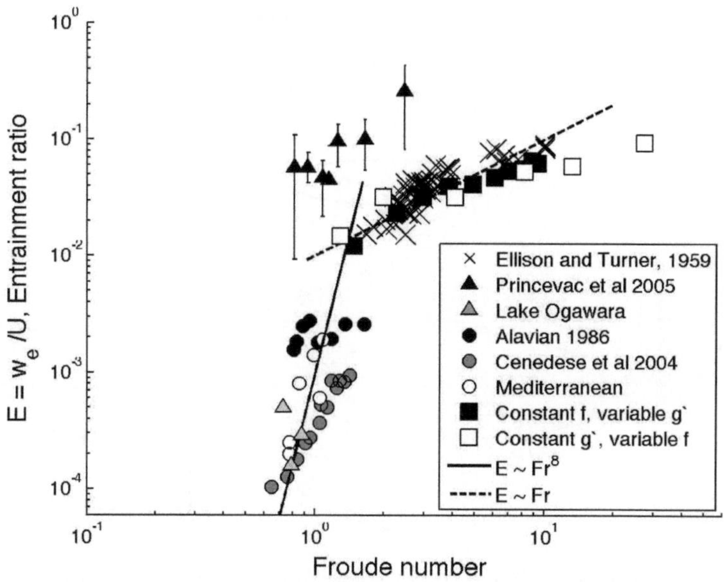

Fig. 4. We plot the new laboratory measurements of entrainment ratio E as a function the Froude number. Previous measurements of entrainment ratio are also plotted.

situation experienced when fresh river-water flows into the salty ocean. In this case a large round bulge initially forms on the river plume, with a radius that scales with the Rossby number (Horner et al. 2006). Eventually water starts to leave this bulge, and is deflected to the right in the Northern hemisphere, to form a coastally trapped gravity current. The significant difference between a buoyant river plume and a dense turbidity current is due to the time-scale of sedimentation of the particles, so that the sedimentation pattern may only show the initial bulge and very little of the sediment may be transported into the geographically steered coastal current.

To make the sediment laden flow, a slurry containing 20g of silicon carbide (density $= 3.15$ g/cm^3) was stirred in a container of volume of 160 cm^3. This produced a turbidity current with initial density 1.08 g/cm^3. The particles we used were Mesh 240 Carbolon SiC. These SiC particles have a mean size of 60 μ m, which implies a Stokes settling velocity of 4 mm s^{-1}. Hence the particles settle rapidly once turbulence stops in these shallow turbidity currents.

Photographs from eight experiments are shown in figure 6 where the deposition patterns of the black silicon carbide stand out against a white background. The photographs are taken from above and show the systematic reduction in deposition radius as the rotation rate is increased. Most of the sediment rapidly falls out of suspension with this radius, however there is a small fraction of much finer sediment that stays in suspension longer, and this

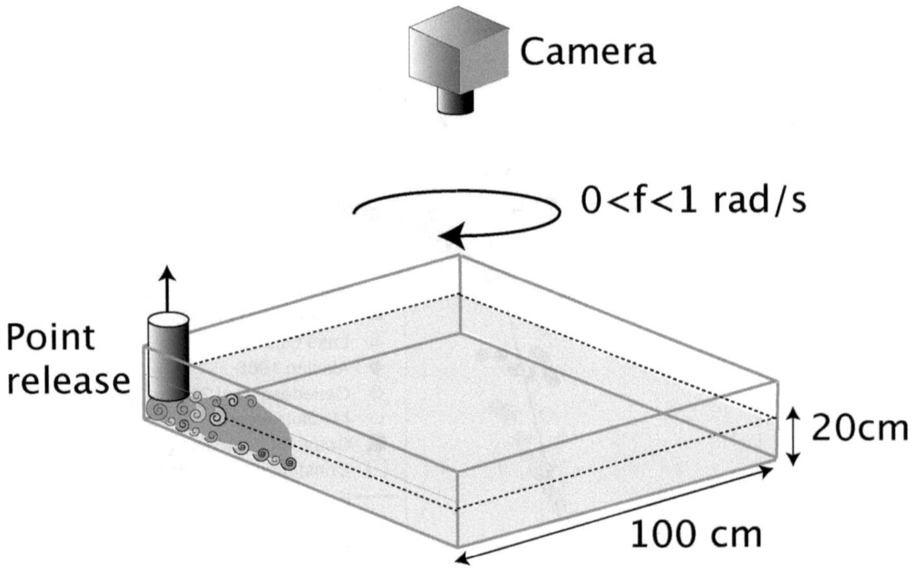

Fig. 5. The laboratory experiment used to determine the influence of Coriolis forces upon sedimentation patterns.

forms a weak turbidity current that is deflected to the right and is seen as the black sediment at the base of the images.

The observed radius L of the sedimentation patterns are plotted in figure 7 and show an inverse dependence upon rotation rate f. In analogy to the radius of the bulge of the buoyant river plume, we assume that the radius of sedimentation on the rotating turbidity current is that which has $Ro = 1$. The Rossby number is defined as $Ro = U/fL$. The initial speed of collapse of the turbidity currents is $U \sim \sqrt{g'h}$, so that the Rossby number is one when $L = \sqrt{g'_o h}/f$. Based upon the low measured values of entrainment for flows where $Fr \sim 1$ in figure 4, we will assume there is little entrainment to change the volume or g'. If we then use conservation of volume of the turbidity current so that $V = hL^2\pi/4$, the radius L of the quarter circle is related to the reduced gravity, the initial volume and the Coriolis parameter by

$$L \sim (4/\pi)^{1/4}(g'_o V/f)^{1/4}. \tag{6}$$

This radius is similar to the scaling of non-sedimenting rotating experiments by Hogg et al. (2001). In figure 7 there is good agreement between the scaling (6) and the observed reduction in L with increasing f.

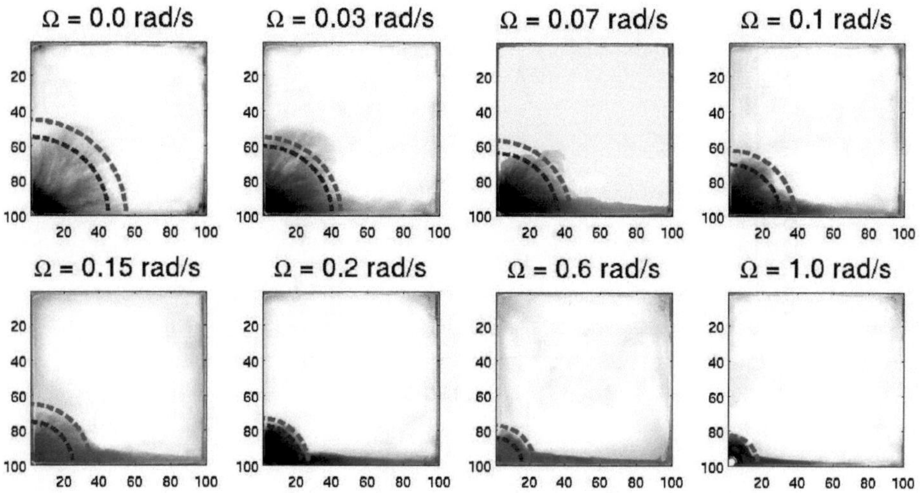

Fig. 6. Eight images of the sedimentation patterns resulting from the release of a black silicon carbide turbidity current in a rotating tank of area 1m^2. The two dashed circles in each picture define the minimum and maximum estimates of the radius L.

4 Applications to the 1929 Grand Banks earthquake

The 1929 earthquake off the Canadian coast of Nova Scotia triggered a turbidity current which spread a 1.5 thick layer of sediment over 280,000 km^2 of the sea floor (Piper et al. 1987). Heezen & Ewing (1952) calculated the speed of the turbidity current based on the times that the trans-Atlantic telegraph service was interrupted, and found that speeds varied from 25 m s^{-1} on the continental slope to under 4 m s^{-1} on the flat abyssal plain. The time for propagation of the current from the shelf to the deepest regions of the flat abyssal plain 800 km away was about 12 hours, comparable to the inertial period, $T_{in} = 2\pi/f$, of about 19 hours (Nof 1996). Thus the EarthÕs rotation should determine the radius that the turbidity current reaches and the resulting sedimentation patterns. A simple estimate on the size of the turbidite is then that $Ro = 1$ or that $L \sim U/f$. If we use the speed estimates based on Heezen & Ewing (1952), that $U = 25$ m s^{-1} and that $f = 9.5 \times 10^{-5} s^{-1}$ at 40° North then this implies that the radius of deposition is $L = 250$ km. In figure 8 we see that this compares favorably with the distribution of sediment observed by Piper et al. (1985).

5 Conclusions

The experiments described in this paper clearly show two strong effects of rotation upon the dynamics of density or turbidity currents. Firstly rotation

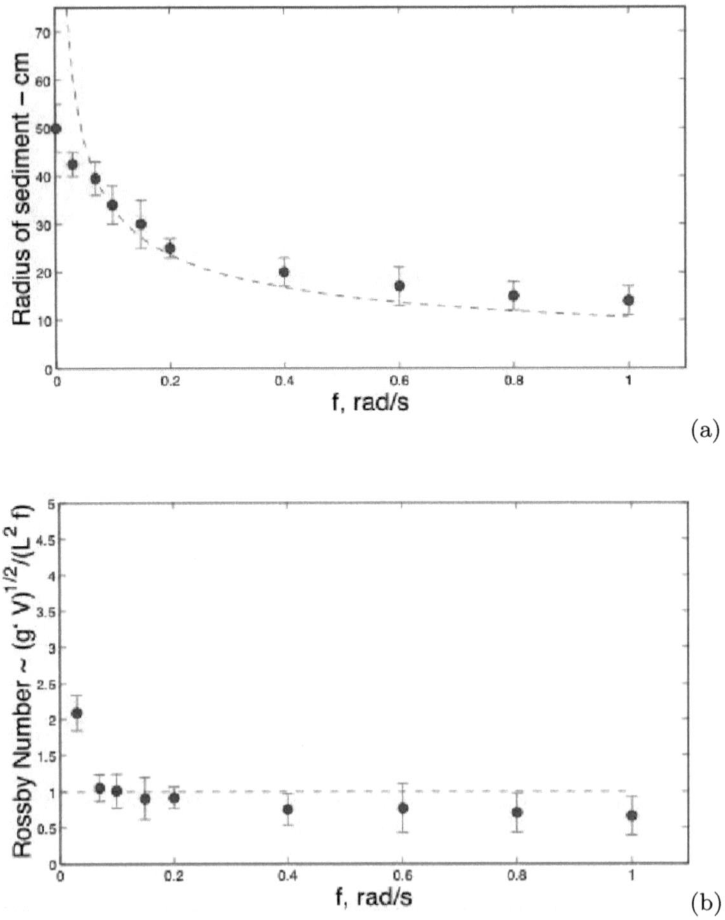

Fig. 7. The experimentally determined radius of sedimentation in figure 6 is plotted as a function of Coriolis frequency f, along with the theoretical predication that $L = 1.06(g'V/f)^{1/4}$ in a). The Rossby number for all the experiments can be seen to be close to one in b) where we plot $\sqrt{g'V}/L^2 f$.

controls the entrainment ratio in such currents, as the velocity is in geostrophic balance. Our theoretical prediction that $E \sim \sqrt{g'}/f\sqrt{h}$ showed good agreement with experimental results in figure 4. Secondly we showed that the radius of a large turbidity current influenced by Coriolis forces is comparable the Rossby radius of deformation, so that the deposition patterns of turbidites should be determined by (6) or $L \sim U/f$. This theoretical prediction again showed good agreement with laboratory experiments.

As there is an inverse dependence of speed and the deposition radius upon the Coriolis parameter, these effects should be most striking for high latit-

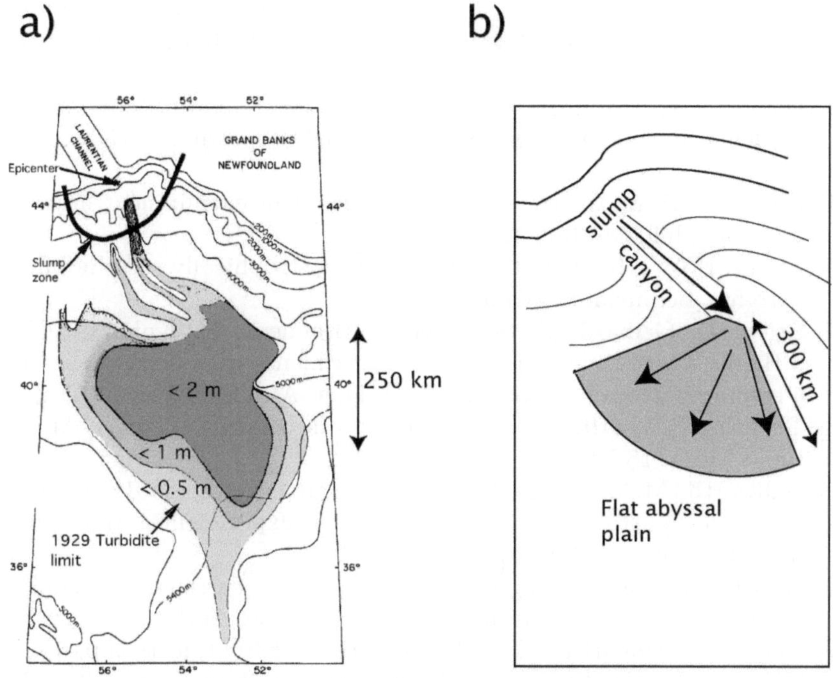

Fig. 8. a) The distribution of turbidites after the 1929 earthquake is shown in grey on this contour map. Most of the sediment lies within 250 km of the canyon mouth, but a small tongue of sediment between 0-50 cm thickness extends south for approximately 600 km. Modified from Piper et al. (1985). b) A simplified conceptual drawing of the sediment distribution, showing a quarter circle of radius 300 km from the point where the turbidity current entered onto the abyssal plain, within this radius lies all of the turbidite between 50-200 cm thickness.

ude turbidity currents and their resulting turbidites. We predict that at high latitudes the turbidites would be of smaller spatial extent and have thicker deposition patterns (assuming similar initial conditions). We found favorable comparisons of the order of magnitude of the spatial extent of 1929 Grand Banks turbidite with the Rossby number scaling. Future work will compare these predictions with a more extensive set of geological observations at different latitudes.

References

[1] Alavian, V. (1986) Behavior of density currents on an incline. J. Hydraulic Eng. 112:27–42

[2] Cenedese, C., Whitehead, J.A., Ascarelli, T.A. & Ohiwa, M. (2004) A dense current flowing down a sloping bottom in a rotating fluid. J. Phys. Oceanog. 34:188–203

[3] Dallimore, C.J., Imberger J., & Ishikawa, T. (2001) Entrainment and turbulence in saline underflow in Lake Ogawara. J. Hydraul. Eng. 127:937–948
[4] Davies, P.A., Wahlin, A.K. & Guo, Y. (2006) Laboratory and analytical model studies of the Faroe Bank Channel deep-water outflow. J. Phys. Ocean. 36:1348-1364
[5] Ellison, T.H. & Turner, J.S. (1959) Turbulent entrainment in stratified flows. J. Fluid Mech. 6:423–448
[6] Emms, P.W. (1999) On the ignition of geostrophically rotating turbidity currents. Sedimentology 46:1049–1063
[7] Etling, D., Gelhardt, F., Schrader, U., Brennecke, F., Kuhn, G., Chabert d'Hieres, G. & Didelle, H. (2000) Experiments with density currents on a sloping bottom in a rotating fluid. Dyn. Atmos. Oceans. 31:139–164
[8] Griffiths, R.A. (1986) Gravity currents in rotating systems. Ann. Rev. Fluid Mech. 18:59–89
[9] Hallworth, M.A., Huppert, H.E. & Ungarsish, M. (2001) Axisymmetric gravity currents in a rotating system: experimental and numerical investigations. J. Fluid Mech. 447:1–29
[10] Heezen, B.C. & Ewing, M. (1952) Turbidity currents and submarine slumps, and the 1929 Grand Banks earthquake. Am. J. Sci. 12:849–873
[11] Hogg, A.J., Ungarish, M. & Huppert, H.E. (2001) Effects of particle sedimentation and rotation on axisymmetric gravity currents. Phys. Fluids 13:3687–3698
[12] Horner-Devine, A.R., Fong, D.A., Monismith, S.G. & Maxworthy, T. (2006) Laboratory experiments simulating a coastal river inflow. J. Fluid Mech. 555:203–232
[13] Huppert, H.E. (1998) Quantitative modelling of granular suspension flows. Proc. Royal Soc. 356:2471-2496
[14] Jacobs, P. & Ivey, G.N. (1998) The influence of rotation on shelf convection. J. Fluid Mech. 369:23–48
[15] Kneller, B. & Buckee, C. (2000) The structure and fluid mechanics of turbidity currents: a review of some recent studies and their geological implications. Sedimentology 47:62–94
[16] Middleton, G.V. (1993) Sediment deposition from turbidity currents. Annu. Rev. Earth Planet. Sci. 21:89–114
[17] Nof, D. (1996) Rotational turbidity flows and the 1929 Grand Banks earthquake. Deep Sea Res. 43:1143–1163
[18] Parker, G., Fukushima, Y. & Pantin, H.M. (1986) Self-accelerating turbidity currents. J. Fluid Mech. 171:145–181
[19] Piper, D.J.W., Shor, A.N., Far're, J.A., O'Connell, S. & Jacobi, R. (1985) Sediment slides and turbidity currents on the Laurentian Fan; sidescan sonar investigations near the epicentre of the 1929 Grand Banks earthquake. Geology 13:538–541
[20] Price, J.F. & Baringer, M.O. (1993) Outflows and deep water production by marginal seas. Prog. Ocean. 33:161–200

[21] Princevac, M., Fernando, H.J.H. & Whiteman, C.D. (2005) Turbulent entrainment into natural gravity driven flows. J. Fluid. Mech. 533:259–268

[22] Shapiro, G.I. & Zatsepin, A.G. (1997) Gravity current down a steeply inclined slope in a rotating fluid. Ann. Geophysicae 15:366–374

[23] Turner, J.S. (1986) Turbulent entrainment–the developement of the entrainment assumption and its application to geophysical flows. J. Fluid. Mech. 173:431–471

[24] Ungarish, M. & Huppert, H.E. (1999) Simple models of Coriolis-influenced axisymmetric particle-driven gravity currents. Int. J. Multi. Flow 25:715-737

[25] Wells, M.G. & Wettlaufer, J.S. (2005) Two-dimensional density currents in a confined basin. Geophys. Astro. Fluid Dyn. 99:199–218

[26] Wells, M.G. & Wettlaufer, J.S. (2006) The long-term circulation driven by density currents in a two-layer stratified basin. J. Fluid. Mech. (accepted)

A stochastic model for large eddy simulation of a particle-laden turbulent flow

Christian Gobert, Katrin Motzet, Michael Manhart

Fachgebiet Hydromechanik, Technische Universität München, Arcisstr. 21, D-80333 München, Germany Ch.Gobert@bv.tum.de

Summary. This paper focuses on the prediction of particle distributions in a flow field computed by large eddy simulation (LES). In an LES, small eddies are not resolved. This gives rise to the question in which cases these eddies need to be reconstructed (modeled) for tracing particles. Therefore the influence of eddies on the particles in dependence on eddy and particle time-scales is discussed. For the case where modeling is necessary, a stochastic model is presented. The model proposed is a model in physical space and not in velocity space, i.e. not the velocities of the unresolved eddies but the effects of these eddies on particle positions are reconstructed. The model is evaluated by an a priori analysis of particle dispersion in turbulent channel flow.

1 Introduction

Particle laden flows in nature often reach Reynolds numbers for which direct numerical simulation (DNS) is not possible on nowadays computers. For detailed numerical predictions of such flows, large eddy simulation (LES) is considered to be an appropriate method. This paper focuses on the simulation of a particle-laden flow by LES.

In a LES, not all length scales of the turbulent fluctuations are resolved. This can be described formally by applying a spatial filter to the velocity field. To solve the Navier-Stokes equations for the filtered velocity fields, a subgrid-scale (SGS) model is required which accounts for the effect of the unresolved scales on the resolved ones (SGS stresses). In the present work, this model is referred to as fluid SGS model. In order to evaluate the performance of a fluid SGS model, an a priori analysis can be conducted. In such an analysis, the SGS-stresses are computed explicitly on the basis of an unfiltered solution and its corresponding filtered one.

In many applications (e.g. prediction of sedimentation processes, dispersion of aerosols in the atmosphere) the dynamics of the carrier fluid is only of secondary interest. It is more important to predict the distribution of the suspended phase. Therefore only the scales in the carrier fluid which have a

Bernard J. Geurts et al. (eds), Particle Laden Flow: From Geophysical to Kolmogorov Scales, 345–358.

significant influence on the suspended phase must be computed. Nevertheless, these scales are often too small to be resolved by LES; the corresponding eddies might be in the subgrid range and a particle-SGS model will be required.

In the present work we focus on such cases. We compute the carrier fluid by a LES and the suspended phase by solving the transport equation of particles in a Lagrangian framework. Effects of the suspended phase on the carrier flow as well as particle-particle interactions are neglected. For the effect of the unresolved eddies of the carrier fluid on the particle motion, a stochastic particle SGS model is developed. This model is validated by an a priori analysis conducted for dispersion in turbulent channel flow. The carrier fluid is computed by DNS and subsequently filtered to eliminate errors that would be introduced by a fluid SGS model.

It will be shown that the SGS eddies are most important for computing particle distributions if the relaxation time of the particles is small. Therefore we restrict the development and validation of the model on inertia free particles.

This paper is organized as follows: In sections 2 and 3, the governing equations and numerical methods for DNS of the carrier flow and the suspended phase are presented. In section 4, we discuss the significance of the subgrid scale (SGS) velocities on the suspended phase. For the case where these velocities are significant, we propose a stochastic model for including their effects on the suspended phase. This model is developed in section 5 and verified by an a priori analysis in section 6.

2 Numerical simulation of the carrier fluid

In order to conduct an a priori analysis, in this study a DNS of the carrier fluid is performed by solving the Navier Stokes equations

$$\operatorname{div} \mathbf{u} = 0 \tag{1}$$

$$\frac{D\mathbf{u}}{Dt} = -\frac{1}{\rho}\nabla p + \nu \Delta \mathbf{u}. \tag{2}$$

Here, $\mathbf{u}$ represents the fluid velocity, ρ the density, ν the kinematic viscosity and p the pressure. $\frac{D}{Dt} = \frac{\partial}{\partial t} + \mathbf{u}.\nabla$ denotes the material derivative.

For solving equations (1) and (2), we used a Finite-Volume method. This method is a modified version of the projection or fractional step method proposed independently by [2, 20]. For spatial discretization a second order scheme (mid point rule) was implemented. For advancing in time, we use a third order Runge-Kutta scheme as proposed by Williamson [22] with constant time step Δt. The continuity equation (1) is satisfied by solving the Poisson equation for the pressure. In this paper, we investigate turbulent channel flow only. Therefore the Poisson equation can be solved by a direct method using Fast-Fourier transformations in the homogeneous streamwise and spanwise directions of the channel flow and a tridiagonal solver in wall-normal direction. For a detailed description of the implemented flow solver the reader is referred to [13]. Please note that in [13] a second-order scheme for advancing in time was used whereas here, we implemented a third order Runge-Kutta scheme.

We use periodic boundary conditions in the two homogeneous directions and no slip conditions at the walls. The flow is driven by a constant pressure gradient that adjusts the Reynolds number based on the half channel height H and the bulk velocity u_{bulk} to $Re = 2817$. This corresponds to a wall units based Reynolds number of $Re_\tau = 180$. In our coordinate system, x is pointing in streamwise, y in spanwise and z in wall-normal direction. The size of our computational domain is $9.6H$ in streamwise, $6.0H$ in spanwise and $2.0H$ in wall normal direction. For all computations staggered Cartesian grids were used.

For the DNS we used $96 \times 80 \times 64$ grid cells. The cell distance in wall units in streamwise and spanwise direction is $\Delta x^+ = 18$ and $\Delta y^+ = 13.5$, respectively. In wall normal direction a stretched grid was used with a stretching factor less than 5%. Here, the cell width is $\Delta z^+ = 2.7$ at the wall and $\Delta z^+ = 9.8$ at the channel center-plane. We compared our results up to second order statistics with the spectral DNS of [8] and found excellent agreement. Further validations of the solver are given in [12, 13]. For evaluating the grid dependency on the suspended phase, further computations were conducted on a refined grid. This grid was obtained by refining the grid mentioned above by a factor of 2 in each direction, i.e. the number of grid cells was incremented by a factor of 8.

For the a priori analysis, the fluid velocity was filtered by top-hat filters using a trapezoidal rule. Most of the results presented in this study are based on a three dimensional filter with a filter width of 4 cell widths in each direction. This filter will be referred to as `fil3d`. Please note that this filter does not correspond to filtering over a cube due to the different cell widths in each direction. For analyzing anisotropic effects we implemented a two dimensional top-hat filter which filters in spanwise and wall normal direction only (`fil2d`). In these directions again the filter width was chosen to be 4 cell widths. For a detailed investigation on the effect of different filters in a particle laden flow, the reader is referred to [1].

3 Numerical simulation of the suspended phase in a DNS

For computing the suspended phase, single particles are traced. In all computations, only effects of the fluid on the particles are considered; effects of the particles on the fluid are neglected (one way coupling). For computing traces of particles other than fluid particles it is assumed that the acting forces on these particles are given by the Stokes drag, fluid acceleration force and gravity. Hence, according to Maxey and Riley [14] the equation of motion for a particle is given by

$$\frac{d\mathbf{v}}{dt} = \underbrace{-\frac{c_D Re_P}{24 t_P}(\mathbf{v}-\mathbf{u})}_{\text{Stokes drag}} + \underbrace{\frac{\rho}{\rho_P}\frac{D\mathbf{u}}{Dt}}_{\text{fluid acceleration}} + \underbrace{\frac{\rho_P-\rho}{\rho_P}\mathbf{g}}_{\text{gravity}} . \tag{3}$$

Here, $\mathbf{v}(t)$ denotes the particle velocity, ρ_P the density of the suspended phase and $\mathbf{g}$ the gravity. t_P is the particle relaxation time, i.e. the timescale for the particle to adopt to the velocity of the surrounding fluid. The particle Reynolds number Re_P is based on particle diameter and particle slip velocity $\|\mathbf{u}-\mathbf{v}\|$ which leads to a nonlinear term for the Stokes drag. The drag coefficient c_D was computed in dependence of Re_P according to the scheme proposed by Clift et al. [3]. $\frac{D\mathbf{u}}{Dt}$ as well as the fluid velocity $\mathbf{u}$ must be evaluated at the particle position $\mathbf{x}_P(t)$, i.e. $\frac{D\mathbf{u}}{Dt} = \frac{D\mathbf{u}}{Dt}(\mathbf{x}_P(t),t)$ and $\mathbf{u} = \mathbf{u}(\mathbf{x}_P(t),t)$. Hence, these values must be interpolated (see below).

In the cases which we considered in this study (for parameters cf. section 4), we found the Stokes drag to be a stiff term whereas fluid acceleration force as well as gravitation are independent of $\mathbf{v}$ and thus not stiff. Therefore it is appropriate to solve equation (3) by a numerical scheme that can treat stiff terms and non stiff terms separately. Such a scheme is given by a Rosenbrock/Wanner method [7]. Here, in each time step the stiff term (i.e. the Stokes drag) is linearized and discretized by an implicit Runge-Kutta scheme. For the other terms an explicit Runge-Kutta method is used.

The stiffness is dependent on particle properties. In order to trace different suspended phases, an adaptive method was chosen. Altogether we decided to implement the adaptive Rosenbrock/Wanner scheme of 4th order together with an error estimation of 3rd order. This scheme can be found in [7].

In the remaining part of this section we will describe how we approximated $\frac{D\mathbf{u}}{Dt}(\mathbf{x}_P(t),t)$ and $\mathbf{u}(\mathbf{x}_P(t),t)$.

Let t_1 and $t_1+\Delta t$ be two instants at which the fluid velocity $\mathbf{u}$ is computed on the given grid by solving the Navier-Stokes equations (1) and (2). $\frac{D\mathbf{u}}{Dt}$ equals the right hand side of the momentum equation (2) and is therefore also computed on this grid at the given instants. Let t be some instant in between two time steps of the flow solver, $t_1 < t < t_1+\Delta t$. For computing the particle velocity according to equation (3), the terms $\mathbf{u}(\mathbf{x}_P(t),t)$ and $\frac{D\mathbf{u}}{Dt}(\mathbf{x}_P(t),t)$ are required. These can be obtained by interpolation in space (at $\mathbf{x}_P$) and in time (at t).

The spatial interpolation uses a second order interpolation in direction of the velocity vector and first order interpolation in the remaining directions. This ensures a conservative interpolation which we found to be important for the particle distributions. A change to a second order interpolation did not affect the results significantly.

In detail, first $\mathbf{u}(\mathbf{x}_P(t_1),t_1)$ and $\frac{D\mathbf{u}}{Dt}(\mathbf{x}_P(t_1),t_1)$ were computed by spatial interpolation. For the fluid acceleration force this was sufficient,

$$\frac{D\mathbf{u}}{Dt}(\mathbf{x}_P(t), t) \approx \frac{D\mathbf{u}}{Dt}(\mathbf{x}_P(t_1), t_1) \; \forall t_1 < t < t_1 + \Delta t. \tag{4}$$

If this was done for $\mathbf{u}(\mathbf{x}_P(t), t)$ as well, this would correspond to a non-continuous fluid velocity along a particle path. Due to the stiffness ($\mathbf{v} \to \mathbf{u}$) this would result in large amplitude higher order terms for $\mathbf{v}$. An adaptive solver would considerably reduce the time step size in such a situation which would render the scheme ineffective. In order to circumvent this problem, we approximated $\mathbf{u}$ during one time step Δt linearly in time by using the flow field of the previous time step $t_1 - \Delta t$.

4 Influence of SGS velocities on the particles

In a LES context, not $\mathbf{u}$ but the filtered velocity $\bar{\mathbf{u}}$ is computed. The question at hand is whether replacing the velocity $\mathbf{u}$ by $\bar{\mathbf{u}}$ in equation (3) has a significant effect on the particle dynamics, i.e. if the non resolved eddies could be neglected or not. This question will be addressed in the present section.

Consider a particle with a relaxation time t_P residing in an eddy with a much larger lifetime t_{EL}, $t_{EL} \gg t_P$. Here, it can be assumed that the particle will eventually adopt to the eddy velocity. On the other hand, if the particle relaxation time is large with respect to the eddy lifetime, $t_{EL} \ll t_P$, the eddy will disappear before the particle can adopt its velocity. Seen on the timescale of the particle, this particle is pushed very slightly by such eddies. Soon (referring to the timescale of the particle), the particle will be located in the next eddy with $t_{EL} \ll t_P$ and the particle will be pushed again. For such a particle, this will result in an effect similar to Brownian motion and can therefore be considered as noise for the particles.

Concluding, the effect of a specific eddy on a particle is dependent on t_{EL}/t_P. This was also found experimentally by Fessler et al. [5]. They investigated the distribution of Lycopodium, glass and copper in air and found a preferential concentration for Lycopodium but not for copper particles. This is due to the different Stokes numbers $St = t_P/t_K$, t_K being the Kolmogorov timescale. For Lycopodium the Stokes number is $St = 0.6$ whereas for copper the Stokes number is $St = 56$.

As shown by Rouson and Eaton [18], the effect under consideration can be shown by DNS at a lower Reynolds number at fixed Stokes number. We did the same computations and found the results depicted in figures 1 and 2. The flow field was computed as described above, i.e. at $Re_\tau = 180$, discretized on $96 \times 80 \times 64$ cells. Recalling that the flow field in the two figures is identical due to one-way coupling, it can be seen that the influence of an eddy varies with the material properties of the particles.

For these computations we took Stokes drag, fluid acceleration and gravity into account. The corresponding parameters were chosen in accordance to the experiment by Fessler et al. Thus, Stokes numbers were chosen as stated above for Lycopodium and copper resp., density ratio was $\rho/\rho_P = 0.0017$

for Lycopodium and $\rho/\rho_P = 0.000136$ for copper. Gravity points in streamwise direction. For scaling gravity with the smallest eddies, particle Froude numbers based on Kolmogorov scales $Fr_P = \sqrt{\rho_P \eta}/\sqrt{(\rho-\rho_P) g t_K^2}$ were held constant. Here, η is the Kolmogorov length scale. Thus, $Fr_P = 1.9151$ for Lycopodium and $Fr_P = 1.9136$ for copper.

In contrast to our computations, Rouson and Eaton neglected fluid acceleration and gravity. We found good agreement between their results and ours. Thus, in this case only Stokes drag affects the particle distribution significantly. Hence, the influence of the eddies varies with the particle relaxation time, the only material inherent parameter appearing in the Stokes drag.

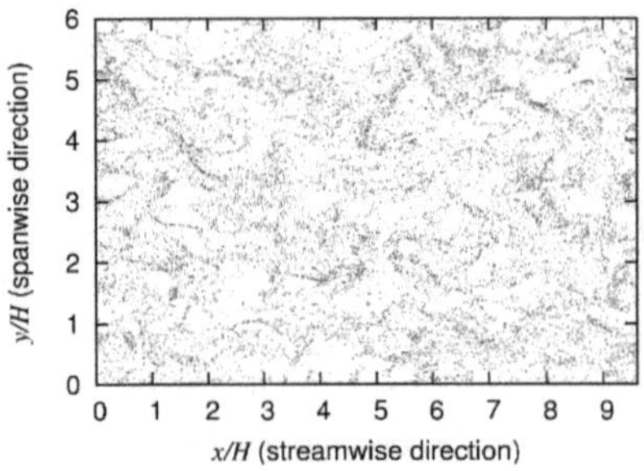

Fig. 1. Lycopodium in air, instantaneous distribution on channel center-plane, $Re_\tau = 180$, $St = 0.6$

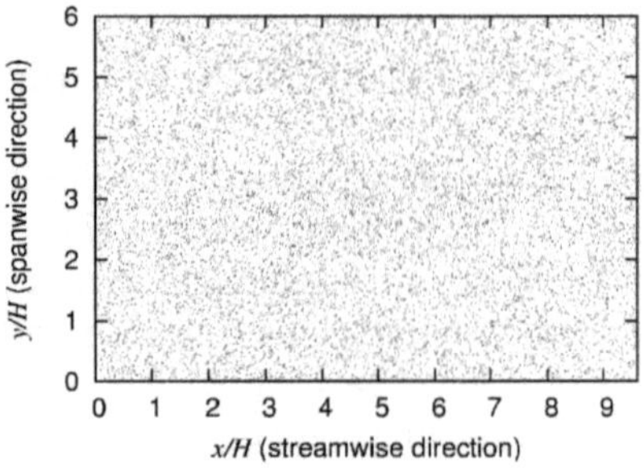

Fig. 2. Copper particles in air, instantaneous distribution on channel center-plane, $Re_\tau = 180$, $St = 56$

In a LES, the size of the resolved eddies depends on the coarseness of the LES grid. Equivalently, the minimal lifetime of the resolved eddies depends on the LES cutoff frequency $1/t_{LES}$. According to this analysis, the SGS terms are not negligible for tracing particles if $t_{LES} \gg t_P$. In such a case, a model is required for recovering these effects. Evidently modeling is most important if $t_P = 0$, i.e. for tracer particles. Therefore a model can be evaluated by applying it on such particles.

5 A stochastic SGS model

For cases in which SGS velocities cannot be neglected, several models were already proposed by different authors. Some of these models are stochastic [15, 19, 21], some are deterministic models [10, 15, 16].

In all the approaches mentioned, modeling is done in velocity space, i.e. the SGS fluctuations $\mathbf{u}'$ are approximated. When modeling $\mathbf{u}'$ as a stochastic variable, time correlations along the particle path must be included. Therefore in many of the models mentioned above, an additional differential equation has to be solved.

In what follows, a SGS model for dispersion of inertia-free particles is derived. Here, we propose to model the SGS-effect in physical space rather than in velocity space. This will be explained as follows. For tracer particles the particle position is given by

$$\mathbf{x}(t) = \mathbf{x}(0) + \underbrace{\int_0^t \bar{\mathbf{u}}(\tau) d\tau}_{=:\,\bar{\mathbf{x}}(t)} + \underbrace{\int_0^t \mathbf{u}'(\tau) d\tau}_{=:\,\mathbf{x}'_t} . \tag{5}$$

Here, $\bar{\mathbf{u}}(\tau) = \bar{\mathbf{u}}(\mathbf{x}(\tau), \tau)$ and $\mathbf{u}'(\tau) = \mathbf{u}'(\mathbf{x}(\tau), \tau)$ are the filtered and SGS velocities on a particle path, resp.

The filtered (i.e. resolved) velocities would result in deterministic particle positions $\bar{\mathbf{x}}(t)$. In our model, the non-resolved (SGS) velocities are considered as random displacements, denoted here by $\mathbf{x}'_t$. Thus, we propose to model $\mathbf{x}'_t$ as a stochastic process. To this end, we consider the moments of $\mathbf{x}'_t$. Here we start with the model proposed by Shotorban and Mashayek [19]. This is a model for the velocity fluctuations $\mathbf{u}'$. Under the assumption of isotropic SGS fluctuations they propose to solve for the SGS fluctuations at each time step in a Lagrangian sense a stochastic differential equation

$$d\bar{\mathbf{u}} + d\mathbf{u}' = d\mathbf{u} = \left(-\frac{1}{\rho} \nabla \bar{p} + \nu \Delta \bar{\mathbf{u}} - \frac{\mathbf{u} - \bar{\mathbf{u}}}{T_L} \right) dt + \sqrt{C_0 \epsilon_{sgs}}\, d\mathbf{W}_t . \tag{6}$$

Here, $\mathbf{W}_t$ is a three dimensional Wiener process, ϵ_{sgs} is the SGS dissipation rate, T_L is the lifetime of a representative SGS eddy and C_0 is a model constant.

According to Deardorff [4], ϵ_{sgs} can be computed by

$$\epsilon_{sgs} = \frac{\nu_t^3}{C_S^4 \Delta^4} \tag{7}$$

with the Smagorinsky constant C_S, filter width Δ and the eddy viscosity ν_t. In most LES models, the eddy viscosity is estimated from the gradients of the resolved velocity field.

Combining the results of Gicquel et al. [6] and Lilly [11], a formula for the SGS relaxation time T_L can be obtained,

$$T_L = \left(\frac{1}{2} + \frac{3}{4} C_0 \right)^{-1} \frac{\Delta^2 C_T}{\nu_t} \tag{8}$$

with another model constant C_T which was set to 0.094 according to [6] and [11].

Now we read the model by Shotorban and Mashayek in a LES context. Therefore the filtered velocity in a Lagrangian framework can be written as

$$d\bar{\mathbf{u}} = \left(-\frac{1}{\rho} \nabla \bar{p} + \nu \Delta \bar{\mathbf{u}} - \operatorname{div} \tau \right) dt \tag{9}$$

with τ being the SGS stress tensor. By subtracting equation (9) from equation (6) a stochastic differential equation for the SGS fluctuations is obtained,

$$d\mathbf{u}' = \left(\operatorname{div} \tau - \frac{\mathbf{u}'}{T_L} \right) dt + \sqrt{C_0 \epsilon_{sgs}} d\mathbf{W}_t. \tag{10}$$

Equation (10) is a linear stochastic differential equation. For such equations Kloeden and Platen [9] give differential equations for the first and second moments of $\mathbf{u}'$:

$$\frac{dE(\mathbf{u}')}{dt} = -\frac{1}{T_L} E(\mathbf{u}') + \operatorname{div} \tau \tag{11}$$

$$\frac{dE(\mathbf{u}'^2)}{dt} = -\frac{2}{T_L} E(\mathbf{u}'^2) + 2E(\mathbf{u}') \operatorname{div} \tau + C_0 \epsilon_{sgs} \tag{12}$$

Again, these equations are linear. Assuming a deterministic velocity for the particles at $t = 0$, i.e. $E(\mathbf{u}')(0) = E(\mathbf{u}'^2)(0) = 0$, the solution reads

$$E(\mathbf{u}') = T_L (1 - e^{-t/T_L}) \operatorname{div} \tau \tag{13}$$

$$\begin{aligned} E(\mathbf{u}'^2) &= (T_L \operatorname{div} \tau)^2 (1 + e^{-2t/T_L} - 2e^{-t/T_L}) \\ &+ \frac{C_0 \epsilon_{sgs} T_L}{2} (1 - e^{-2t/T_L}). \end{aligned} \tag{14}$$

Hence, the variance of the SGS fluctuations can be computed as

$$\operatorname{Var}(\mathbf{u}') = E(\mathbf{u}'^2) - E^2(\mathbf{u}') = \frac{C_0 \epsilon_{sgs} T_L}{2} (1 - e^{-2t/T_L}). \tag{15}$$

Equation (15) gives the variance for the SGS fluctuations in velocity space. This information will be used now in order to compute the variance of the SGS fluctuations in physical space, $\operatorname{Var}(\mathbf{x}'_t)$. To this end, several assumptions will be taken in the remaining part of this section. We will not present validations for each assumption individually; instead, in section 6 results will be presented which support the correctness of the resulting function $\operatorname{Var}(\mathbf{x}'_t)$.

We start by integrating the model equation (10) under the assumption that T_L and ϵ_{sgs} vary little[1]:

---

[1] for a rigorous deduction the mean value theorem can be applied instead

$$\int_0^t d\mathbf{u}' = \int_0^t \operatorname{div} \tau \, dt - \frac{\mathbf{x}'_t - \mathbf{x}'_0}{T_L} + \sqrt{C_0 \epsilon_{sgs}} \mathbf{W}_t \tag{16}$$

Solving for $\mathbf{x}'_t$ gives

$$\mathbf{x}'_t = \mathbf{x}'_0 + T_L \left(-\mathbf{u}'(t) + \mathbf{u}'(0) + \int_0^t \operatorname{div} \tau dt + \sqrt{C_0 \epsilon_{sgs}} \mathbf{W}_t \right) . \tag{17}$$

$\mathbf{x}'_0$, $\mathbf{u}'(0)$ and $\int_0^t \operatorname{div} \tau dt$ are deterministic. Thus,

$$\begin{aligned} \operatorname{Var}(\mathbf{x}'_t) = & - \operatorname{Var}\left(T_L \mathbf{u}'(t)\right) + \operatorname{Var}\left(T_L \sqrt{C_0 \epsilon_{sgs}} \mathbf{W}_t\right) \\ & - 2T_L^2 \sqrt{C_0 \epsilon_{sgs}} \operatorname{Cov}\left(\mathbf{W}_t, \mathbf{u}'(t)\right) . \end{aligned} \tag{18}$$

Now we compute the covariance of the fluctuations and the Wiener process generating the fluctuations, $\operatorname{Cov}(\mathbf{W}_t, \mathbf{u}'(t))$. For this we assume that the fluctuation velocities are unbiased, $E(\mathbf{u}'(t)) = 0$. Multiplication of (17) by $\mathbf{u}'(t)$ and computing the expectation gives

$$\int_0^t E\left(\mathbf{u}'(t)\mathbf{u}'(s)\right) ds = -T_L E(\mathbf{u}'^2) + T_L \sqrt{C_0 \epsilon_{sgs}} E\left(\mathbf{u}' \mathbf{W}_t\right) \tag{19}$$

Now we assume that the autocorrelation of the fluctuations decays exponentially with the Lagrangian correlation timescale t_L,

$$\operatorname{Cor}(\mathbf{u}'(t), \mathbf{u}'(t+\tau)) = e^{-\tau/t_L} . \tag{20}$$

Furthermore in many applications one is interested in the long time behavior. Therefore in the following we will consider large t only. According to equation (14), $E(\mathbf{u}'^2)$ is constant for large t. Thus, by substituting (20) into (19) we obtain

$$E(\mathbf{u}'^2)(t_L + T_L - t_L e^{-t/t_L}) = T_L \sqrt{C_0 \epsilon_{sgs}} E\left(\mathbf{u}' \mathbf{W}_t\right) . \tag{21}$$

Substituting this into equation (18) gives

$$\operatorname{Var}(\mathbf{x}'_t) = C_0 \epsilon_{sgs} T_L^2 (\alpha + t_L e^{\frac{-t}{t_L}})(1 - e^{\frac{-2t}{T_L}}) + T_L^2 C_0 \epsilon_{sgs} t \tag{22}$$

$$\alpha = -\frac{3}{2} T_L - t_L \tag{23}$$

The first terms in equation (22) are exponentially decaying functions whereas the last term is linear. This means that for large t the first terms are negligible. We assume that these terms appear due to the suppression of fluctuations at $t = 0$, $\mathbf{x}'_0 = \mathbf{u}'(0) = 0$. If the latter terms were random, the exponential and constant terms in equation (22) might disappear.

Therefore we neglect these terms and model the SGS fluctuations according to

$$\mathrm{Var}(\mathbf{x}'_t) = T_L^2 C_0 \epsilon_{sgs} t. \tag{24}$$

One possibility to model $\mathbf{x}_t$ with this variance is to model $\mathbf{x}_t$ as a scaled Wiener process. This is not only consistent with the derivation given but can also be interpreted physically as dispersion. Armenio et al. [1] found that neglecting SGS velocities results in a lack of dispersion. We want to compensate this by modeling $\mathbf{x}_t$ as scaled Wiener process:

$$d\mathbf{x}'_t = T_L \sqrt{C_0 \epsilon_{sgs}}\, d\mathbf{W}_t. \tag{25}$$

This process can be implemented by adding in each time step of the flow solver the appropriate stochastic term on the position of each particle:

$$\mathbf{x}(t + \Delta t) = \mathbf{x}(t) + \int_t^{t+\Delta t} \bar{\mathbf{u}}\, d\tau + T_L \sqrt{C_0 \epsilon_{sgs} \Delta t}\, \mathbf{Z} \tag{26}$$

$\mathbf{Z}$ is a Gaussian distributed random variable with expectation 0 and variance 1. T_L, C_0 and ϵ_{sgs} can be obtained according to the model of Shotorban et al. [19]. In our computations we implemented an explicit Euler scheme for solving equation (26).

6 A priori analysis

In this section we present an a priori analysis of the SGS dispersion model developed in section 5. For this analysis we distributed particles on the center-plane of the channel described in section 2. Therefore each particle has the same wall distance and by sampling over the particles, statistics in wall normal direction can be obtained. All results will be displayed in wall units.

We traced the particles with velocities computed by DNS using the two different grids described in section 2. During the simulations we stored the particle positions on hard disk. In a post-processing step we computed the variance of the wall normal coordinate of the particles which develops as predicted by Pope [17] and computed by Armenio et al. [1] (cf. figure 3). We compared both, coarse and fine grid simulations and decided that the resolution of the coarse grid was sufficient for our purposes.

For conducting the a priori analysis, we filtered the DNS-velocity field by top-hat filters as described in section 2. We traced particles with the filtered velocities and found that the dispersion is reduced by filtering. This was already found by Armenio et al. [1]. In contrast to their work, we filtered also in wall normal direction. The corresponding variance of the particle position in wall normal direction (dispersion) is depicted in figure 3. Here, for both filters mentioned in section 2 the dispersion is depicted. Evidently dispersion in wall normal direction is not very sensitive to filtering in streamwise direction.

In a LES context, we are interested in the SGS dispersion, i.e. the difference between dispersion computed by DNS and by filtered velocities. For `fil3d` this SGS dispersion is plotted in figure 4. Please note that in such an analysis the dependency of $\mathbf{x}'$ on $\bar{\mathbf{x}}$ is neglected whereas in our model this dependency is respected due to adding the SGS dispersion subsequently during the simulation. In order to validate the shape of the theoretically derived SGS dispersion (equation (22)), we fitted the SGS dispersion according to equation (22). The fit for $t_L = 0.19, T_L = 0.001, C_0 \epsilon_{sgs} T_L^2 = 1100$ is also shown in figure 4. We find very good agreement between fitted function and numerically computed SGS dispersion. This justifies the assumptions of section 5.

In the next step, we computed the particle positions with the filtered velocities only and added subsequently the modeled SGS dispersion as given by equation (26). For small time steps the stochastic term in equation (26) (i.e. the contribution to the SGS dispersion) is so low that roundoff errors become significant. In our simulations we used a time step for the flow solver of $\Delta t = 0.01 \frac{H}{u_{bulk}}$. We found that when using this time step in combination with the filter `fil3d` these roundoff errors become dominant. In order to compensate for this, we added the term corresponding to the SGS dispersion at every 50 time steps only. The corresponding result is shown in figure 5.

It can be seen that for short times the deviation between DNS and modeled result is still large whereas this becomes somewhat better for large times. This was to be expected since we developed our model for long term behavior. The difference occurs due to the neglecting of the constant and exponentially decaying terms in equation (22). Therefore it is more appropriate to validate the model on the dispersion rate, i.e. the time derivative of the dispersion. This is plotted in figure 6. According to these results we are very satisfied with the performance of our model.

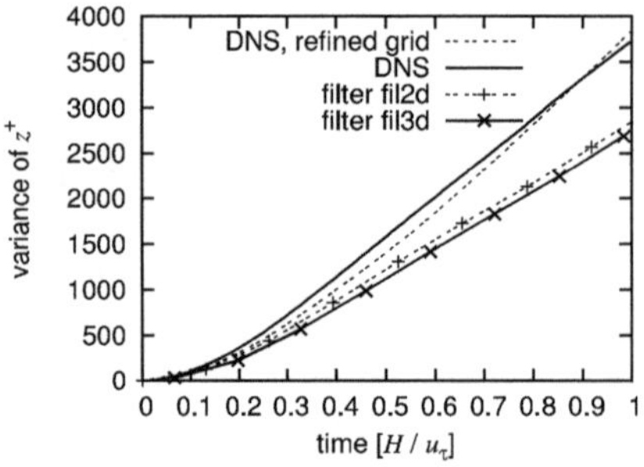

Fig. 3. Dispersion computed by DNS and filtered velocities, displayed in wall units.

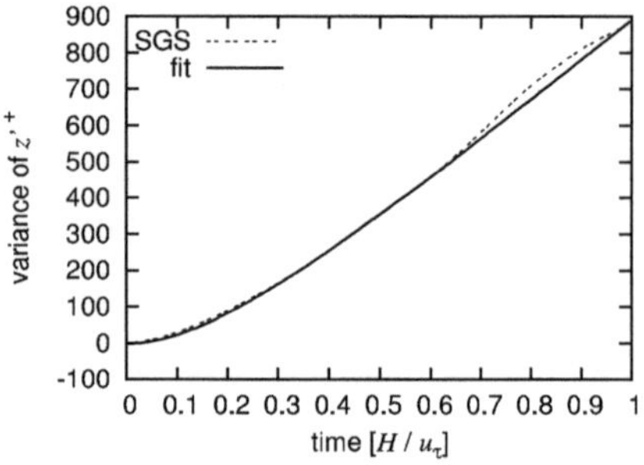

Fig. 4. SGS dispersion and fit according to equation (22).

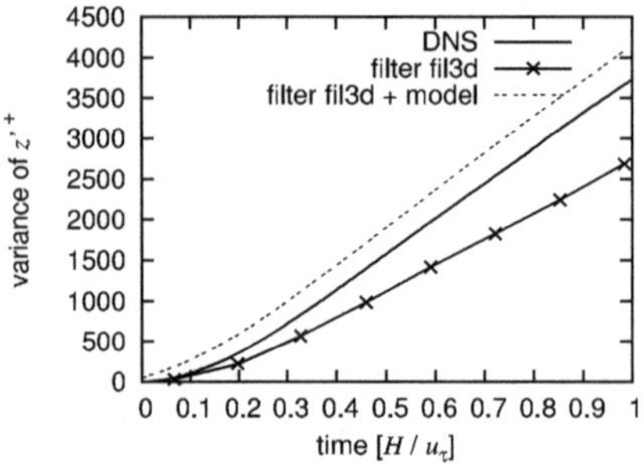

Fig. 5. Dispersion computed by DNS and filtered velocities as well as filtered velocities plus modeled SGS dispersion

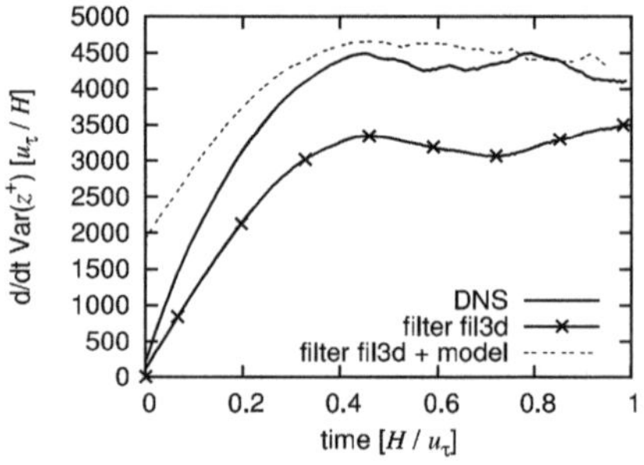

Fig. 6. Rate of dispersion computed by DNS and filtered velocities as well as filtered velocities plus modeled SGS dispersion

7 Conclusions

In this work we developed a stochastic SGS model for computing particle dispersion in turbulent flows. The model was developed for Eulerian-Lagrangian simulations, i.e. where the fluid phase is computed by an Eulerian method and the suspended phase is computed by tracing single particles in a Lagrangian view. For such simulations the effect of the non resolved scales in the carrier fluid must be modeled for tracing the suspended particles.

Modeling is done in physical space by subsequently adding a stochastic term on the particle position which can be seen as dispersion caused by unresolved scales. For developing the model simplifications were taken which are only valid when focusing on the long term behavior of the suspended particles; the model is only capable of predicting the dispersion of particles when a certain time after particle injection has passed. Since the model is of stochastic nature, it can only be used for prediction of statistical properties.

In order to maximize the effect of subgrid scale influence, we focused on tracer particles only in this study. An a priori analysis was conducted in turbulent channel flow. Particles were released on the channel center-plane and the evolution of particle dispersion was computed. In order to get a reference solution, particles were traced by velocities computed from DNS. In another simulation particles were traced by using filtered velocities and adding LES dispersion as given by the model proposed. We found good agreement between both simulations.

References

[1] V. Armenio, U. Piomelli, and V. Fiorotto. Effect of the subgrid scales on particle motion. Physics of Fluids, 11(10):3030-3042, 1999

[2] A. J. Chorin. Numerical solution of the Navier Stokes equations. Math. Comput., 22:745-762, 1968

[3] R. Clift, J. R. Grace, and M. E. Weber. Bubbles, Drops and Particles. Academic Press, New York, 1978

[4] J. W. Deardorff. A numerical study of three-dimensional turbulent channel flow at large Reynolds numbers. J. Fluid Mech., 41:453-480, 1970

[5] J. R. Fessler, J. D. Kulick, and J. K. Eaton. Preferential concentration of heavy particles in a turbulent channel flow. Phys. Fluids, 6:3742-3749, 1994

[6] L. Y. M. Gicquel, P. Givi, F. A. Jabei, and S. B. Pope. Velocity filtered density function for large eddy simulation of turbulent flows. Phys. Fluids, 14:1196-1213, 2002

[7] E. Hairer and G. Wanner. Solving Ordinary Differential Equations II. Stiff and Differential-Algebraic Problems. Springer Series in Computational Mathematics. Springer-Verlag, 1990

[8] J. Kim, P. Moin, and R. Moser. Turbulence statistics in fully developed channel flow at low Reynolds number. J. Fluid Mech., 177:133-166, 1987

[9] P. E. Kloeden and E. Platen. Numerical solution of stochastic differential equations. Applications of mathematics. Springer Verlag, Berlin, 3 edition, 1999

[10] J. G. M. Kuerten. Subgrid modeling in particle-laden channel flow. Phys. Fluids, 18:025108, 2006

[11] D. K. Lilly. The representation of small-scale turbulence in numerical simulation experiments. In Proceedings of the IBM Scientific Computing Symposium on Environmental Sciences, pages 195-210, 1967

[12] M. Manhart. Rheology of suspensions of rigid-rod like particles in turbulent channel flow. Journal of Non-Newtonian Fluid Mechanics, 112(2-3):269-293, 2003

[13] M. Manhart. A zonal grid algorithm for DNS of turbulent boundary layers. Computers and Fluids, 33(3):435-461, 2004

[14] M. R. Maxey and J. J. Riley. Equation of motion for a small rigid sphere in a nonuniform flow. Phys. Fluids, 26:883-889, 1983

[15] N. A. Okong'o and J. Bellan. A priori subgrid analysis of temporal mixing layers with evaporating droplets. Phys. Fluids, 12:1573-1591, 2000

[16] N. A. Okong'o and J. Bellan. Consistent large-eddy simulation of a temporal mixing layer with evaporating drops. Part 1. Direct numerical simulation, formulation and a priori analysis. J. Fluid Mech., 499:1-47, 2004

[17] S. B. Pope. Lagrangian pdf methods for turbulent flows. Annu. Rev. Fluid Mech., 26:23-63, 1994

[18] D. W. I. Rouson and J. K. Eaton. On the preferential concentration of solid particles in turbulent channel flow. J. Fluid Mech., 428:149-169, 2001

[19] B. Shotorban and F. Mashayek. A stochastic model for particle motion in large-eddy simulation. JoT, 7:N18, 2006

[20] R. Temam. On an approximate solution of the Navier Stokes equations by the method of fractional steps (part 1). Arch. Ration. Mech. Anal, 32:135-153, 1969

[21] Q. Wang and K. D. Squires. Large eddy simulation of particle-laden turbulent channel flow. Phys. Fluids, 8:1207-1223, 1996

[22] J. H. Williamson. Low-storage Runge-Kutta schemes. Journal of Computational Physics, 35:48-56, 1980

Aggregate formation in 3D turbulent-like flows

A. Domínguez, M. van Aartrijk, L. Del Castello and H.J.H. Clercx

Fluid Dynamics Laboratory, Eindhoven University of Technology, Den Dolech 2, 5612 AZ, Eindhoven, The Netherlands. A.Dominguez@tue.nl

Summary. Aggregate formation is an important process in industrial and environmental turbulent flows. Two examples in the environmental area, where turbulent aggregate formation takes place, are raindrop formation in clouds and Marine Snow (aggregate) formation in the upper layer in the oceans.

The dispersion of inertial particles differs from that of (passive) fluid particles and is dominated by particle-turbulence interaction. This is especially important when the particle scales match the small-scale turbulent flow scales. Our motivation to study turbulent aggregate formation comes from the need to describe aggregate formation in small-scale turbulence in the oceans.

For a proper description, the study of aggregate formation in turbulent flows requires a particle-based model, i.e. following single particles. Therefore, three main processes should be modeled: the turbulent flow, the motion of the particles, and the collision between particles and subsequent aggregate formation. In this study we use 3D kinematic simulations to model the turbulent flow. A simplified version of the Maxey-Riley equation is used to describe the motion of the particles. For the collision and aggregate formation a geometrical collision check is used: when the distance between two particles is smaller than the sum of their radii a collision takes place. All the particles that collide stay together to form an aggregate, i.e. 100% coagulation efficiency. To account for the porosity of the aggregates a Fractal Growth Model is used.

In this study the importance of the Stokes number and the fractal dimension of the aggregates on collision rates and aggregate formation has been explored, finding that the preferential concentration plays a very important role in aggregate formation by creating regions of high particle concentration. Other results are: the net effect of fractal growth is to increase the aggregate Stokes number and to decrease the density of the aggregate.

In order to determine the performance and applicability of 3D-KS models on aggregate formation processes, DNS simulations and supplementary laboratory experiments are planned.

Bernard J. Geurts et al. (eds), Particle Laden Flow: From Geophysical to Kolmogorov Scales, 359–371.

1 Introduction

Phytoplankton populations in oceans are immersed in a turbulent environment. Turbulence increases encounter rates among plankton cells and other organic and inorganic particles, promoting the formation of aggregates. These aggregates are known as marine snow, and they are a vehicle for transporting organic matter (CO_2) to the bottom of the ocean [1]. The spectrum of spatial and temporal scales in the ocean is extremely broad, ranging from kilometers to micrometers and from days to seconds. This makes direct numerical simulation (DNS) of this system not feasible with the present computational capacity. Implying that the large-scale modeling requires the use of empirical models to be able to represent the effects of small-scale processes on large-scale phenomena, i.e. to represent the gap between small- and large-scale phenomena. Phytoplankton dispersion in large-scale simulations is an example.

In models the phytoplankton is usually represented as a passive tracer, but under some conditions this representation fails. The transport of organic material to the bottom of the ocean, also called the Carbon pump, is an example of a process that cannot be described by considering plankton and particulate matter as passive tracers. The net transport depends on the settling velocity of individual particles which in turn depend in the particle-flow interaction. The settling velocity of particles in turbulent flows is modified from the settling velocity in quiescent fluids [2, 3] and particle properties can change by particle-particle interaction (e.g. aggregation), or by particle-fluid interaction (e.g. breaking up of aggregates). Therefore Lagrangian-based models following trajectories of single particles are an ideal method to study aggregate formation, where particle-particle and particle-flow interaction are important. These models capture in simplified ways these interactions, but they are computationally expensive.

In general, when a few solid particles have formed an aggregate, the shape of this aggregate is complex (although for many constituting particles the aggregate can become more or less spherical). The aggregate is porous and its density is different of that of the constituting particles. Aggregate growth is known to be of fractal nature [4], and this fractal behavior should be taken in account, although with a rather simplified model.

The inertial effects of plankton in turbulent flows have been studied previously by Squires and Yamazaki [5], finding that plankton can experience preferential concentration. There are some works on aggregate formation: turbulent aerosol aggregate formation using DNS [6]; polymer flocculation in turbulent pipe flows [7] and aggregate formation in the ocean has been explored with coagulation theory [8].

In the first part of this paper a brief theoretical introduction is presented. The model is described in the second part, in the third section the results are presented and conclusions are presented in the final section.

2 Turbulence-particle interaction

The interaction between a particle and a (turbulent) flow is complex. When the volume fraction of the solid phase (particles), denoted by $\phi = NV_p/V_d$ (with N the total number of particles, V_p the particle volume, and V_d the total volume of the system), is very small, say $\phi \leq 10^{-4}$, the particles do not influence the flow properties but the flow influences the particle trajectories. This is known as one-way coupling. Under different conditions particles can act as passive tracers, showing a homogeneous distribution of particles or, on the contrary, behave in "resonance" with the flow, then showing preferential concentration [9, 10]. This flow-particle interaction depends on the combination of the particle and flow scales.

The smallest scales of the turbulent flow are the Kolmogorov time and length scales,

$$\tau_k = \sqrt{\frac{\nu}{\varepsilon}} \quad \text{and} \quad \eta = \left(\frac{\nu^3}{\varepsilon}\right)^{1/4}. \tag{1}$$

An important quantity characterizing the response of a particle of diameter d_p to the turbulent flow is known as the Stokes number and is given by

$$St = \frac{\tau_p}{\tau_k} \tag{2}$$

where τ_p, the particle response time. It is the time the particle needs to adapt to a change in velocity and is given by

$$\tau_p = \gamma \frac{d_p^2}{18\nu} \tag{3}$$

where $\gamma = \rho_p/\rho_f$. This response time is valid for small spherical particles with $d_p \ll \eta$. The particle-based Reynolds number $Re_p = \frac{|\mathbf{u}-\mathbf{v}|d_p}{\nu} \ll 1$, where $\mathbf{v}$ is the particle velocity and $\mathbf{u}$ is the velocity of the fluid at the particle position $\mathbf{x}$. When $St \ll 1$ particles behave approximately as passive tracers and they are homogeneously distributed in the flow domain. When $St \simeq 1$ particles respond to the smallest scales of the flow ($\tau_p \approx \tau_k$). The particles are then ejected from the high-vorticity regions and accumulate in the high-strain regions, thus giving rise to preferential concentration. Note that for $St \gg 1$ the particles are virtually insensitive to the small-scale turbulent velocity fluctuations.

The equation of motion for a small, rigid spherical particle in the limit of zero particle Reynolds number is given by the Maxey-Riley equation [11]

$$\begin{aligned} m_p \frac{d\mathbf{v}}{dt} &= 6\pi a\mu(\mathbf{u} - \mathbf{v} - \frac{1}{6}\nabla^2\mathbf{u}) + (m_p - m_f)\mathbf{g} + m_f \frac{D\mathbf{u}}{Dt} \\ &+ \frac{m_f}{2}\frac{d}{dt}(\mathbf{u} - \mathbf{v} - \frac{1}{10}\nabla^2\mathbf{u}) \\ &+ 6\pi a^2\mu \int_0^t \frac{d}{d\tau}(\mathbf{u} - \mathbf{v} - \frac{1}{6}\nabla^2\mathbf{u}) \frac{d\tau}{\pi\nu\sqrt{(t-\tau)}}, \end{aligned} \tag{4}$$

where a is the particle radius and μ is the dynamic viscosity of the fluid. Furthermore, $\mathbf{g}$ is the gravitational acceleration, $m_f = \rho_f V_p$ and $m_p = \rho_p V_p$, with V_p the particle volume, and ρ_f and ρ_p are the fluid and particle densities, respectively. The forces appearing in the right hand side of (4) are: the Stokes drag, the buoyancy force, the pressure gradient force, the added mass and Basset history force, respectively. For particles with $\gamma \gg 1$, often called heavy particles, the only important forces are the Stokes drag and the buoyancy force, and the resulting equation is:

$$m_p \frac{d\mathbf{v}}{dt} = 6\pi a \mu (\mathbf{u} - \mathbf{v}) + (m_p - m_f)\mathbf{g}. \tag{5}$$

This study is restricted to turbulent dispersion and aggregation of heavy particles.

3 Flow and aggregation model

The particle-based aggregation model described here consists of three parts: the turbulent flow, the equation of motion for the particles and the collision-aggregation process. The basic assumptions of the model are that at the smallest scales of the flow the turbulence is homogeneous and isotropic, that $d_p \ll \eta$ and that $\phi < 10^{-4}$ to assume only one-way coupling and to consider particle-pair collisions as the dominant process for aggregate formation.

3.1 3D Kinematic Simulations

Three-dimensional kinematic simulations (3D-KS) of turbulent flows [12] are models of turbulent flows whose velocity field is constructed as a sum of random Fourier modes with a prescribed Eulerian energy spectrum. The velocity at position $\mathbf{x}$ and time t is given by:

$$\mathbf{u}(\mathbf{x}, t) = \sum_{n=1}^{N_k} [(\mathbf{A}_n \times \hat{\mathbf{k}}_n) \cos(\mathbf{k}_n \cdot \mathbf{x} + \omega_n t) + (\mathbf{B}_n \times \hat{\mathbf{k}}_n) \sin(\mathbf{k}_n \cdot \mathbf{x} + \omega_n t)], \tag{6}$$

where N_k is the number of modes in the simulation, $\mathbf{A}_n \times \hat{\mathbf{k}}_n$ and $\mathbf{B}_n \times \hat{\mathbf{k}}_n$ are spatial Fourier amplitudes, $\hat{\mathbf{k}}_n = \mathbf{k}_n / |\mathbf{k}_n|$ and $\mathbf{k}_n$ is the wavenumber. By construction, the velocity field satisfies the incompressibility condition $\nabla \cdot \mathbf{u} = 0$. The positive amplitudes of $\mathbf{A}_n$ and $\mathbf{B}_n$ are chosen according:

$$\frac{3}{2}|\mathbf{A}_n| = \frac{3}{2}|\mathbf{B}_n| = E(k)\Delta k_n, \tag{7}$$

where $E(k)$ is the Eulerian energy spectrum, $\Delta k_n = (k_{n+1} - k_{n-1})/2$ for $2 \leq n \leq N_k - 1$, $\Delta k_1 = k_2 - k_1$ and $\Delta k_{N_k} = k_{N_k} - k_{N_k - 1}$. The distribution of wave numbers is geometric,

$$k_n = k_1 \left(\frac{L}{\eta}\right)^{\frac{n-1}{N_k-1}}, \tag{8}$$

where $k_1 = 2\pi/L$ and $k_\eta = k_{N_k} = 2\pi/\eta$, the wave number associated with the largest scale in the flow and the Kolmogorov length scale, respectively. We are interested in modeling the so-called inertial range, where the energy spectrum is given by:

$$\begin{aligned} E(k) &= C_\varepsilon \varepsilon^{2/3} k^{-5/3} \qquad k_1 < k < k_\eta \\ &= 0, \qquad\qquad\qquad\quad \text{otherwise} \end{aligned} \tag{9}$$

where ε is the energy dissipation rate, k is the wavenumber, C_ε is an empirical constant which value is chosen to be 1.5, according to Kraichnan [13]. The time dependence of a turbulent flow is determined by the non-linear interaction between different modes and by advection of the vorticity field by the velocity field. However, in $3D$-KS there is no dynamics and therefore the time evolution of the flow is determined by the unsteadiness frequencies ω_n in equation (6). Following Fung [12], we choose the frequency of the mode n to be proportional to the eddy turnover time associated with the wave-vector $\mathbf{k}_n$:

$$\omega_n = \lambda \sqrt{k_n^3 E(k_n)}, \tag{10}$$

where λ is the unsteadiness parameter, and in this study $\lambda = 0.5$. The resulting velocity field incorporates turbulent-like flow structures, eddying, straining, and streaming regions. Moreover, given the exact way in which the velocity field is calculated we avoid interpolation errors calculating the velocities of the flow at the particle position. One of the disadvantages of $3D$-KS models is the impossibility of having periodic boundary conditions.

3.2 Reduced equation of motion for the particle

The motion of the particles is calculated from equation (5) without the buoyancy term, in order to understand as a first step solely the effects of inertia. This assures that the collision rates are not affected by other processes like differential sedimentation. So, the final equation used in this study is

$$m_p \frac{d\mathbf{v}}{dt} = 6\pi a \mu (\mathbf{u} - \mathbf{v}). \tag{11}$$

This equation is solved using a fourth order Adams-Bashforth (predictor)-Adams-Moulton (corrector) method.

3.3 Collision-Aggregation

To explore aggregate formation we chose to start with a population of spherical particles all having the same properties: all have diameter d_p and density

ρ_p. After each collision a new aggregate is formed and to keep the background population of particles constant a new particle is added at a random position. This would represent a steady state for the background particle number density. However, this will slowly increase the volume fraction ϕ and might have some effect in increasing the collision rate, but the effect is expected to be small compared to other processes involved. A geometric check for the collision is used: a collision takes place if the distance of separation between two particles or aggregates is smaller than the sum of their radii in the interval $[t, t+\Delta t]$. If the diameters of the colliding particles (or aggregates) are d_i and d_j then this condition for the collision to take place is

$$r_{ij} \leq \frac{d_i + d_j}{2}, \tag{12}$$

with $r_{ij} = |\mathbf{r}_{ij}|$ their mutual distance. To check if the particles may collide in a (given) time interval Δt, the relative velocity between the particles i and j, $V_{ij} = (\mathbf{V}_i - \mathbf{V}_j)\cdot\mathbf{r}_{ij}$, is calculated at the beginning of the time-step. If $V_{ij} > 0$, i.e. the particles are approaching, then the collision time, $\Delta t^{(ij)}$, defined by

$$\Delta t^{(ij)} = \frac{r_{ij} - (d_i + d_j)/2}{V_{ij}}, \tag{13}$$

is calculated. If $\Delta t^{(ij)} < \Delta t$ the particles collide and aggregate. The colliding particles are not allowed to collide again in the same time-step, an assumption based on the observation that the probability of two collisions within the time interval Δt scales approximately with ϕ^2. This probability is extremely small, due to $\phi \ll 1$.

We consider a Fractal Growth Model (FGM), which implies that the aggregate properties are different from those of the initial particles. An aggregate formed by n of the original particles is assumed to be spherical with diameter d_a given by,

$$d_a = d_p n^{1/D_f}, \tag{14}$$

where D_f is the fractal dimension, a parameter that accounts for the porosity of the aggregate [8]. To first approximation, considering that $\gamma \gg 1$ the density of the aggregate is changed as

$$\rho_a = \rho_p n^{1-3/D_f}. \tag{15}$$

From this equation it can be seen that in the case of coagulation of droplets, where there is no porosity and $D_f = 3$, the density is conserved. For solid particles where $D_f < 3$ (the aggregates are porous) and if $\rho_p > \rho_f$, the density of the aggregate will always be smaller than the density of the particles forming it. It is important to mention that we only model the effects of the fractal growth (porosity) on the particle size and density. Using equations (2), (14) and (15) the Stokes number of the aggregate composed by n particles with St_p will be given by:

$$St_a = St_p n^{1-1/D_f}. \tag{16}$$

In order to model the effects of breaking-up up particles, the aggregates are not allowed to grow indefinitely. We choose to allow them to reach a maximum size equal to the Kolmogorov length scale. In the experiments performed, when the aggregates reach that size then $St_a \gg 1$ thus the effect of the small scales is expected to be reduced.

4 3D-KS of particle dispersion and aggregation

4.1 Encounter rates of non-aggregating particles

The encounter or collision rates among non-interacting ("ghost") particles is a measure of the probability of interaction between real particles. If the particles collect in particular regions of the flow domain because of their interaction with the structures in the flow (see Fig. (1)), the collision rate will be increased as can be seen in Fig. (2). To explore the effects of St on the encounter or collision rates of "ghost" particles four numerical experiments were performed with properties given in table 1.

Table 1. Parameter of the numerical experiments exploring the effects of St on the collision rates of non-aggregating particles. For all runs $Re_\lambda = 70$ and $\phi = 10^{-5}$

	$C1$	$C2$	$C3$	$C4$
St_p	0.14	0.91	2.29	9.17

The Stokes number of the particles depends on the ratio $\gamma = \rho_p/\rho_f$ and on the particle size, and is varied from $O(\frac{1}{10})$ to $O(10)$. This is done by changing the density of the particles keeping the size and the volume fraction (ϕ) constant. In the long term an increased encounter rate of non-interacting particles is found to be higher for particles with $St \approx 1$ as can be seen in figure (2). This increase can be explained by the preferential concentration effect, a phenomenon that has been observed in several numerical and experimental works [9, 2, 10]. When $St \approx 1$ the distribution of particles is inhomogeneous, thus increasing the encounter probability of particles in areas with high (particle) concentrations.

It is important to point out that for the initial stage the collision rates are much higher for particles with $St \ll 1$. In this Stokes number regime the particles follow the flow and the preferential concentration effects are nearly absent. The collisions are expected to be caused by relative motion generated by the turbulent small-scale shear flows.

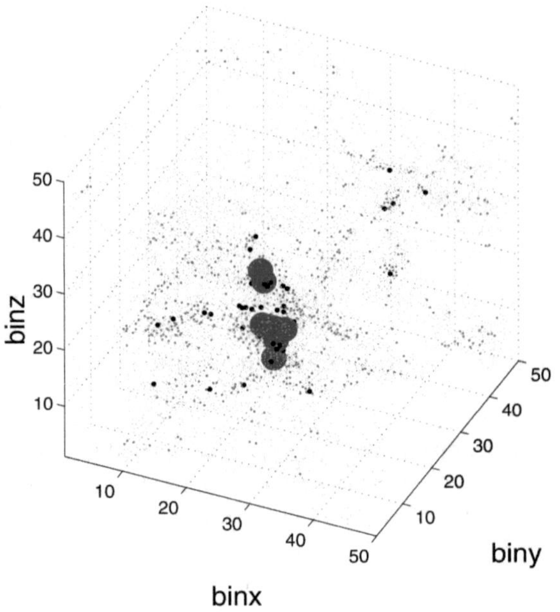

Fig. 1. Preferential concentration for particles with $St \simeq 1$. Snapshot of particle distribution in domain emphasizing the regions of high particle concentration. The big dots are the regions of the flow were the number of particles is 50 times larger than the homogeneous case. The smallest dots denote the regions with a number of particles between 1 and 5, and the mid-sized dots are regions with a number of particles between 5 and 20.

4.2 Effect of initial Stokes number on aggregate formation rates

To explore in more detail the effect of the initial Stokes number of the particles (St_p) on the collision rate five experiments were performed. The details are given in the table 2. In these experiments $Re_\lambda \sim 70$, $D_f = 2.5$ and St_p is varied from 0.56 to 4.30. This is done by keeping the same d_p in all experiments and changing ρ_p.

Table 2. Experiment parameters exploring the effects of initial Stokes number St_p and $D_f = 2.5$ on aggregate formation. For all runs $Re_\lambda = 70$ and $\phi = 10^{-5}$.

	$S1$	$S2$	$S3$	$S4$	$S5$
St_p	0.56	0.70	1.00	2.10	4.30

In the figure (3) three different regimes can be identified. The first concerns the initial stage of the simulation, where a peak in the collision rate is found and a subsequent decrease is observed. This transient behavior can be ex-

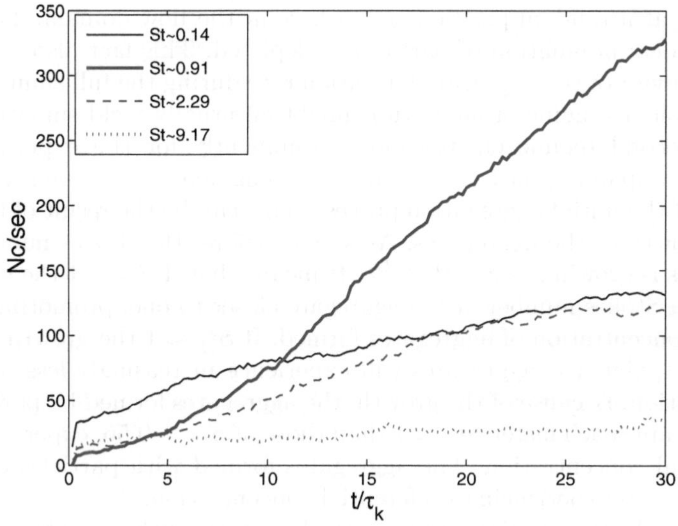

Fig. 2. Effects of St on the encounter (collision) rates.

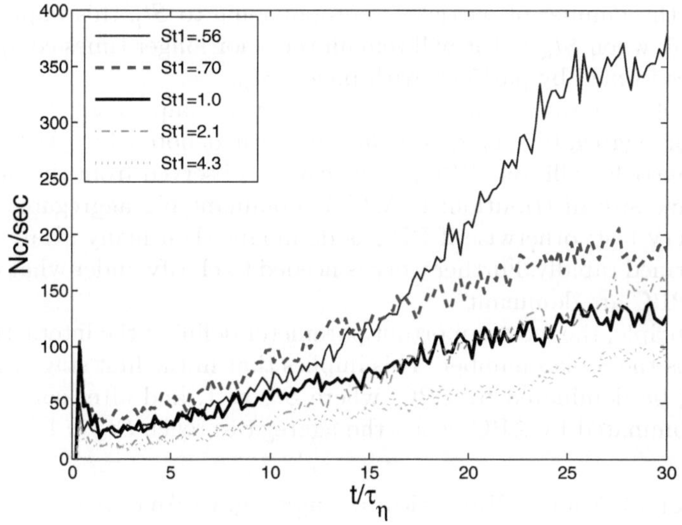

Fig. 3. Effects of the Stokes number of the initial population of particles on aggregate formation rates

plained as an effect of initial inhomogeneities in the particle distribution and preferential concentration. After a period of initialization the collisions are turned on, and that is considered to be $t = 0$. In a high concentration region there will be an initial stage of high collision rates, due to the collision model: when two particles collide one bigger particle is formed and another particle is created at a random position somewhere in the flow domain. This implies that the local population of particles is depleted. This fact also explains the spiky behavior of the aggregate formation rate during the full simulation (and an ensemble averaging of many runs might eventually yield smooth curves).

The second regime emerge most prominently for the experiment with $St_p = 0.7$, apparently just in a regime where augmented collision rates occur as a result of the initial aggregation process. This can be the result of the change in the growth of the aggregates. As stated before, the Stokes number of the aggregates is growing as $St_p n^{1-1/D_f}$. It means that if $St_p < 1$ the growth will result in a Stokes number of the aggregate closer to one, promoting the preferential concentration of aggregates formed. If $St_p \approx 1$ the growth will result in $St_a > 1$, thus the aggregates will experience increasingly less preferential concentration. Because of the growth, the aggregates formed by particles with $St_p = 0.7$ approach faster $St_a \approx 1$ than those of $St_p = 0.56$, experiencing then preferential concentration. The aggregates formed with particles of $St_p \approx 1$ gradually stop experiencing preferential concentration.

Finally, the third regime concerns the long-term evolution where the higher collision rates are found when $St_p = 0.56$. This can be explained as the result of the rate of change in St_a. Equation (16) shows that the rate of change of St_a with the number of particles is proportional to St_p this implies that for $St_p = 0.56$, when $St_a \sim 1$ it will remain there for longer times compared with aggregates formed by particles with bigger St_p.

The collision rates can be divided in two components: the aggregate-particle or aggregate-aggregate collision, both denoted by (APC), and the particle-particle collision (PPC). This can be observed from the evolution of the particle size distribution: if APC is dominant big aggregates should be formed very fast; otherwise if PPC is dominant then many small aggregates will be formed rapidly. Further work is needed to clarify under what conditions APC or PPC are dominant.

In principle, the most important parameter defining the interaction among particles is the Stokes number. This implies that in the first stage the collision rates will be dominated by PPC, where $St_1 \approx 1$, and after some time they will be dominated by APC, when the aggregates reach $St_a \approx 1$.

4.3 Effect of fractal dimension on aggregate formation

To explore the effects of the fractal dimension of the aggregates (D_f) three experiments were performed, see table 3.

The numerical experiments are performed with $Re_\lambda \sim 70$, $\phi = 10^{-5}$ and particle Stokes number $St_p = 1.1$. Only the porosity, accounted for by D_f, is changed. The results are shown in figure (4).

Once again figure (4) shows the initial peak that can be explained as before as an effect of the preferential concentration of the particles with $St_p = 1.1$. The long-term evolution where the experiment with $D_f = 2.2$ dominates in the number of collision can be explained by two factors. The size of the

Table 3. Experiment parameters exploring the effects of D_f on aggregate formation. For all runs $Re_\lambda = 70$, $\phi = 10^{-5}$ and $St_p = 1.1$.

	$D1$	$D2$	$D3$
D_f	3.0	2.5	2.2

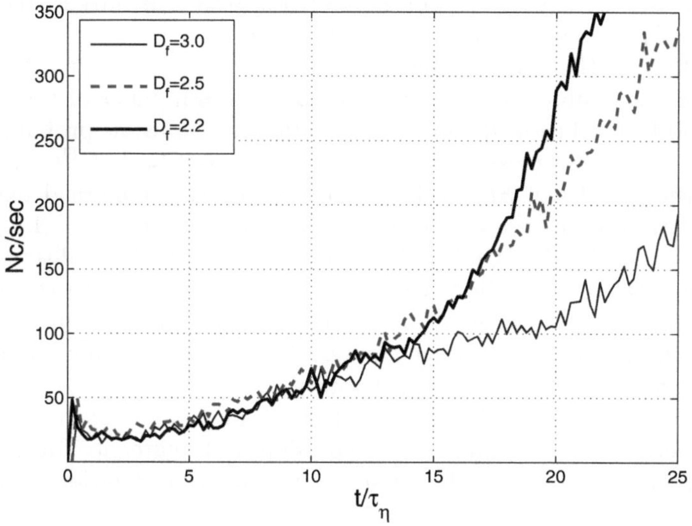

Fig. 4. Effect of D_f on collision rates and aggregate formation.

aggregates grows faster when D_f is decreased, bigger particles have more probability to interact with other particles. The second factor is the rate of growth of $St_a \sim n^{1-1/D_f}$, i.e. given St_p, decreasing D_f (increasing porosity) will result in a slower increase rate of St_a, thus resulting in the aggregates experiencing preferential concentration for larger times.

5 Conclusions

In this work we report on the effects of the Stokes number of the particles on the encounter rates of non-interacting particles, on the effects of the initial Stokes number of aggregating particles and the effects of porosity (fractal growth) on the collision rates.

In spite of the fact that the volume fraction is very small, the number of collisions between particles is high for particles with $St \simeq 1$. This can be explained by the phenomena of preferential concentration: it collects the particles in localized regions of the domain promoting collisions between them.

The effects of St_p on aggregate formation rates present three regimes: In the initial regime the higher collision rates are found in particles whose

$St_p \simeq 1$; in the long term, the higher collision rates are found when the aggregates are formed by particles with $St_p \simeq 0.5$; and in the intermediate stage the collision rates are dominated by the particles whose $St_p \simeq 0.7$. All these results are explained by the preferential concentration effects when either St_p or St_a are close to one. When $St_p < 1$ the Stokes number of the aggregates will approach one and will experience preferential concentration. When $St_p \simeq 1$ the aggregate Stokes number will become larger than one and the effect of preferential concentration is reduced.

The effect of the fractal growth (porosity) on the aggregate formation is indirect, in particular a modification of the Stokes number of the aggregates and a modified collision diameter (also affecting collision probability). The result for the Stokes number is that St_a grows as $St_p n^{1-1/D_f}$. If D_f is decreased, the growth of St_a is decreased. In our the numerical experiments performed to explore the effects of D_f, particles with $St_p = 1.1$ are used, and three values for D_f. Under these conditions the higher collision rates are found when $D_f = 2.2$. This is the result of small St_a growth rates near St_p implying that St_a stays longer near one, experiencing the effects of preferential concentration for prolonged times. Additionally, higher d_a (lower D_f) promotes an increased collision probability.

This work only considered the effect of inertia on encounter rates of non-interacting particles and on the collision and aggregate formation rates of interacting particles. Next step is to modify the equation of motion of the particles adding buoyancy and added mass forces, see equation (4). This will help us to consider realistic conditions for plankton systems where the density of the particles is close to the density of the fluid, $\rho_p \sim \rho_f$.

Acknowledgments

The program is founded by the Netherlands Organization for Scientific Research (NWO) and Technology Foundation (STW) under the Innovational Research Incentives Scheme grant ESF.6239.

References

[1] Kiørboe, T. (1997) Scientia Marina 61:141-158.
[2] Wang, L-P. and M.R. Maxey, (1993) J. Fluid Mech. 256:27-68.
[3] Fung, JCH (1990) Kinematic Simulation of Turbulent Flow and Particle Motions. PhD Thesis, University of Cambridge, Cambridge
[4] Meakin, P. (1988) Adv. Colloid Interface Sci. 28:249-331.
[5] Squires, K. D. and H. Yamazaki, (1995) Deep Sea Res. 42:1989-2004.
[6] Reade, K. D. and L. R. Collins, (2000) J. Fluid Mech. 415:45-64.

[7] Heath A.R, P .A. Bahri, P. D Fawell and J. B. Farrow (2006) AIChE J 52:1987-1994
[8] Jackson G. A. (1990) Deep-Sea Res 37:1197-1211
[9] Squires KD, Eaton JK (1991) Phys Fluids 3:1169-1178.
[10] Fessler, J. R., J. D. Kulick, and J. K. Eaton (1994) Phys. Fluids 6:3742-3749.
[11] Maxey M. R. and Riley J. (1983) Phys Fluids 26:883-889
[12] Fung, J. C. H., Hunt, J. C. R., Malik, N. A. and Perkins, R. J. (1992) J Fluid Mech 236:281-318.
[13] Kraichnan, R. H. (1970) Phys Fluids 13:22-31.

Influence of the turbulence structure on the particle sedimentation in wall-bounded flows

M. Cargnelutti and L.M. Portela

Delft University of Technology, MSP, Prins Bernhardlaan 6, 2628 BW Delft, The Netherlands. M.F.Cargnelutti@TNW.TUDelft.nl

Summary. In this paper we analyze the influence of the particle Froude number in the settling velocity of particles in wall-bounded turbulence. We consider dilute situations using point-particle direct numerical simulations, neglecting the modifications of the flow by the particles (one-way coupling). We compare our results with the cases of particles settling around a single vortex and a vortex array. The behavior of the settling velocity in the wall-bounded turbulence case is found to be similar to the one of particles settling in a vortex array with a "large" inter-vortex distance. We analyze how the particle settling velocity is related to the particle-fluid two-point velocity correlations.

1 Introduction

The transport of sediment in wall-bounded turbulent flows is important in many engineering applications. This situation happens in numerous scenarios, from the motion of sand in the ocean, to the mixing of microscopic particles in a pipe of a chemical plant. A common situation involves the transport and sedimentation of sand-like particles in turbulent water-flows. In this case, the sedimentation of the particles depends strongly on the interaction between the particles and the turbulence, and there is a lack of good understanding about it. It is important to get more physical insight of the situation, using both numerical simulations and experiments [1]. This physical insight can be used to develop physically-based models for sediment transport.

It is known that in a turbulent flow the non-dimensional particle settling velocity (V_s) is different than in a stagnant fluid. In the past, some work has been done in simplified situations, trying to understand this phenomenon. Hunt and Davila [3] considered the motion of particles around a single vortex, and they found that V_s can be described by two dimensionless parameters: V_t, the non-dimensional particle settling velocity in stagnant fluid, and F_p, the particle Froude number. F_p can be interpreted as the ratio of the inertial forces experienced by a particle to the buoyancy forces. They showed that,

Bernard J. Geurts et al. (eds), Particle Laden Flow: From Geophysical to Kolmogorov Scales, 373–384.

depending on F_p, V_s can be higher (up to 80% for $F_p \sim 1$) or lower (up to 20% for $F_p \sim 10$) than V_t.

Some authors extended the analysis of Hunt and Davila for a group of vortices. De Ronde [2] found that in a symmetric vortex array, the settling velocity depends not only on V_t and F_p, but also on the vortex separation, D. He found that, for $F_p \sim 0.1$, depending on the value of D, V_s can be higher or lower than the settling velocity in a stagnant medium. For $F_p < 0.01$ and $F_p > 1$, he found the value of V_s is approximately the same regardless of the value of D. Eames and Gilbertson [4] analyzed the settling and dispersion of particles by spherical vortices. They showed that the use of V_t and F_p is not enough to properly describe the physical situation, and that other parameters are needed, in order to take into account the distribution of the vortices.

The work of De Ronde and Eames and Gilbertson suggest that the turbulence structure can play an important role in determining V_s. In their work, the influence of F_p and other parameters on the value of V_s was considered for simplified situations. In this work, we study how the falling velocity of particles in wall-bounded turbulent flows is affected by V_t, F_p and the turbulence structure, using DNS simulations of a particle-laden turbulent open-channel flow. We use a standard point-particle approach, where the forces acting on the particle are determined from the fluid velocity interpolated at the center of the particle [6]. We consider dilute situations and neglect the modification of the flow by the particles (one-way coupling).

In this paper, we present a quantitative systematic analysis of the variation of V_s with V_t and F_p, and its relation with the turbulence structure. In section 2 there is a description of the particle equation of motion, with the forces considered in the simulations. In section 3 there is a discussion about the particle Froude number. In section 4 and 5 there is an analysis of the sedimentation of particles around a single vortex and a vortex array. In section 6, the sedimentation of particles in an open-channel flow is presented. Finally, in section 7 we give some concluding remarks.

2 Particle equation of motion

The forces acting on a particle immersed in a flow were described by Maxey and Riley[5]. In this work, in order to keep the situation as simple as possible, we consider only Stokes drag, gravity, added mass and the surrounding fluid stress. Then, the motion of a particle is described by:

$$\frac{d\tilde{\mathbf{v}}}{d\tilde{t}} = \underbrace{\frac{18\,\tilde{\mu}}{\tilde{\rho}_p\,\tilde{d}_p^2}(\tilde{\mathbf{u}}-\tilde{\mathbf{v}})}_{drag} + \underbrace{\tilde{\mathbf{g}}}_{gravity} + \underbrace{\frac{1}{\tilde{\rho}_p}\left(\tilde{\nabla}\cdot\tilde{\mathbf{T}}\right)}_{stresses} + \underbrace{\frac{1}{2}\frac{\tilde{\rho}_f}{\tilde{\rho}_p}\left(\frac{D\tilde{\mathbf{u}}}{D\tilde{t}} - \frac{d\tilde{\mathbf{v}}}{d\tilde{t}}\right)}_{added\,mass} \tag{1}$$

were, $\tilde{\mathbf{v}}$ is the particle velocity, $\tilde{\rho}_p$ and $\tilde{\rho}_f$ are the particle and fluid density, respectively, $\tilde{d}_p$ is the particle diameter, $\tilde{\mu}$ is the fluid viscosity, $\tilde{\mathbf{g}}$ is the gravity acceleration, $\tilde{\mathbf{u}}$ and $D\tilde{\mathbf{u}}/D\tilde{t}$ are the surrounding fluid velocity and acceleration, respectively, at the particle position, and $\tilde{\mathbf{T}}$ is the fluid stress tensor (throughout this paper, the superscript $\sim$ is used to denote dimensional quantities). Neglecting the influence of the particles on the fluid, the acceleration of the surrounding fluid is given by:

$$\frac{D\tilde{\mathbf{u}}}{D\tilde{t}} = \frac{\partial \tilde{\mathbf{u}}}{\partial \tilde{t}} + \tilde{\mathbf{u}} \cdot \left(\tilde{\nabla}\tilde{\mathbf{u}}\right) = \frac{1}{\tilde{\rho}_f}\left(\tilde{\nabla}\cdot\tilde{\mathbf{T}}\right) + \tilde{\mathbf{g}} \tag{2}$$

Finally, we can express the particle equation of motion as:

$$\frac{d\tilde{\mathbf{v}}}{d\tilde{t}} = \frac{\beta-1}{\beta+\frac{1}{2}}\frac{1}{\tilde{\tau}_p}\left(\tilde{\mathbf{u}}-\tilde{\mathbf{v}}\right) + \frac{\beta-1}{\beta+\frac{1}{2}}\tilde{\mathbf{g}} + \frac{3}{2}\frac{1}{\beta+\frac{1}{2}}\frac{D\tilde{\mathbf{u}}}{D\tilde{t}} \tag{3}$$

where $\beta = \tilde{\rho}_p/\tilde{\rho}_f$ is the particle-fluid density ratio, and $\tilde{\tau}_p = (\beta-1)\,\tilde{d}_p^2/(18\tilde{\nu})$ is the particle relaxation time.

3 Particle Froude number

Using the equations from section 2 we can define two parameters which are important for the following analysis: the particle terminal velocity in a stagnant fluid and the particle Stokes number. From equation 3, the terminal velocity in a stagnant fluid is $\tilde{V}_t = \tilde{\tau}_p\,\tilde{g}$. Considering a relevant fluid velocity scale, $\tilde{U}_f$, we define the dimensionless terminal velocity in a stagnant fluid as $V_t = \tilde{V}_t/\tilde{U}_f$. The second parameter is the particle Stokes number, which is the ratio between the particle relaxation time $\tilde{\tau}_p$, and some representative time scale of the fluid, $\tilde{\tau}_f$: $St = \tilde{\tau}_p/\tilde{\tau}_f$.

As can be seen from equation 3, the motion of the particles depends on the flow field, the density ratio, the particle relaxation time and the acceleration of gravity. For heavy particles ($\beta >> 1$), besides the flow field, the motion of the particles depends only on the particle relaxation time and the acceleration of gravity; i.e., for a given flow field, the particle motion is completely determined by two non-dimensional parameters: St and V_t. Davila and Hunt [3] suggested that the motion of particles near a vortex can be better understood by combining these two parameters into a single parameter, the particle Froude number, F_p, defined as $F_p = V_t^2 St$. F_p can be interpreted as a rescaled Stokes number, and is a measure of the ratio of the inertial forces experienced by a particle to the buoyancy force. In the following sections we present an analysis of the influence of the particle Froude number on the settling velocity of particles for different flow fields.

4 Single vortex

Davila and Hunt performed an asymptotic analysis of the influence of the particle Froude number in the settling velocity of particles around a single

Rankine-like vortex, like the one presented in figure 1. They showed that V_s can be higher or smaller that V_t, depending on the value of F_p.

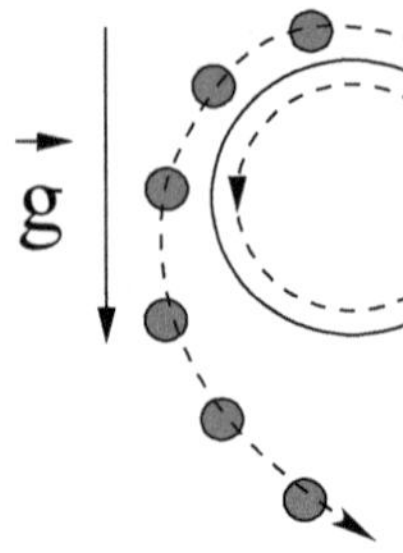

Fig. 1. Single vortex schematic representation. The gray circles represent the trajectory of a particle settling down around the vortex center.

Similarly, de Ronde [2] analyzed the settling of particles around a Rankine-like using numerical simulations in a finite domain. As in the case of Davila and Hunt, the flow field was constant in time. The maximum circumferential velocity of the vortex, $\tilde{U}_{max}$, and the vortex strength $\tilde{\Gamma}$ were used to make the different quantities dimensionless. The non-dimensional settling velocity was defined as $V_t = \tilde{V}_t/\tilde{U}_{max}$, and the Stokes number as $St = \tilde{\tau}_p/\tilde{\tau}_f$, were $\tilde{\tau}_f = \tilde{\Gamma}/\tilde{U}^2_{max}$ is the fluid time-scale. As explained in section 3, the particle Froude number is $F_p = V_t^2 St$.

In the simulations, De Ronde used a constant non-dimensional particle terminal velocity, equal to $V_t = 0.1$, and a constant particle diameter equal to $d_p = 10^{-3} R_v$, where $\tilde{R}_v = \tilde{\Gamma}/\tilde{U}_{max}$ is the vortex radius. The particles were released at the top of the vortex, distributed in a line, with an equal number of particles on each side of the vortex center. This line of particles was released at two different distances d from the vortex center: $d = 2R_v$ and $d = 4R_v$. The values of d were chosen in order to compare the results of a single vortex with a vortex array (see section 5). The situation is presented in figure 2.

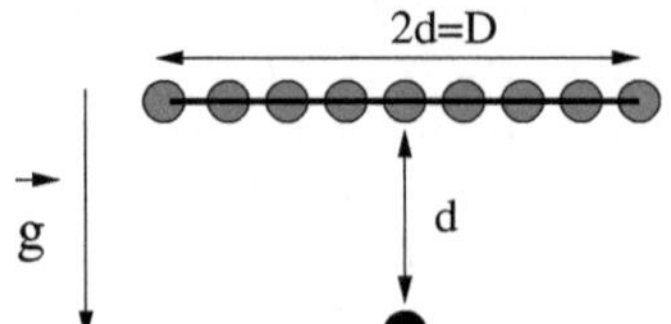

Fig. 2. Initial particle distribution for the single-vortex cases. The black circle represents the center of the vortex. The gray circles represent the particles at their initial positions.

The particles were released with an initial downward velocity equal to V_t. As the particles fell down trough the vortex field, if a particle was on the the left hand side of the vortex (see figure 2), both the gravity and the vertical fluid-velocity pointed in the same direction, and the particle vertical velocity increased. If a particle was on the right hand side of the vortex, the vertical velocity of the fluid was opposite to gravity, and the particle velocity decreased. Therefore, if a particle was on the right hand side of the vortex, it could reach a position where its vertical velocity was zero. At this point, its acceleration was opposite to gravity, and the particle was ejected to the top of the vortex, a phenomenon called the "over-shooting effect".

The average particle settling velocity V_s was determined by computing the time it took for a particle to fell down a fixed distance L from the vortex center. The distance L was set equal to $L = 50\,R_v$, large enough to capture the "over-shooting effect". The average value of V_s, as a function of F_p, is presented in figure 3.

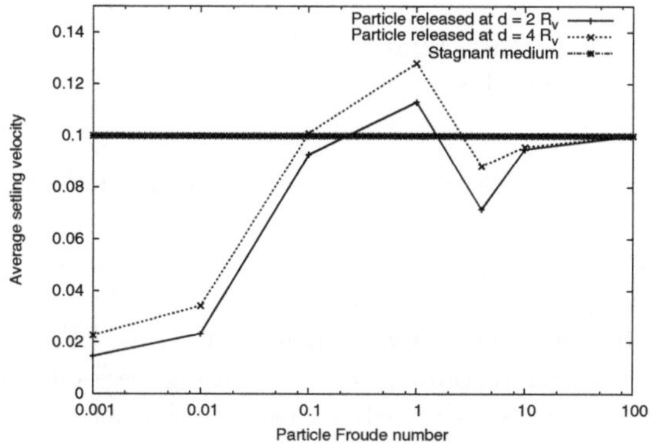

Fig. 3. Average settling velocity for a single vortex as a function of the particle Froude number.

The qualitative behavior of both cases presented in figure 3 is the same, but the values are a bit different. For $F_p << 1$, the particle settling velocity was smaller than V_t, because the particles found regions where the gravity was in equilibrium with the flow velocity, and then they "over-shooted" to the top of the vortex. As a consequence, the particles followed a longer trajectory, and the average settling velocity was smaller that V_t.

For $F_p >> 1$ the particles ignored the presence of the flow, and $V_s \sim V_t$. Finally, when F_p was of the order of unity, V_s could be either higher or slower than V_t.

5 Vortex array

De Ronde analyzed also the particle sedimentation in a vortex array. A domain of counter-rotating vortices was constructed, where each vortex rotated in an opposite direction to its closest neighbor. The situation is plotted in figure 4.

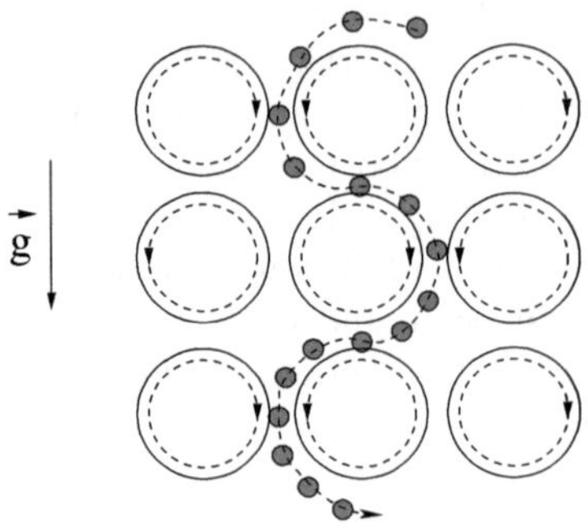

Fig. 4. Vortex array schematic representation. The gray circles represent the trajectory of a particle settling down.

Similarly to the single vortex case, the non-dimensional parameters were defined using the maximum velocity $\tilde{U}_{max}$, and the strength $\tilde{\Gamma}$ of one vortex.

In the simulations, the particles were distributed in a line in the central region between two rows of vortices (see figure 5). Two different values for the inter-vortex distance, D, where considered: $D = 4\,R_v$ and $D = 8\,R_v$. In both cases, the non-dimensional particle terminal velocity was set equal to $V_t = 0.1$. Similarly to the single vortex case, the diameter of the particles was set equal to $d_p = 10^{-3}\,R_v$.

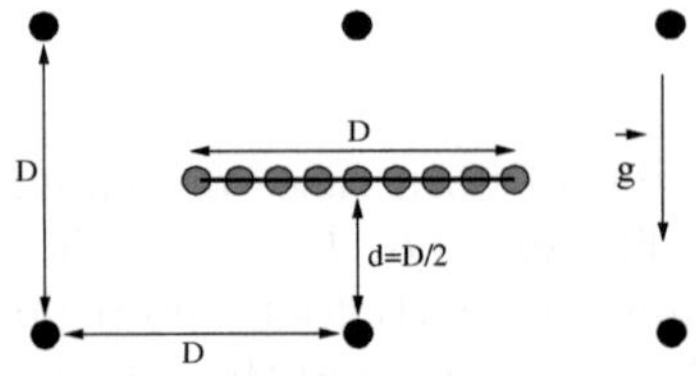

Fig. 5. Initial particle distribution for the vortex array cases. The black circles represent the center of the vortices. In the middle of a vortices row, the gray circles represent the particles at their initial positions. An equal number of particles was used on each side of the middle vortex center.

The particles were released with an initial downward velocity equal to V_t. The average settling velocity V_s was determined as in the previous section, by computing the time it took for a particle to travel a distance $L = 50\,R_v$. In order to make sure that the particles felt an infinite vortex array, the domain was big enough such that each particle was always surrounded by at least five vortices in each direction. The results for different values of F_p are presented in figure 6.

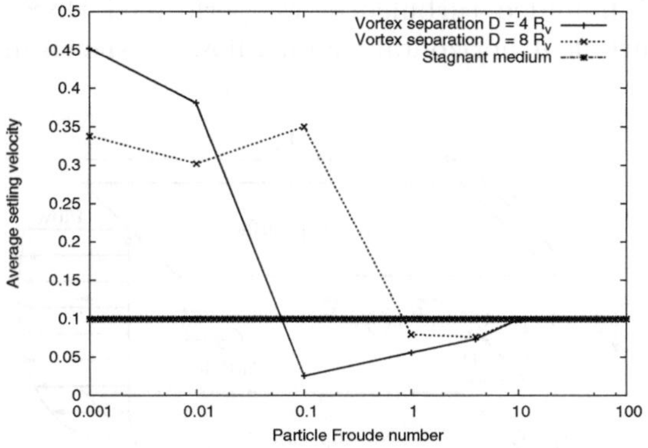

Fig. 6. Average settling velocity for a vortex array as a function of the particle Froude number.

When the particles were released, they started to fell down due to gravity. The trajectories they followed depended on the particle Froude number. For $F_p << 1$ the particles were ejected to the vortex edges and they tended to follow the flow streamlines. However, since they had a fixed terminal velocity, they could feel the effect of gravity, and they did not followed exactly the fluid streamlines. When the particles reached an stagnation point, they jumped to the next flow streamline and kept on traveling downward, as shown in figure 4. In this way, the trajectories they followed were longer than in a stagnant medium, however, the speed they gained was big enough to make the average settling velocity higher than V_t, for both values of the vortex separation D.

For $F_p >> 1$, as in the single vortex case, the particles ignored the presence of the flow and fell down at a velocity almost equal to V_t.

When F_p was of the order of one, the particles had more inertia, and they did not follow exactly the streamlines. Instead, they were driven mainly to regions of higher flow velocity. In this particular configuration, the velocity of the fluid was higher in between the vortices, either upward or downward, at an angle of 45 degree. Therefore, the particles preferentially followed regions of fluid either going upward or downward. The important parameter here was the distance between the vortices D. For $F_p = 0.1$, if $D = 8\,R_v$, the particles moved mainly in regions were the flow was going down. On the other hand, when $D = 4\,R_v$, the particles moved either up or down [2]. As a result, the average particle settling velocity could be either higher ($D = 8\,R_v$) or smaller ($D = 4\,R_v$) than V_t, as can be observed in figure 6.

6 Channel flow simulations

In order to analyze the influence of the particle Froude number in the settling velocity for wall-bounded turbulence, we performed DNS simulations of a particle-laden turbulent horizontal channel flow, as shown in figure 7.

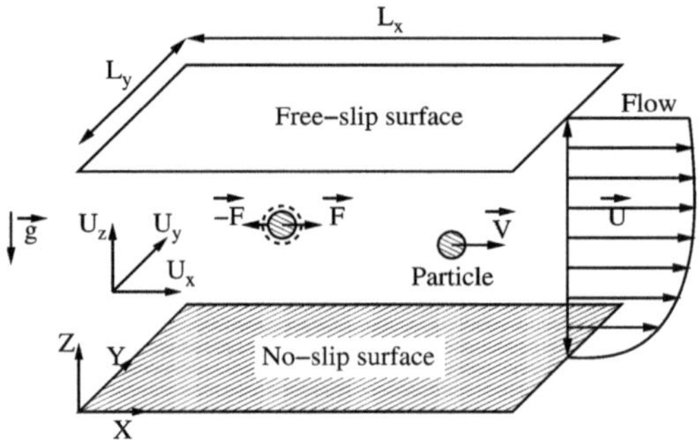

Fig. 7. Scheme open-channel flow.

We imposed a free-slip boundary condition at the top wall, and a no-slip condition at the bottom-wall. In the streamwise and spanwise directions we used periodic boundary conditions. The flow was driven by a streamwise pressure gradient. The code used a standard finite-volume approach on a staggered grid, were the continuous phase was solved using a predictor-corrector scheme, with a second-order Adams-Bashforth time discretization. The time step was obtained using the Courant stability criterion. For the discrete phase, we integrated the particle equation of motion using a explicit method. To calculate the fluid velocity at the particle position we used a tri-linear interpolation. Further details can be found in [6].

The equations of motion were made dimensionless using the friction velocity $\tilde{u}_\tau$ and the height of the channel, $\tilde{H}$. With this parameters, the Reynolds number was set equal to $Re_\tau = \frac{\tilde{u}_\tau \tilde{H}}{\tilde{\nu}} = 500$, where $\tilde{\nu}$ is the kinematic viscosity of the fluid. The non-dimensional particle terminal velocity was defined as $V_t = \tilde{V}_t/\tilde{u}_\tau$, and the Stokes number number as $St = \tilde{\tau}_p/\tilde{\tau}_f$, where the fluid time scale $\tilde{\tau}_f$ was defined as $\tilde{\tau}_f = \tilde{u}_\tau^2/\tilde{\nu}$

The size of the computational domain was $5\,H$ and $2\,H$ in the streamwise (x) and spanwise (y) directions, respectively. We used a grid of 256 x 192 x 128 nodes, which gave us a domain of $x^+ = 2500$, $y^+ = 1000$ and $z^+ = 500$, where the + superscript denotes wall units. The grid was uniform in the x and y directions, with $\Delta x^+ \sim 10$ and $\Delta y^+ \sim 5$. A hyperbolic-tangent stretching was used for the normal direction, with $\Delta z^+ \sim 0.9$ at the wall, and $\Delta z^+ \sim 7$ at the center of the channel.

The particles were released homogeneously distributed in a plane at a distance $z = 0.9\,H$ from the bottom of the channel, which corresponds to $z^+ = 450$, with an initial vertical velocity equal to $V_t = 0.1$. For each particle, we computed the time it took to travel: (*i*) from $z^+ = 450$ to $z^+ = 250$ (center of the channel), (*ii*) from $z^+ = 250$ to $z^+ = 50$ (buffer region), and (*iii*) from $z^+ = 50$ to $z^+ = 3$.

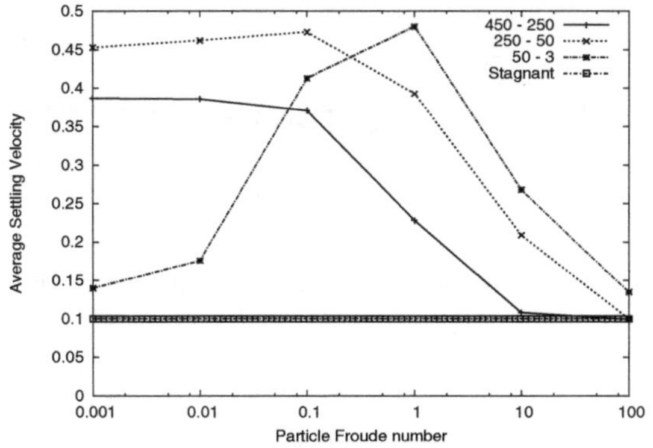

Fig. 8. Average settling velocity for an open-channel as a function of the particle Froude number.

The results for different particle Froude numbers, are presented in figure 8. When the particle Froude number was smaller than 1, and when the particles were falling down between $z^+ = 450$ and $z^+ = 250$, and between $z^+ = 250$ and $z^+ = 50$, the average settling velocity V_s was higher than V_t. In this case, the relation between V_s and F_p is somehow similar to the case of a vortex array where the vortex distance is "large" ($8R_v$), with an almost monotonic decrease in the average settling velocity as F_p increases. On the other hand, in the near-wall region, there is a maximum in the average settling velocity at $F_p \sim 1$. In the vortex array case we saw that for "intermediate values" of F_p, the average settling velocity had a strong dependence on the vortex spacing, with a more complex behavior when the vortex spacing was smaller. Near the wall the streamwise vortices play an important role and their spacing is smaller than further away from the wall [6]. This could be a possible explanation for the behavior near the wall. However, the behavior is quite different from the "compact vortex array" ($D = 4\,R_v$), and contrary to the vortex array V_s is always higher than V_t. Clearly, the turbulence structure appears to play an important role in determining the settling velocity.

In order to quantify the importance of the turbulence structure on the particle motion, we analyzed the particle-fluid two-point velocity correlations.

In figures 9 and 10 are plotted, respectively, the spanwise and normal-wise particle-fluid velocity correlation.

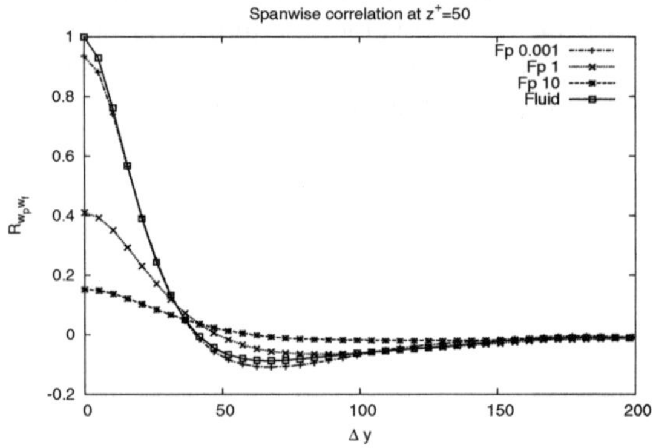

Fig. 9. Particle-fluid vertical velocity two-point spanwise correlation.

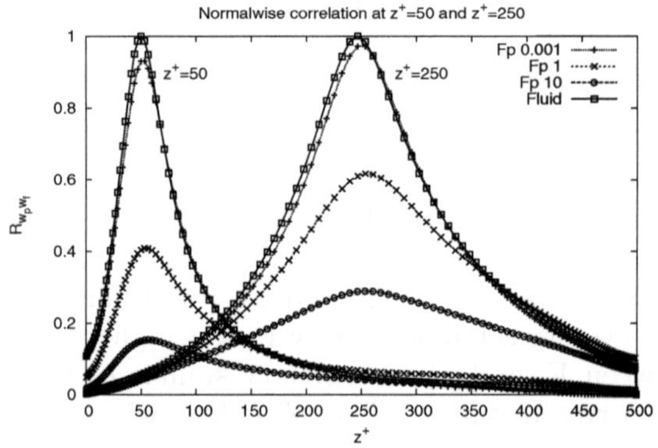

Fig. 10. Particle-fluid vertical velocity two-point normal-wise correlation.

In the spanwise correlation plots, for the fluid auto-correlation at $z^+ = 50$, there is a minimum around $\Delta y^+ = 60$, which can be seen as a measure of the vortices diameter. Even though the particle-fluid correlation is in general smaller than the fluid auto-correlation, for the smallest values of F_p we notice than the particle-fluid correlation is higher at $\Delta y^+ \sim 60$. This seems to indicate than the effect of the fluid structures on the spanwise direction persist in time. On the other hand, when $F_p >> 1$, the velocity correlation is almost zero for all values of Δy^+, which means that the particles ignored the presence of the turbulence and fell down with a velocity equal to V_t.

In the normal-wise velocity correlations (figure 10) it can be seen that the loss of correlation is not the same in the central part of the channel as in the near-wall region. For example, for $F_p = 1$ the correlation is larger at $z^+ = 250$ than at $z^+ = 50$. This seems to indicate that the particles tend to follow in a stronger way the larger fluid structures at the center of the channel than the smaller structures closer to the channel wall.

In figure 10 we can also note that in both regions (center of the channel and near wall region), there is an asymmetry in the correlations. The particles seem to correlate more with the structures close to the top of the channel than with those structures close to the bottom. This effect is more pronounced for $F_p < 1$, where the particle-fluid correlation at $z^+ = 250$ can be even higher in the top part of the channel than the fluid auto-correlation. This seems to indicate that the particles feel more the presence of the fluid structures from the top of the channel than from below, and that they keep a "memory" of the fluid structure above them.

7 Conclusions

Clearly, the turbulence structure appears to play an important role in determining the settling velocity in wall-bounded turbulence. Far from the wall the behavior is somehow similar to a vortex array with a "large" vortex spacing. Near the wall, the behavior is more complex and a maximum in the settling velocity is found for $F_p \sim 1$.

The precise mechanisms through which the turbulence structure influences the settling velocity are still not clear. However, a preliminary analysis of the two-point fluid-particle correlation shows that the particles "feel" the normal-wise and spanwise velocity correlation and appear to keep a "memory" of the fluid structure above them.

Acknowledgments

We gratefully acknowledge the financial support provided by STW, WL—Delft Hydraulics and KIWA Water Research. The numerical simulations were performed at SARA, Amsterdam, and computer-time was financed by NWO.

References

[1] W.A. Breugem and W.S.J. Uijttewaal. Sediment transport by coherent structures in a horizontal open channel flow experiment. Proceedings of the Euromech-Colloquium 477, to appear

[2] W.H. de Ronde. Sedimenting particles in a symmetric array of vortices. BSc Thesis, Delft University of Technology, 2005

[3] J. Davila and J.C. Hunt. Settling of small particles near vortices and in turbulence. Journal of Fluid Mechanics, 440:117-145, 2001

[4] I. Eames and M.A. Gilbertson. The settling and dispersion of small dense particles by spherical vortices. Journal of Fluid Mechanics, 498:183-203, 2004

[5] M.R. Maxey and J.J. Riley. Equation of motion for a small rigid sphere in a nonuniform motion. Physics of Fluids, 26(4):883-889, 1983

[6] L.M. Portela and R.V.A. Oliemans. Eulerian-lagrangian dns/les of particle-turbulence interactions in wall-bounded flows. International Journal of Numerical Methods in Fluids, 9:1045-1065, 2003.

Mean and variance of the velocity of solid particles in turbulence

Peter Nielsen

Dept Civil Engineering, The University of Queensland, Brisbane Australia
p.nielsen@uq.edu.au

Summary. Even the simplest velocity statistics, i. e., the mean and the variance for particles moving in turbulence still offer challenges. This paper offers simple conceptual models/explanations for a couple of the most intriguing observations, namely, the enhanced settling rate in strong turbulence and the reduced Lagrangian velocity variance for even the smallest of sinking particles. While simultaneous experimental observation of the two effects still do not exist, we draw parallels between two classical sets of experiments, each exhibiting one, to argue that they are two sides of the same phenomenon: Selective sampling due to particle concentration on fast tracks like those illustrated by Maxey & Corrsin (1986).

1 Settling in strong turbulence

Figure 1 shows comprehensive experimental data on mean vertical velocity $\overline{w}$, i. e., the settling or rise velocity of particles with still water settling/rise velocity w_o in turbulence with vertical rms velocity w'.

The settling/rise delay at moderate turbulence strength, $0.3 < w'/w_o < 3$, can be understood in terms of vortex trapping. Vortex trapping was shown experimentally by Tooby et al. (1977), see their magnificent stroboscopic photo showing a heavy particle and bubbles trapped in the same vortex. The trapped particles move in closed orbits analogous to those of the fluid but offset horizontally. Heavy particles thus move predominantly in the upward moving fluid while light particles and bubbles move predominantly in the downward moving fluid. Closed sediment/bubble paths result from the simple superposition law $\mathbf{u}_p = \mathbf{u}_f + \mathbf{w}_o$ which is a good approximation as long as the flow accelerations are small compared with g, see, e. g., Nielsen (1992) p 182. Non-linear drag may also cause a settling delay. However, this effect is very weak. It's magnitude A may be estimated as

$$A < \frac{w_o}{4}\left(\frac{du_p}{dt}/g\right)^2$$

Bernard J. Geurts et al. (eds), Particle Laden Flow: From Geophysical to Kolmogorov Scales, 385–391.

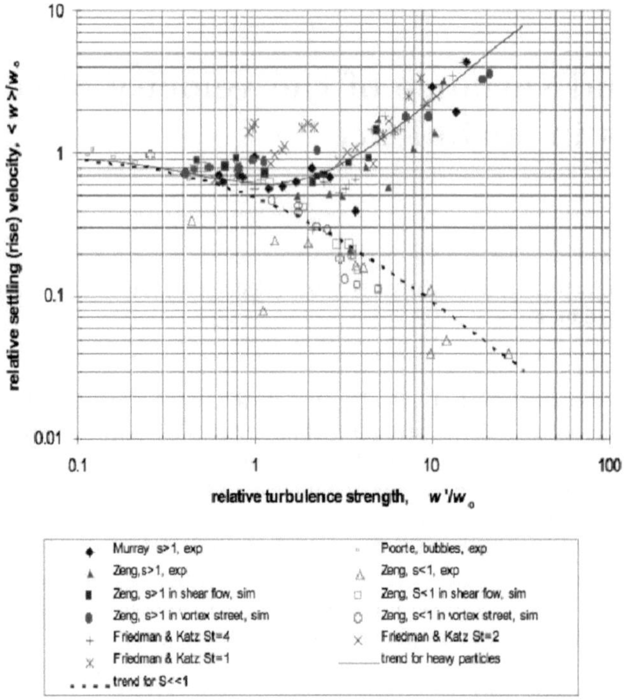

Fig. 1. Measured (exp) and simulated (sim) settling velocities of dense particles (solid symbols) and rise velocities of light particles and bubbles (open symbols), and rise velocities of diesel droplets ($+, *, \times$) in water.

in most natural scenarios. To measure the non-linear drag effect one must thus use a 'flow' free of trapping vortices like the vertically oscillating jar of Ho (1964).

2 Accelerated or delayed settling/rise in strong turbulence

While the data in Figure 1 indicate that light and heavy particles are similarly delayed by turbulence of moderate strength, $0.3 < w'/w_o < 3$, the effects of strong turbulence are qualitatively different depending on particle density. Broadly speaking, heavy particles are accelerated asymptotically for $w'/w_o \rightarrow \infty$, while light particles are increasingly delayed by stronger turbulence. The intriguing thing is that the critical particle density separating delay from acceleration is not $\rho_p = \rho_f$. That is, the diesel droplets of Friedman & Katz (2002) while lighter than the surrounding water are accelerated by strong turbulence like the heavy particles of Murray (1970) and others.

To get a qualitative understanding of the accelerated settling of heavy particles in turbulence it is helpful to consider the cellular flow field in Fi-

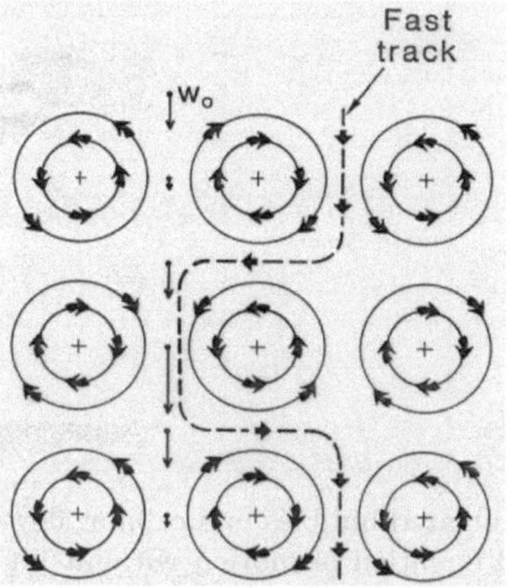

Fig. 2. In a field of vortices, heavy particles will, spiral outwards and become concentrated on the 'fast tracks' along the vortex boundaries.

gure 2. Maxey & Corrsin (1986) showed that dense particles initially uniformly distributed in such a velocity field, will after a while, end up on the 'fast track' and experience enhanced settling.Based on this scenario, Nielsen (1993) suggested the asymptotic relation:

$$\overline{w} \approx 0.4 w_o \quad for \quad w' \gg w_o \tag{1}$$

While heavy particles spiral out, light particles and bubbles will generally spiral towards the neutral or stationary point given by $\mathbf{u}_f = -\mathbf{w}_o$. This inward spiraling and ensuing stable trapping corresponds to the descending curve in Figure 1, i.e., stronger rise-delay with increasing turbulence intensity. This inward spiraling might thus lead to the expectation that all light particles and bubbles plot along the descending curve in Figure 1. However, curiously, the 'diesel droplets in water results' of Friedman & Katz show an increasing trend similar to (1) except that they recommended the factor 0.25 in stead of 0.4.

An explanation for this enhanced rise velocity for some light particles might be the 'rising fast tracks' in Figures 5 and 6 of Maxey (1990). Based on a simplified equation of motion, excluding lift forces and the Basset history term, Maxey found that bubbles, which were initially uniformly distributed on the cellular flow field, would after a long time, either spiral into the stationary points or move along rising fast tracks. The rising fast tracks are in fact, within each cell, pieces of inward spirals towards the stationary points, see Figure 3.

A set of unique rising fast tracks like those in Figure 3 probably exist within a certain domain of the $(U_{max}/w_o, g/(w_o\omega))$-plane, where U_{max} is the maximum velocity in the flow field and ω its angular velocity. Determining this domain by further simulations (or analysis) might lead to an understanding of the parameter ranges within which accelerated rise of light particles like the

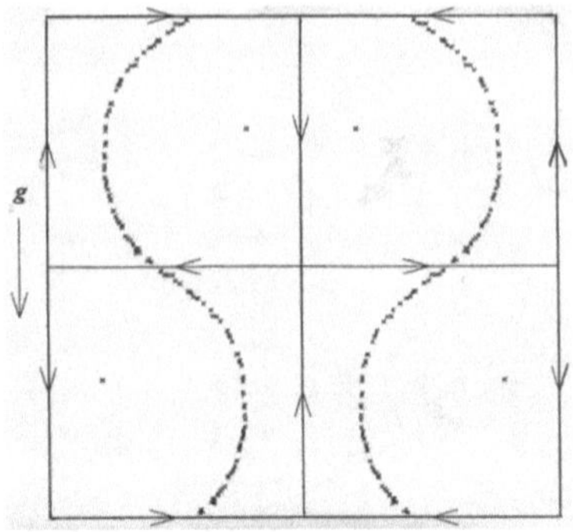

Fig. 3. Pattern of concentrated bubbles in a cellular flow field calculated by Maxey (1990) using a simplified equation of motion without lift forces and Basset history term. The bubbles were initially uniformly scattered. The isolated 'bubble' in each cell is at the stable neutral point, where $\mathbf{u}_f = -\mathbf{w}_o$, into which a great number of particles have actually converged. The curves are rising fast tracks which are pieced together from arcs, which within each cell are inward spirals towards the neutral point.

diesel droplets of Friedman & Katz may occur. A complete understanding may also require consideration of lift forces although the fast tracks predominantly occupy areas of low velocity shear and correspondingly weak lift forces.

3 Velocity variance for suspended particles

The velocity variance offers a long standing conundrum raised by Snyder & Lumley (1971) (S&L). After carefully designing their smallest particle to follow the fluid perfectly (for all practical purposes), they still found

$$Var(w_p) \approx 0.6 Var(w_{Eulerian}) \tag{2}$$

see Figure 4. That is, the particle's Lagrangian velocity variance was significantly smaller than the fluid velocity variance observed by a fixed probe. S&L were at a loss to explain this reduction. Apparently, they expected the Lagrangian variance from the particles to be the same as the Eulerian one from the fixed probe. However, while that identity would hold for any pair of point statistics for fluid particles in an incompressible fluid, there should be no such expectation, where disperse suspended particles are con-cerned. Disperse particles do not behave as an incompressible fluid, and their one point statistics need not be the same as those of the fluid.

A qualitative explanation for $Var(w_p) \approx 0.6Var(w_{Eulerian})$ can again be based on the tendency for heavy particles to become concentrated in certain parts of the flow and hence sample fluid velocities with a reduced range/variance. Particles on the fast tracks in Figure 2 only see downward fluid velocity and hence only half the fluid velocity range:

$$w_{fluid,min} < w_p < 0 \quad (3)$$

A probe which 'sweeps' this velocity field at random sees the full range of fluid velocities, i.e.,

$$w_{fluid,min} < w_{Eulerian} < w_{fluid,max} \quad (4)$$

Correspondingly, particles on the fast track see a smaller velocity variance than a fixed probe. The precise relation depends on exactly how the particles turn the corners on the fast track, but a value which agrees with the observation of S&L can be obtained with reasonable estimates.

A possible objection to explaining the reduction of $Var(w_p)$ for the smallest of S&L's in terms of the fast tracks in Figure 2 is that these small particles had too little inertia or velocity bias to actually get onto the fast tracks. Unfortunately, the necessary experimental information about w_p is not available to settle this question on direct evidence. What is available, is indirect evidence in the form of accelerated settling data from Murray (1970).

Like Snyder & Lumley, Murray also used a set of low inertia particles, which had been designed to follow the fluid perfectly. These particles were observed to experience very significantly accelerated settling: In strong turbulence ($10 < w'/w_o < 20$) they settled two to four times faster than in still water, see Figure 1. This is taken as evidence that Murray's particles did get on to the fast tracks.

Whether the particles have enough inertia to get onto the fast tracks may be measured by the time scale ratio

$$\frac{T_p}{T_L} = \frac{w_o/g}{T_L} \quad (5)$$

This time scale ratio also measures the particles' ability to respond to fluid velocity oscillations and hence also the expected velocity variance ratio $Var(w_{particle})/Var(w_{fluid})$. In the absence of coherent flow structures and fast tracks, i.e., in what might be termed structure-less turbulence, a plausible frequency response function is

$$\frac{Var(w_{particle})}{Var(w_{fluid})} = \frac{1}{\left(1 + 0.3(\frac{T_P}{T_L})^2\right)^2} \quad (6)$$

However in order to get a good match with Snyder & Lumley's data in Figure 4 an 0.6 reduction is required. That is, the trend of Snyder and Lumley's data is mimicked very nicely by

$$\frac{Var(w_{particle})}{Var(w_{fluid})} = \frac{0.6}{\left(1 + 0.3(\frac{T_P}{T_L})^2\right)^2} \quad (7)$$

in Figure 4.

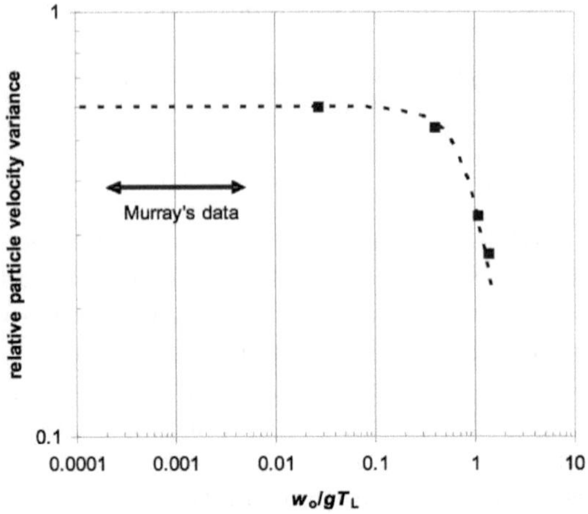

Fig. 4. Larger, more inert particles will have smaller velocity variance in a given flow. The solid squares correspond to the data of Snyder & Lumley and dashed shows Equation (6). The range of T_P/T_L for Murray's data is also indicated.

The suggestion that the 0.6-factor is due to S&L's particles moving along fast tracks is supported by Murray's observations in the following way: as indicated on Figure 4, Murray's particles were significantly smaller than those of S&L in terms of $w_o/(gT_L)$. Murray's particles clearly experienced fast tracking, see Figure 1, so they moved along fast tracks. If Murray's particles were big enough to get onto the fast tracks, so were those of S&L.

4 Conclusions

We argue that the accelerated settling of heavy and the accelerated rise of some moderately buoyant particles in turbulence can be seen as analogous with the fast-racking in cellular the flow fields initially explored by Maxey & Corrsin (1986).

Since particles on the fast tracks sample a subset of fluid velocities with a reduced variance one should expect a smaller Lagrangian velocity variance from particles in a flow with coherent eddy structures than from an Eulerian probe which samples the eddies at random.

This applies in particular to the smallest particles used by Snyder & Lumley (1971). The variance reduction by 40%, which was unexpected at the time, can be explained in terms of the particles moving along the turbulence equivalent of the fast tracks in the cellular flow field in Figure 2. Even the smallest of S&L's particles were big enough to spiral onto the fast tracks because they were, in terms of T_P/T_L, more than one order of magnitude bigger than MurrayÕs (1970) smallest particles which showed clear signs of fast tracking via strongly enhanced settling.

References

[1] Friedman, P D & J Katz (2002): Mean rise rate of droplets in isotropic turbulence. Physics of Fluids, Vol 14, No 9, pp 3059-3073

[2] Ho, H W, (1964): Fall velocity of a sphere in an oscillating fluid. PhD Thesis, University of Iowa

[3] Maxey, M R & S Corrsin (1986): Gravitational settling of aerosol particles in randomly orientated circular flow fields. J. Atmospherical Sci, Vol 43, pp 1112-1134

[4] Maxey, M R (1990): On the advection of spherical and non-spherical particles in a non uniform flow. Phil Trans Roy Soc Lond, Vol 333, pp 289-307

[5] Murray, S P (1970): Settling velocity and vertical diffusion of particles in turbulent water. J Geophys Res, vol 75, No 9, pp 1647-1654

[6] Nielsen, P (1992); Coastal bottom boundary layers ands sediment transport. World Scientific, Singapore, 324pp

[7] Nielsen, P (1993): Turbulence effects of the settling of suspended particles. J Sed Petrology, Vol 63, No 5, pp 835-838

[8] Snyder, W H & J L Lumley (1971): Some measurements of particle velocity autocorrelation functions in a turbulent flow. J Fluid Mech, Vol 48, pp 41-71

[9] Tooby, P F, G L Wick & J D Isacs (1977): The motion of a small sphere in a rotating velocity field: A possible mechanism for suspending particles in turbulence. J Geophysical Res, Vol 82, No 15C, pp 2096-2100

[10] Zeng, Q (2001): Motion of particles and bubbles in turbulent flows. PhD Thesis, The University of Queen-sland, Brisbane, 191pp

The turbulent rotational phase separator

J.G.M. Kuerten and B.P.M. van Esch

Dept. of Mechanical Engineering, Technische Universiteit Eindhoven, The Netherlands j.g.m.kuerten@tue.nl

Summary. The Rotational Phase Separator (RPS) is a device to separate liquid or solid particles from a lighter or heavier fluid by centrifugation in a bundle of channels which rotate around a common axis. Originally, the RPS was designed in such a way that the flow through the channels is laminar in order to avoid eddies in which the particles become entrained and do not reach the walls. However, in some applications the required volume flow of fluid is so large, that the Reynolds number exceeds the value for which laminar Poiseuille flow is linearly stable. Depending on the Reynolds numbers the flow can then be turbulent, or a laminar time-dependent flow results. In both cases a counter-rotating vortex is present, which might deteriorate the separation efficiency of the RPS. This is studied by means of direct numerical simulation of flow in a rotating pipe and particle tracking in this flow. The results show that the collection efficiency for larger particles decreases due to the combined action of the vortex and turbulent velocity fluctuations, while it is unchanged for smaller particles.

1 Introduction

The Rotational Phase Separator (RPS) is a separation device built around a rotating filter element consisting of a large number of narrow parallel channels (see Fig. 1 for a schematic drawing). Usually, the RPS is applied in addition to a conventional tangential or axial cyclone in order to decrease the cut-off particle diameter by one order of magnitude [1, 2]. In the original design of the RPS, the flow in the channels of the filter element is kept laminar to prevent capture of particles or droplets in turbulent eddies. In case of the tangential design, mainly used to separate droplets or particles from a gas flow, it is normally not a problem to design within this limit as the throughput is low compared to the flow area of the cyclone and filter element.

The opposite is true for the axial version which is mainly used for in-line (offshore) separation of condensed droplets from another liquid or gas flow. In such applications the pressure and required volume flow lead to higher Reynolds numbers and the conditions for stable Poiseuille flow might become

Bernard J. Geurts et al. (eds), Particle Laden Flow: From Geophysical to Kolmogorov Scales, 393–405.

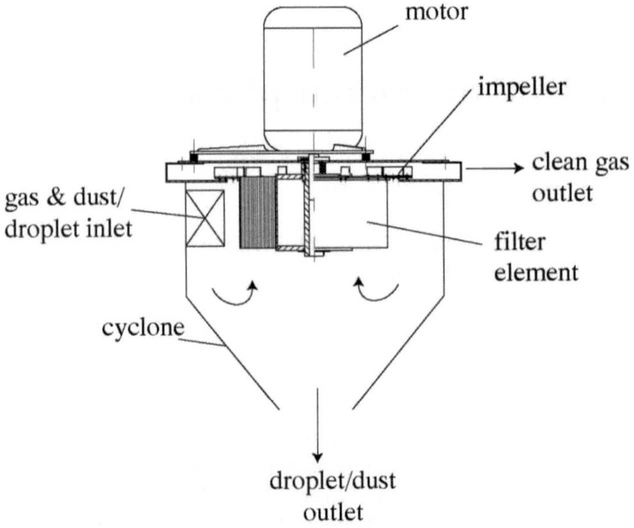

Fig. 1. Schematic drawing of tangential RPS mounted in a cyclone.

too restrictive. However, in many cases the Reynolds number is low enough for the flow and particle behavior to be studied in detail by means of direct numerical simulation of the fluid flow (DNS) and Lagrangian particle tracking. In this study, the consequences of allowing conditions in the channels of the filter element for which Poiseuille flow is unstable, are investigated.

In section 2 of this paper an analytical model for the calculation of particle collection efficiency will be presented briefly. Section 3 provides the governing equations and numerical method for the computation of particle-laden flow in a rotating pipe and in section 4 results are presented. Finally, in section 5 the conclusions of the paper are given.

2 Analytical model

Brouwers [3] derived the elementary particle collection efficiency of the RPS for channels with circular, triangular and sinusoidal shape in case the flow in the channels is laminar and stationary. At entry of the filter element, or soon after, the fluid co-rotates with the filter element. As we are concerned with particles in the micrometer range, inertial forces are neglected. Whether a particle reaches the outer wall depends on the radial distance to be traveled by the particle, the centrifugal force, the axial velocity profile and the length of the channel. The centrifugal force depends on the angular velocity of rotation, the difference in mass density between particle and fluid, the particle diameter and the distance between particle and axis of rotation. The velocity at which the particles move radially can be calculated using Stokes' law for drag force.

Assuming a constant axial fluid velocity U_b and a uniform distribution of the particles over the cross-sectional area, an expression can be derived for the smallest particle which is collected with 100% probability in a channel at a radial location R [3]:

$$d_{p,100}^2 = \frac{18\mu U_b D}{(\rho_p - \rho_f)\Omega^2 RL}. \tag{1}$$

Here μ denotes the dynamic viscosity of the fluid, D the pipe diameter, ρ_p and ρ_f the mass density of the particle and fluid, Ω the angular velocity and L the length of the pipe.

To derive an expression for the particle collection efficiency in the presence of a Hagen-Poiseuille velocity profile, the circular cross section is divided into a system of parallel planes within which the movement of particles takes place. Equation (1) can be used for the local conditions. Subsequent integration gives the particle collection efficiency η in circular channels subject to Hagen-Poiseuille flow

$$\eta = \begin{cases} \frac{4}{\pi}x^2 a - \frac{4}{3\pi}\sqrt{(1-a^2)}a(\frac{5}{2} - a^2) - \frac{2}{\pi}\arcsin(a) + 1 & \text{if } x < \sqrt{4/3} \\ 1 & \text{if } x \geq \sqrt{4/3}. \end{cases} \tag{2}$$

Here $a = [1 - (3x^2/4)^{2/3}]^{1/2}$ and $x = d_p/d_{p,100}$. Although current production methods produce sinusoidal or rectangular channel geometries, the case of a circular geometry is adopted for this study as there are more reference cases available and numerical simulation is easier.

3 Numerical method

In this paper the flow in a pipe rotating with angular velocity $\mathbf{\Omega}$ around an axis parallel to its own axis is studied by solving all relevant scales of motion. To this end the three-dimensional Navier-Stokes equation for incompressible flow is solved in a cylindrical geometry in the vorticity formulation. The equation is solved in a rotating frame of reference and reads:

$$\frac{\partial \mathbf{u}}{\partial t} + \boldsymbol{\omega} \times \mathbf{u} + \mathbf{\Omega} \times \mathbf{\Omega} \times \mathbf{r} + 2\mathbf{\Omega} \times \mathbf{u} = -\frac{1}{\rho}\nabla P + \nu \Delta \mathbf{u}. \tag{3}$$

Here, P denotes the total pressure, $P = p + \frac{1}{2}u^2$, $\mathbf{u}$ the fluid velocity, $\boldsymbol{\omega}$ the fluid vorticity, ρ the fluid density, ν the kinematic viscosity and $\mathbf{r}$ the position vector with respect to the rotation axis. Compared to the Navier-Stokes equation in a stationary frame of reference, two additional terms appear. The centrifugal acceleration, $\mathbf{\Omega} \times \mathbf{\Omega} \times \mathbf{r}$, can be incorporated in the pressure [1]. The Coriolis acceleration, $2\mathbf{\Omega} \times \mathbf{u}$, does not depend on the distance to the rotation axis. Hence, the fluid velocity does not depend on this distance, which implies that in one calculation the flow in all pipes in the bundle can be simulated. Note however, that the pressure field does depend on this distance; only the sum of the pressure and centrifugal pressure is independent of the distance to the rotation axis.

In the calculations a pipe of a finite length equal to five times its diameter is taken with periodic boundary conditions in the axial direction. Since the tangential direction is periodic by definition, a spectral method with a Fourier-Galerkin approach in the two periodic directions is a natural choice. In the radial direction a Chebyshev-collocation method is applied, but, in order to avoid a large number of collocation points near the axis of the pipe, the radial direction is divided into five elements with a Chebyshev grid in each element [5]. The coupling between the elements is continuously differentiable.

For integration in time a second-order accurate time-splitting method is chosen. In the first step the nonlinear terms, including the Coriolis force, are treated in an explicit way. The nonlinear terms are calculated pseudo-spectrally by fast Fourier transform, where the 3/2-rule is applied to prevent aliasing errors. In the second step the pressure is calculated in such a way that the velocity field at the new time level is approximately divergence free. Finally, in the last step the viscous terms are treated implicitly. The wall of the pipe acts as a no-slip wall. The correct boundary conditions at the pipe axis follow from the property that the Cartesian velocity components and pressure are single-valued and continuously differentiable.

The mean axial pressure gradient is chosen in such a way that the volume flow remains constant. The simulations are started from an arbitrary initial solution. After a large number of time steps a state of statistically stationary flow is reached. In [5] it is shown that for turbulent flow in a non-rotating pipe the DNS results for mean flow, velocity fluctuations and terms in the kinetic energy balance agree well with results of other DNS codes and experimental results.

Particle-laden flows can be described in two different ways. In Lagrangian methods an equation of motion for each particle is solved, whereas in Eulerian methods the particles are described as a second phase for which conservation equations are solved. We chose a Lagrangian approach for two reasons. First, the number of particles is limited and the particle mass loading small, so that a Lagrangian method with one-way coupling is possible. Second, the length of an actual channel of an RPS is much larger than the length of the pipe used in the calculations. In Eulerian approaches a particle concentration field for the whole channel length and for each particle diameter would be needed, which leads to huge memory and computational resources. Hence, particles are tracked by solving an equation of motion for each particle.

If $\mathbf{x}$ is particle position and $\mathbf{v} = d\mathbf{x}/dt$ its velocity, the equation of motion reads in general:

$$m\frac{d\mathbf{v}}{dt} = \sum \mathbf{f}. \tag{4}$$

Here, m denotes the mass of the particle and the right-hand side contains all (effective) forces acting on the particle. In the simulations considered here, we restrict to cases where particles are small and have a large mass density compared to the fluid mass density. As a result the only forces which cannot be neglected are the drag force and centrifugal force. This leads to an equation of motion of the form:

$$\frac{d\mathbf{v}}{dt} = \frac{\mathbf{u}(\mathbf{x},t) - \mathbf{v}}{\tau_\mathrm{p}}(1 + 0.15 Re_\mathrm{p}^{0.687}) + \Omega^2(R\mathbf{e}_x + \mathbf{r}_2), \tag{5}$$

where τ_p is the particle relaxation time, $\mathbf{e}_x$ is the unit vector in the direction from the rotation axis to the pipe axis and $\mathbf{r}_2$ the position vector of the particle in the two-dimensional plane perpendicular to the pipe axis. The standard drag correlation for particle Reynolds number, Re_p, between 0 and 1000 is used. Note that in contrast to the fluid velocity, the particle equation of motion depends on the distance between the pipe axis and axis of rotation through the centrifugal force. Since the particle relaxation times of the particles considered are very small, the inertia term on the left-hand side of Eq. (5) could be neglected. However, since the equation is nonlinear in the particle velocity due to the particle Reynolds number, it is easier to solve it in this way. A partially implicit two-step Runge-Kutta method, in which the particle velocity appearing in Re_p is treated explicitly, is used to this end. Finally, the fluid velocity at the particle position, which appears in Eq. (5) is found from fourth-order accurate interpolation from its values at grid points.

The particle simulations start from a fully-developed velocity field with a homogeneous distribution of particles over the entire pipe. The initial particle velocity is chosen in such a way that its initial acceleration equals zero. In a real RPS the length of a channel is much larger than the length of the computational domain. Therefore, if a particle reaches the end of the computational domain in the axial direction, it is re-inserted at the corresponding position at the pipe entrance until it has traveled an axial distance equal to the length of the real pipe. If a particle reaches the wall of the pipe before it travels the whole length it is considered as being collected.

In an actual experiment where the particles are homogeneously distributed over the total flow domain, the number of particles that enter a channel of the RPS at a certain radial position per unit of time, is proportional to the axial velocity at that position. Therefore, in the calculation of the collection efficiency, each particle has a weight proportional to its exact initial axial velocity.

4 Results

In this section results will be presented. The fluid flow is determined by two non-dimensional parameters, the bulk Reynolds number $Re = U_b D/\nu$ and the rotation Reynolds number $Re_\Omega = \Omega D^2/(4\nu)$, where U_b is the bulk velocity and D the diameter of the pipe. Without rotation the laminar Hagen-Poiseuille flow is unstable for large perturbations if $Re > 2300$ approximately. Rotation reduces the stability of the laminar flow considerably as shown by

Mackrodt [7]. In order to study the resulting flow and the effects on particle motion, we will consider three typical test cases.

4.1 Turbulent flow at $Re = 5300$

For the first test case with $Re = 5300$ and $Re_\Omega = 980$, the flow without rotation is already turbulent. Flows without particles in this regime have been studied by means of direct numerical simulation before by Orlandi and Fatica [6]. As a second non-dimensional parameter they used the rotation number defined as the ratio of the rotation Reynolds number and the bulk Reynolds number. The rotation number in our simulations equals 0.37. The DNS is performed with 106 collocation points in the wall-normal direction and 128 Fourier modes in both the axial and tangential direction. In the following, results of the fluid calculations will be presented and analyzed first, and then the results of the particle simulations will be discussed.

For rotating pipe flow time-averaged quantities depend on the radial coordinate only and from the continuity equation it follows that the mean radial velocity component equals zero, but in contrast to the non-rotating case, the mean tangential velocity is not equal to zero.

In Fig. 2 the mean tangential velocity component in wall units is plotted as a function of the radial coordinate. In this figure also the result for the same bulk Reynolds number and $Re_\Omega = 490$ is included. It can be seen that the mean tangential velocity is almost exactly linearly dependent on Re_Ω when scaled with the friction velocity,

$$u_\tau = \sqrt{\nu \left. \frac{d\bar{u}_z}{dr} \right|_{r=D/2}} \tag{6}$$

The non-zero mean tangential velocity can be understood from the equation for the radial-tangential component of the Reynolds stress tensor, which reads after disregard of the very small viscous terms:

$$\begin{aligned} \bar{u}_\phi \left(2\overline{u_\phi'^2} - \overline{u_r'^2} \right) - r\overline{u_r'^2}\frac{d\bar{u}_\phi}{dr} = &-2\Omega r \left(\overline{u_\phi'^2} - \overline{u_r'^2} \right) + \frac{1}{r}\frac{d}{dr}\left(r^2 \overline{u_r'^2 u_\phi'} \right) \\ &- \overline{u_\phi'^3} + \frac{1}{\rho}\left(\overline{r u_\phi' \frac{\partial p'}{\partial r}} + \overline{u_r' \frac{\partial p'}{\partial \phi}} \right) . \end{aligned} \tag{7}$$

In this expression primes denote the fluctuating part of a quantity, subscripts r and ϕ refer to the radial and tangential component and bars denote mean quantities. The third order moments appearing in Eq. (7) turn out to be very small throughout the pipe, whereas the last term on the right-hand side is only significant close to the wall of the pipe. Furthermore, due to the behavior of the tangential velocity component near the pipe axis $r d\bar{u}_\phi/dr \cong \bar{u}_\phi$ there. Therefore, Eq. (7) simplifies to $\bar{u}_\phi \cong -\Omega r$ close to the axis of the pipe. The

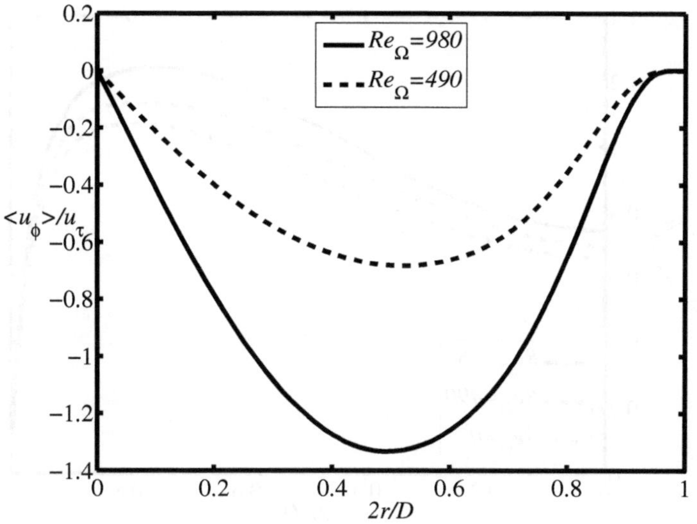

Fig. 2. Mean tangential velocity component in wall units, for rotating pipe flow with $Re = 5300$. The brackets have the same meaning as the overbar in the text.

results presented in Fig. 2 indeed agree with this behavior close to the axis of the pipe.

A further flow property which is important for the understanding of particle behavior is the fluctuating part of the fluid velocity in the plane perpendicular to the pipe axis. In Fig. 3 the root-mean-square of the tangential velocity component is plotted as a function of the radial coordinate in wall units. Included are results at $Re_\Omega = 490$ and for a non-rotating pipe. It can be seen that the rotation slightly increases these velocity fluctuations. Moreover, it appears that the magnitude of the velocity fluctuations is almost equal to the mean tangential velocity component in case $Re_\Omega = 980$. The increase in velocity fluctuations with increasing Re_Ω occurs for all three velocity components.

Particle behavior in turbulent rotating pipe flow can be understood from a simplified equation of motion in the plane perpendicular to the pipe axis. To this end all forces on the particle are disregarded except the linearized drag force and the centrifugal force. If r and ϕ are the radial and tangential coordinate of a particle, the equations of motion are:

$$\begin{cases} \frac{dr}{dt} = u'_r + \tau_p \Omega^2 (r + R\cos(\phi)) \\ r\frac{d\phi}{dt} = \bar{u}_\phi + u'_\phi - \tau_p \Omega^2 R \sin(\phi) \end{cases} \tag{8}$$

The equations of motion contain three different terms: the mean tangential fluid velocity, which has Ωr as order of magnitude, the fluctuating velocity with the friction velocity u_τ as order of magnitude and the last term on the right-hand sides of Eq. (8), which represents the centrifugal velocity. For the

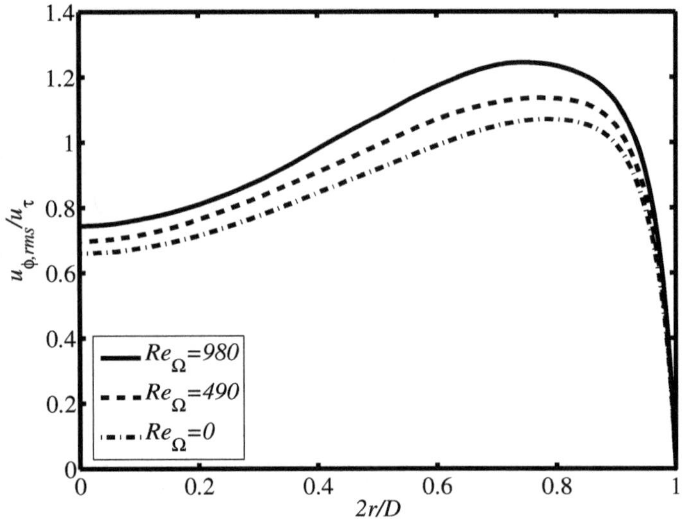

Fig. 3. Root-mean square of tangential velocity component in wall units, for rotating pipe flow at $Re = 5300$.

smallest particles which are completely separated in uniform laminar flow, the order of magnitude of the centrifugal velocity equals $U_b D/L$ with L the length of the pipe. For situations relevant in practice, the centrifugal velocity is always smaller than the fluctuating velocity. In our example the mean tangential velocity is only slightly smaller than the fluctuating velocity.

We first consider a hypothetical velocity field with a mean tangential velocity, but without velocity fluctuations. In Fig. 4 the collection efficiency for this flow is compared with that for laminar Hagen-Poiseuille flow. The particle diameter is non-dimensionalized with the smallest diameter which is collected with 100% probability for uniform laminar flow. Fig. 4 shows that the collection efficiency is reduced dramatically by the presence of the axial vortex. Particles are trapped in this vortex and follow a path which differs only slightly from the path they would follow without centrifugal force. Only those particles which are initially close to the wall are collected. This situation is similar to the one obtained for laminar flow in a slightly tilted rotating pipe, which was studied by Brouwers [4]. Also in that case particles are trapped in the secondary flow perpendicular to the pipe axis, which results in a reduced collection efficiency.

Next, we return to particle behavior in turbulent rotating pipe flow. In the simulation particles with diameters ranging between $0.1d_{p,100}$ and $1.6d_{p,100}$ are inserted in the flow, where $d_{p,100}$ is the smallest particle collected with 100% probability in a uniform laminar flow. For each diameter 25,000 particles are initially uniformly distributed over the pipe and their motion is subsequently tracked by solving their equation of motion until they either reach the wall

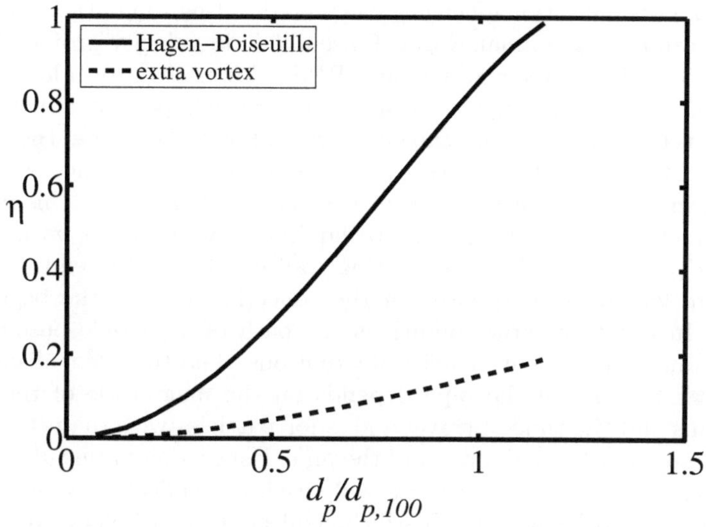

Fig. 4. Collection efficiency for laminar flow with and without extra tangential velocity.

of the pipe or travel over an axial distance larger than the length of the pipe, which equals $133.5D$. The mass density of the particles equals 22.5 times the mass density of the fluid and only one pipe is considered with its axis at a distance of $26.7D$ from the rotation axis.

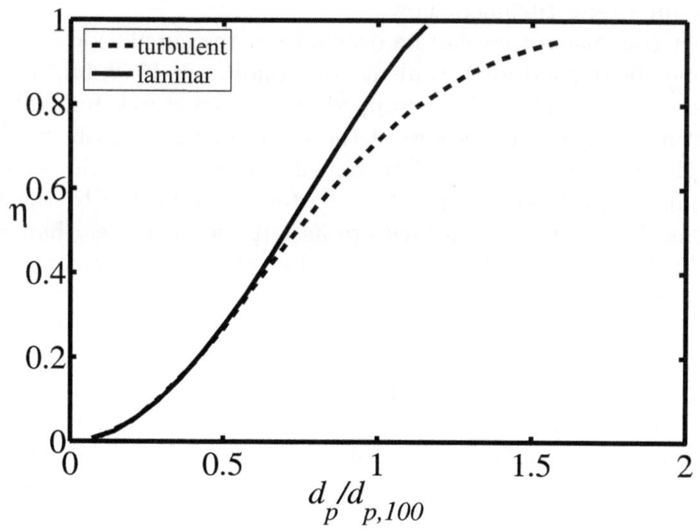

Fig. 5. Collection efficiency for laminar and turbulent flow.

In Fig. 5 the collection efficiency calculated in this simulation is compared with the result for a laminar Hagen-Poiseuille fluid velocity profile. Although the collection efficiency for a turbulent RPS is lower than for a laminar RPS, the reduction obtained is not as dramatic as for the hypothetical flow without turbulent velocity fluctuations shown in Fig. 4. Fig. 5 shows that the collection efficiency of the smallest particles is hardly affected by turbulence, whereas the reduction in efficiency for particles near $d_{p,100}$ is almost 30%. This result can be explained in the following way. In laminar flow, particles with diameter equal to $d_{p,100}$ will reach the collecting wall exactly at the end of the pipe if they are located just opposite of the collecting wall at the beginning of the pipe. In turbulent flow conditions the path of a particle becomes more irregular due to turbulent velocity fluctuations. The turbulent dispersion of particles at the end of the pipe depends on the magnitude of the velocity fluctuations and the time of travel and is for particles with diameter equal to $d_{p,100}$ on the order of the diameter of the pipe. Hence, due to turbulent velocity fluctuations some of these particles will reach the collection wall at a more upstream axial position and will still be collected in turbulent flow, whereas other particles would reach the collection wall at a more downstream position and will not be collected in turbulent flow. Hence, the collection efficiency of particles with diameter close to $d_{p,100}$ will decrease in turbulent flow. On the other hand, some of the particles with a diameter much smaller than $d_{p,100}$ that are collected in laminar flow conditions, will not be collected in turbulent flow, whereas some of these small particles that are not collected in laminar flow, will be collected due to turbulent velocity fluctuations in turbulent flow. The effects of both phenomena on the total collection efficiency approximately cancel, so that the collection efficiency for small particles is approximately the same in laminar and turbulent flow.

Another conclusion that can be drawn from the simulation results is that $d_{p,100}$ cannot be defined for turbulent flow conditions. Even for large particle diameters, some particles will be trapped in flow structures and will not reach the collecting wall before the end of the pipe. The results shown in Figs. 4 and 5 imply that the presence of turbulent velocity fluctuations counteracts the trapping of particles in the axial vortex. This is due to the fact that the tangential velocity fluctuations are as large as or larger than the mean tangential velocity. Two extra simulations have been performed to verify this. In the first the Coriolis force in the Navier-Stokes equation has been set to zero, but the centrifugal force in the particle equation of motion remained unaffected. Hence, in this simulation the mean tangential fluid velocity in Eq. (8) equals zero, but the velocity fluctuations are almost the same. The resulting collection efficiency is only slightly higher than the turbulent result in Fig. 5. On the other hand, a simulation with a mean tangential velocity artificially increased by a factor of 3 resulted in a substantial reduction in collection efficiency.

4.2 Turbulent flow at *Re* = 1400

In the second test case $Re = 1400$ and $Re_\Omega = 980$, so that the flow without rotation would be laminar and stationary. The rotation, however, leads to a turbulent flow state very similar to the one at higher bulk Reynolds number. The calculation is performed at the same parameter values as the one in the previous subsection. The flow differs from the one at $Re = 5300$, but the important features are more or less the same. The flow is still turbulent, in spite of the low bulk Reynolds number. The mean axial velocity and the velocity fluctuations are of the same order of magnitude as at the higher Reynolds number. The mean tangential velocity is higher than at $Re = 5300$, but still of the same order of magnitude as the tangential velocity fluctuations. This explains the result for the collection efficiency shown in Fig. 6, which differs only slightly from the $Re = 5300$ result.

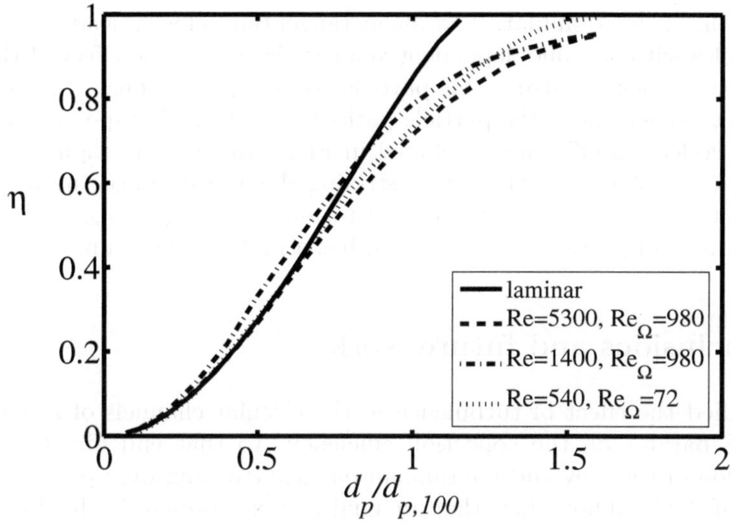

Fig. 6. Collection efficiency for laminar and turbulent flow.

4.3 Laminar time-dependent flow

According to Mackrodt [7] laminar Poiseuille flow already becomes unstable at low bulk Reynolds numbers if a small rotation rate is applied. However, he remarked that this instability will probably lead to another stable flow rather than to turbulence. Indeed, the work of Sanmiguel-Rojas and Fernandez-Feria [8] indicates that a time-dependent laminar flow results. In order to study this and the effect of this flow on particle behavior, we calculated the flow at $Re = 540$ and $Re_\Omega = 72$ with the present DNS code starting from a state of turbulence at a higher Reynolds number. After a transition period the resulting flow turns out to be very different from the turbulent flows discussed in the previous sections. Most of the Fourier modes of the expansion are negligibly small and the solution is in good approximation given by

$$u_i(r, \phi, z, t) = u_i^0(r) + \hat{u}_i(r) \exp(i(\phi - 2\pi z/L + \omega t)) + c.c., \tag{9}$$

where u_i denotes one of the three velocity components. Moreover, the mean axial velocity $u_z^0(r)$ is in very good approximation given by the quadratic Poiseuille profile at this Reynolds number and, because of incompressibility, $u_r^0(r) = 0$. The mean tangential velocity is unequal to zero, like at the higher Reynolds numbers. This solution is not a solution of the linear stability equations: nonlinear interactions play a role as well and the amplitude of the disturbance, $\hat{u}_i$, is determined by the balance of some of the nonlinear terms in the Navier-Stokes equation and the pressure and Coriolis terms.

Although the flow in these conditions is not turbulent, the particle motion in this flow is similar. The mean tangential velocity leads to trapping of particles without time-dependent velocity. However, the effect of the time-dependent velocity destroys this particle trapping mechanism. In contrast to the real turbulent flow, the particle paths have a helical shape. The resulting particle collection efficiency, included in Fig. 6 differs only slightly from the one at $Re = 5300$, but there is a striking difference. Since the flow is not turbulent and the particles follow a deterministic path, $d_{p,100}$, the diameter above which all particles are collected, has a finite value again.

5 Conclusions and future work

We studied the effect of turbulence in the circular channels of a Rotational Phase Separator on the collection efficiency. To that end direct numerical simulation of the flow and Lagrangian particle tracking were performed. The results of the fluid flow show that an axial vortex is present in the flow, caused by the rotation, but, in contrast to the secondary flow in laminar flow in a slightly tilted pipe, this vortex hardly influences the collection efficiency for the parameter settings of the simulated test cases. However, turbulent velocity fluctuations have a negative influence on the collection efficiency, especially for larger particles. One of the consequences is that $d_{p,100}$, the diameter of the smallest particles which are all collected, cannot be defined for turbulent flow conditions. At low Reynolds number, the laminar Poiseuille flow is unstable, but evolves into a time-dependent laminar flow. The resulting particle collection efficiency is comparable to the one at higher Reynolds numbers.

Acknowledgment

This work was sponsored by the National Computing Facilities Foundation, NCF for the use of supercomputer facilities, with financial support from the Netherlands Organization for Scientific Research, NWO.

References

[1] Brouwers JJH, (2002) Exp Thermal Fluid Science 26:325–334
[2] Kuerten JGM, Van Kemenade HP, Brouwers JJH (2005) Powder Technol 154:73–82
[3] Brouwers JJH (1997) Powder Technol 92:89–99
[4] Brouwers JJH (1995) Appl. Scientific Research 55:95–105
[5] Walpot RJE, Kuerten JGM, Van der Geld CWM (2006) Flow, Turbulence and Combustion 76:163–175
[6] Orlandi P, Fatica M (1997) J0 Fluid Mech 343:43–72
[7] Mackrodt P-A (1976) J Fluid Mech 73:153–164
[8] Sanmiguel-Rojas E, Fernandez-Feria R (2005) Phys. Fluids 17:014104

Particle laden geophysical flows: from geophysical to sub-kolmogorov scales

H.J.S. Fernando and Y.-J. Choi

Department of Mechanical and Aerospace Engineering, Environmental Fluid Dynamics Program, Arizona State University, Tempe, AZ 85287-9809
j.fernando@asu.edu

Summary. A brief review of natural particle-laden flows is given, paying particular attention to the wide range of scales of motions and particles found in the environment. Some fundamental concepts underlying particle-turbulence interactions are discussed and their application to a few selected flow configurations (particle laden jets and particle settling through vortices and turbulence) is exemplified. Examples of the application of a sophisticated modeling system to predict particle concentration and visibility in the atmosphere are also illustrated.

1 Introduction

Suspended particles are ubiquitous in environmental flows. They produce interesting and important effects such as dust storms, avalanches, spectacular sunsets and visibility reduction in the atmosphere as well as turbidity currents and thin layers that act as biological hot-spots in oceans. Particulate matter (PM) is a key atmospheric pollutant, and it spans a wide variety of sizes (Figure 1). Small particles (aerodynamic diameter $d_p < 10$ microns or PM_{10}) have a tendency to remain suspended for extended periods of time without being deposited, but larger particles settle out within minutes of entering the atmosphere. Atmospheric aerosols (solid and liquid PM) are of two kinds: primary (directly emitted by anthropogenic and natural sources) and secondary (formed by chemical reactions, typically $d_p < 1$ micron). Much attention has been focused on these aerosols because of their proven association with severe health problems and other quality of life issues (e.g., reduction in visibility and welfare of animal/plants). Also shown in Figure 1 are the different types of aerosols and their sources. Of these, particles of size less than 2.5 microns ($PM_{2.5}$) are known to be most detrimental to human health. Atmospheric visibility is mainly affected by still smaller particles ($d_p < 1$ micron), which are responsible for the appearance of a brown cloud over polluted cities in the morning. Aerosols, irrespective of the nature of their sources [e.g., point (e.g., chimney stacks), line (e.g., unpaved roads) or area (e.g., cities) sources]

Bernard J. Geurts et al. (eds), Particle Laden Flow: From Geophysical to Kolmogorov Scales, 407–421.

are transported by airflow, undergo physical and chemical transformations and can deposit over surfaces. Deep convection and flow convergence can lift aerosols up to several *km*, which are then picked up by large-scale flow and transported over long distances, covering intercontinental (global) and regional (synoptic) scales. These upper atmospheric aerosols are fumigated (mixed down) over land by convection and subsidence, and then transported and dispersed off by mesoscale, urban/rural and local turbulent flows. A continuum of scales are responsible for particle dispersion, the larger scales being responsible for advection while turbulence down to the Kolmogorov scale plays a critical role of dispersing, mixing and suspending the particles. The role of sub-Kolmogorov (micron) scales is yet to be delineated. For suspended ultra-fine particles (< 0.1 microns), the Knudsen number is small, the continuum hypothesis fails and the usual fluid dynamics concepts may become invalid.

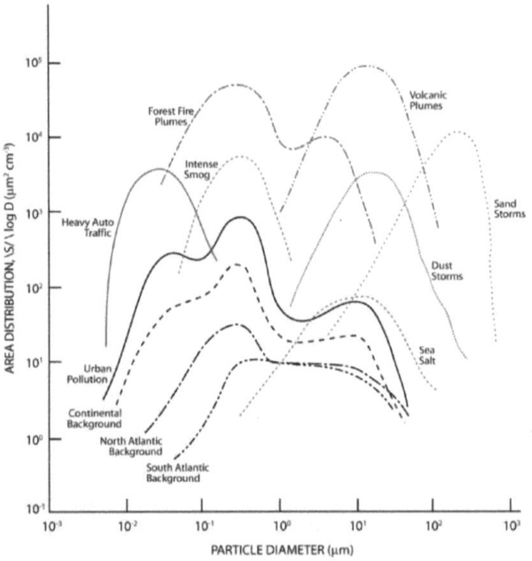

Fig. 1. A typical surface area distribution of atmospheric aerosols associated with different source regions (from Bridgeman, 1990)

Particles found in lakes and oceans (hydrosols) also span a large range of scales, depending on the nature of existing sources and particles, physical and biological processes, aggregation and age. Most oceanic particles are organic material, including living plants and animals, pieces of dead/dying organisms (aggregates of which produce 'marine snow') and their waste products. Dust depositing from the air or pieces of the continents carried by rivers or waves is also common. Oceanic particles thus are more diverse and complex, and only a few measurements exist on suspended-particle size spectra. A typical hypothetical spectrum is shown in Figure 2. As in the atmosphere, bio-geochemical processes are abundant amongst ocean particles, and the modeling of ensuing complicated physiological functions poses intricate challenges (Ghosh et al. 2005). In all, particle-laden flows in the environment spans geophysical (km) to sub-Kolmogorov (10-100 nm) scales, where the Kolmogorov scales in the atmosphere and oceans/lakes are on the order of 1 mm.

The aim of this paper is to present a brief overview of the types of environmental particle-laden flows and their underlying dynamics in the realm of fluid-particle interactions, modeling and applications. Given the preponderance of scales (100 nm to 10^4 km; seconds to weeks) and the multitude of processes involved, a detailed review is beyond the scope of this paper. Therefore, only a few selected topics are discussed. Of particular interest is the application of fluid-dynamics based knowledge to real environmental flows, which is exemplified using visibility and PM modeling for the atmosphere.

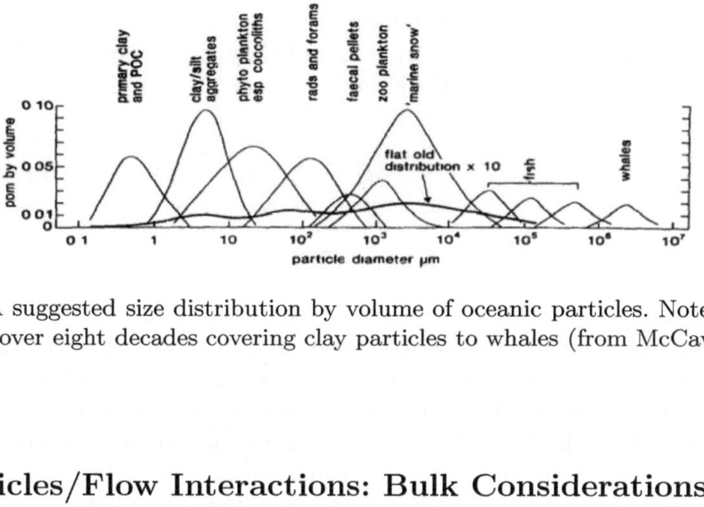

Fig. 2. A suggested size distribution by volume of oceanic particles. Note the distribution over eight decades covering clay particles to whales (from McCave 1984)

2 Particles/Flow Interactions: Bulk Considerations

One of the earliest attempts to describe particle-turbulence interaction was made by Tchen (1947; also see Hinze 1975). By assuming that: (*i*) the turbulence is homogeneous and stationary; (*ii*) the flow has an infinite domain; (*iii*) the particles are spherical; (*iv*) the particles are small compared to the smallest wave length present in turbulence (i.e., Kolmogorov scale); (*v*) the neighborhood of the particles is always formed by the same fluid; and (*vi*) the body forces are conservative, Tchen derived the equation of motion for a single particle as

$$\frac{\pi}{6}d_p^3\rho_p\frac{d\mathbf{v}_p}{dt} = 3\pi\mu_f d_p(\mathbf{v}_f - \mathbf{v}_p) + \frac{\pi}{6}d_p^3\rho_f\frac{d\mathbf{v}_f}{dt} + \frac{1}{2}\frac{\pi}{6}d_p^3\rho_f\left(\frac{d\mathbf{v}_f}{dt} - \frac{d\mathbf{v}_p}{dt}\right)$$

$$+\frac{3}{2}d_p^2\sqrt{\pi\rho_f\mu_f}\int_{t_0}^{t} dt'\,\frac{\left(\frac{d\mathbf{v}_f}{dt'} - \frac{d\mathbf{v}_p}{dt'}\right)}{\sqrt{t-t'}} + \mathbf{F}_b \tag{1}$$

where $\mathbf{v}_p$ is the velocity of the particle, μ_f the dynamical viscosity of the fluid, t the time, $\mathbf{F}_b$ the body force on the particle, and $\mathbf{v}_f$ the fluid velocity in the neighborhood of the particle but at a sufficiently large distance not to be disturbed by the relative motion. The terms in this equation, from left to right, represent the particle inertia, the drag due to relative motion between the fluid and the solid, the local pressure gradient, the virtual inertia, forces due to unsteadiness of the particle (Basset forces), and the body forces. The total derivative d/dt for the particle and the fluid phases should be based on their respective velocities.

As pointed out by Maxey and Riley (1983), among others, (1) is valid only under restrictive conditions, and thus (1) can only give a qualitative description of the motion of a single particle or a set of particles at low concentrations. Currently, more precise expressions exist to describe particle motions (e.g., Maxey and Riley 1983), but they also lead to the same non-dimensional numbers as those based on (1). Thus, in the following bulk analysis we proceed with (1). The vertical component of (1) can be written by including the buoyancy term $\mathbf{F}_b = -\pi d_p^3(\rho_p - \rho_f)\mathbf{g}/6$ and the settling velocity of individual particles $w_s = d_p^2(\rho_p - \rho_f)g/(18\mu_f)$ as

$$\frac{dw_p}{dt} = \frac{18\nu_f}{d_p^2}(w_f - w_p) - \frac{18\nu_f}{d_p^2}w_s + \frac{dw_f}{dt} + \frac{1}{2}\left(\frac{dw_f}{dt} - \frac{dw_p}{dt}\right) + \frac{9}{\sqrt{\pi}}\sqrt{\frac{\nu_f}{d_p^2}}\int_{t_0}^{t} dt' \, \frac{(dw_f/dt' - dw_p/dt')}{t - t'} \qquad (2)$$

where w_f and w_p are the vertical velocities of the fluid and the particle, respectively, and g the gravitational acceleration. In the spirit of the Boussinesq approximation, it is assumed that $\rho_p \approx \rho_f$, except for the buoyancy term. Accordingly, the dimensional variables governing the motion of a solid particle in homogeneous turbulence are ν_f/d_p^2 and w_s as well as the parameters that characterize the velocity field. These governing variables can be utilized in bulk analyses, for example, in deriving non-dimensional numbers. If ρ_p is much different from ρ_f, then ρ_p/ρ_f also should enter as a parameter.

Of course, in the presence of velocity shear, (2) should be modified to include the velocity gradient terms as well as the lift on the particle (Hunt et al. 1994). However, gradient terms can be shown to be small when $d_p \ll |\nabla U|/|\nabla^2 U|$ and $|\mathbf{v}_p - \mathbf{v}_f| \gg d_p|\nabla \times \mathbf{v}_f|$, where U is the characteristic flow velocity. The inclusion of lift force $C_L(\mathbf{v}_f - \mathbf{v}_p) \times (\nabla \times \mathbf{v}_f)$, where C_L is the lift coefficient, does not introduce additional governing parameters.

As an example, consider the evolution of particle-laden jets, a problem motivated by its application to karstic lakes (in particular, Lake Banyoles, Catalonia), wherein resuspension of argillaceous and marly material near the lake bottom by subterranean springs is a common phenomenon (Colomer and Fernando 1996; Casamitjana et al. 2000). An idealized configuration that mimics this flow is a heavy particle-laden jet discharging upward from a point source, as shown in Figure 3a (in some cases, it is also appropriate to use multiple jets; Colomer et al. 1999). It is possible to expect that the resulting velocity field to be determined by the initial momentum of the jet M_0, initial (jet) particle concentration c_0 and parameters governing the fluid-particle interactions, i.e., ν_f/d_p^2 and w_s. For the problem at hand, a plausible set of governing variables is M_0 , ν_f/d_p^2, c_0 and w_s, and thus the maximum height of rise of particles z_m within the jet can be written as

$$\frac{z_m}{(M_0^{1/2}/w_s)} = f_1\Big(\frac{w_s^2 d_p^2}{\nu_f M_0^{1/2}}, c_0\Big) \tag{3}$$

where f_1 is a function.

Likewise, an expression for the ground concentration can be obtained. If the rate of accumulation of particles per unit area on the ground is $\dot{N}$ and the number of particles per unit volume at the discharge is n_0 (which is directly related to c_0), then the expression for the conservation of particles under steady conditions takes the form (Neves and Fernando 1995)

$$\int_0^\infty \dot{N}(r) 2\pi r \, dr = n_0 Q_0 = F \tag{4}$$

where Q_0 is the volumetric rate and F the total flux of particles at the jet discharge. Thus, in ground deposition studies, F can be selected as a governing parameter alternative to c_0, viz.,

$$\dot{N} = f_2(F, M_0, w_s, \nu_f/d_p^2, r) \tag{5}$$

or

$$\dot{N}^* = \frac{\dot{N}}{\left(\frac{F w_s^2}{M_0}\right)} = f_3\Big(\frac{r}{M_0^{1/2}/w_s}, \frac{w_s^2 d_p^2}{\nu_f M_0^{1/2}}\Big) \tag{6}$$

In the laboratory experiment depicted in Figure 3a, a particle-laden jet was injected vertically into a laboratory tank. The maximum height of particle rise and the radial deposition rate of particles were measured. Figure 3b shows the variation of $z_m/(M_0^{1/2}/w_s)$ with c_0. Several trends of the results are evident. For $c_0 < 5\ 10^{-3}$, the normalized z_m is independent on the concentration, but at larger concentrations there is a slight dependence on c_0. The normalized z_m in both regimes appears to be independent on $w_s^2 d_p^2/(\nu_f M_0^{1/2})$. The normalized particle deposition rate is shown in Figure 3c, which shows dependence on both $r/(M_0^{1/2}/w_s)$ and $w_s^2 d_p^2/(\nu_f M_0^{1/2})$. Further analysis (not shown) shows that an excellent collapse of data is possible, if the curves are normalized by the maximum deposition rate, which is a function of $w_s^2 d_p^2/(\nu_f M_0^{1/2})$.

3 Particle Settling in Turbulent Flows

Particles are subjected to a myriad of physical effects as they travel in a turbulent fluid. For example, the so-called crossing trajectory effect arises as a

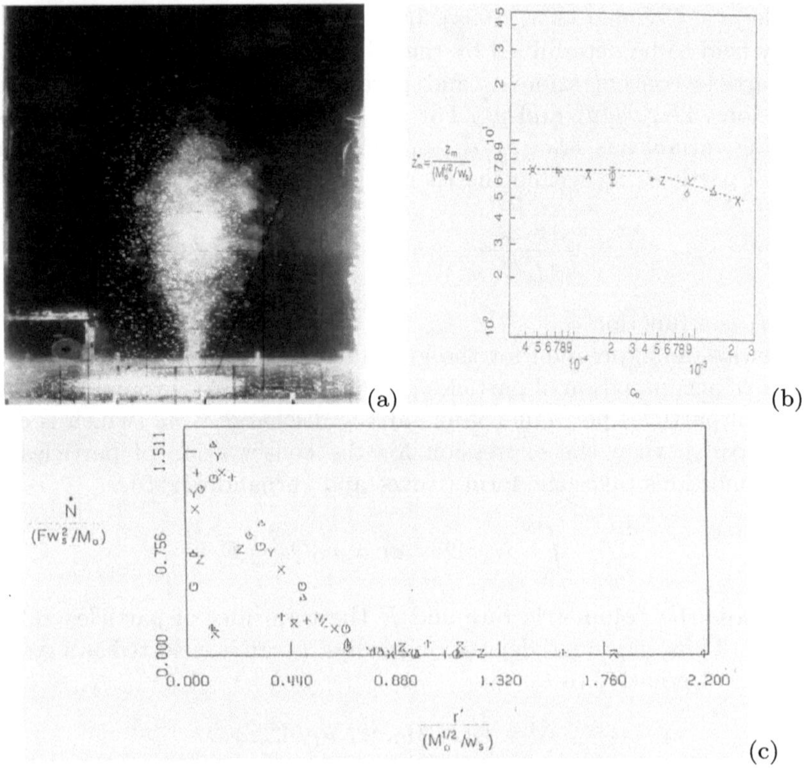

Fig. 3. (a) A laboratory particle driven jet emerging out of a circular slot, (b) The normalized maximum particle rise height (measured using a particle tracking technique) as a function of the concentration at jet discharge, (c). The normalized particle distribution rate as a function of normalized effective radius ($r' = r - d_0/2$, d_0 being the jet orifice diameter). Different symbols indicate different experimental runs conducted with widely varying experimental parameters.

result of the change of surroundings when a particle settles with a deterministic velocity in the direction opposite to the gravity. It moves from a region where flow velocity is highly correlated to another that is de-correlated from the previous region. Settling particles therefore lose their velocity correlation more rapidly than neutrally buoyant particles. Another is the continuity effect, where a back-flow is generated surrounding the particle to satisfy fluid continuity. The inertia effect, which is signified by the time necessary for a particle to adjust to the background flow, is also of critical importance; the particle response time takes the usual form $\tau_p = (\beta - 1)d_p^2/(18\nu_f)$. If $\tau_p = 0$, the particle responds to all changes in the fluid flow immediately, and in this case the velocity of the particle can be expressed as $\mathbf{v}_p = \mathbf{v}_f + \mathbf{v}_T$, where $\mathbf{v}_T$ is the settling velocity, and $|\mathbf{v}_T| = w_s$.

The interaction of settling particles and flow ought to be determined by d_p^2/ν_f, w_s , β and characteristic flow parameters, as discussed in Section 2. Note that τ_p includes the d_p^2/ν_f term, and hence these governing parameters can be recast as τ_p, w_s, and β. For simplicity, as was done in Davila and Hunt (2001), consider particle settling in a vortex characterized by its radius R, velocity U and circulation $\widetilde{A} = UR$. Important parameters for the problem therefore become the Stokes number, $St = \tau_p/(R/U)$, the normalized settling velocity $\overline{V}_T = V_T/U$ and β. Other related parameters have also been used, for example, the Froude number of the flow ($Fr = V^2/(gR)$; the ratio of fluid inertia to gravitational forces) or the particle Froude number $F_p = V_T^3/(g\widetilde{A})$ (inertial forces exerted by particles to gravitational forces).

As an illustration, consider an inertia-less particle entering into a vortex described by the sample velocity field $(u_1, u_2) = (\omega x_2, -\omega x_1)$. It is clear that a particle may stay trapped inside the vortex or escape from it, depending on the value of $\overline{V}_T$. If $\overline{V}_T < 1$, then a particle can be permanently trapped inside the vortex moving in a circle with center $(-V_T/\omega, 0)$. If there is finite inertia, then the particles can break closed orbits without following the fluid particles. Owing to a centrifugal force, particles with finite inertia may spiral inward or outward, depending on the signs of the settling velocity and buoyancy.

Davila and Hunt (2001) considered the instantaneous position and velocity of small spherical particles with inertia ($St \neq 0$) settling through a Rankine vortex having the velocity field

$$\mathbf{u} = \left(\frac{-2x_2}{1 + x_1^2 + x_2^2}, \frac{2x_1}{1 + x_1^2 + x_2^2},\right) \quad ; \quad \max|\mathbf{u}| = 1 \tag{7}$$

For convenience, heavy particles $\beta \gg 1$ were considered, for which the equation of motion becomes

$$\frac{d\mathbf{v}_p^*}{dt} = \frac{1}{St}\left(\overline{\mathbf{v}}_T + \mathbf{u}^* - \mathbf{v}_p^*\right) \tag{8}$$

where $\mathbf{u}_*$ is the background flow velocity and superscript $*$ denoted quantities normalized by flow variables U and R. Equation (8) points to the equilibrium points ($\mathbf{v}_p = d\mathbf{v}_p/dt = 0$) for particles satisfying $\mathbf{u}^*(x_E) + \mathbf{v}_T = 0$. They showed that for the vortex described by (7), there are two equilibrium (E) points

$$(x_1^*)_{E1} = \frac{1 - \sqrt{1 - \overline{V}_T^2}}{\overline{V}_T} \quad ; \quad (x_1^*)_{E2} = \frac{1 + \sqrt{1 - \overline{V}_T^2}}{\overline{V}_T} \tag{9}$$

with $(x_2^*)_{E1} = (x_2^*)_{E2} = 0$, where $(x_1)_{E1}$ is a node and $(x_1)_{E2}$ is a saddle point. It is clear that stationary points are possible only when $\overline{V}_T = V_T/U \leq 1$, and that particles with $\overline{V}_T > 1$ do not show stagnation behavior.

If $\overline{V}_T \leq O(1)$ and if the inertia is small enough, only those particles in regions with $|u| \sim V_T$ ought to be affected by the flow. This points to the possibility of using V_T as the scaling velocity. For the general case of

$|u^*| \sim 1/|x^*|^a$, Davila and Hunt (2001) suggested an alternative scaling, where V_T, $\widetilde{A}/(V_T U^{a-1})^{1/a}$ and $\widetilde{A}/(V_T^{a+1} U^{a-1})^{1/a}$ were the velocity, length and time scales, respectively, which yielded a new non-dimensional parameter (or a rescaled Stokes number)

$$F_p = \overline{V}_T^{1+1/a} St = \frac{\tau_p}{\widetilde{A}/[V_T^{(a+1)} U^{(a-1)}]^{1/a}} \tag{10}$$

as a measure of inertia effects. F_p represents the ratio of distance over which a particle accelerates ($V_T \tau_p$ and the maximum radius of curvature of particle trajectories around line vortices. For the line-vortex case with $a \approx 1$, small F_p implies quick response of particles to flow variation and vice versa. At large F_p (contributed by large St and $\overline{V}_T$), particles cannot maneuver changing courses of fluid trajectories well and overshooting into nearby trajectories occurs.

The above simple model of particle settling in line vortices (for which the governing parameters are F_p, $\overline{V}_T$ and β) can be used to explain observations of heavy particle settling in turbulent flows. Of course, there are clear differences between the two cases, for example, even the simplest case of isotropic turbulence consists of time-dependent random distribution of vortices. Vortices therein are separated by a distance on the order of the integral scale L_x, with each having a radius on the order of Taylor microscale λ_T. Assuming that (*i*) $L_x \gg \lambda_T$, (*ii*) those vortices having axes that are horizontal contribute most to the particle settling, and (*iii*) that the persistence time of vortices L_x/U is greater than the particles penetration time through them, it is possible to extrapolate (qualitatively) the results of Davila and Hunt (2001) to understand particle settling through turbulent fluids.

It should be noted that there are several definitions for particle settling velocity. The Eulerian settling velocity V_E is the average of the particle settling velocity at a given point weighted by the particulate concentration distribution. The Lagrangian settling velocity V_L is defined in terms of the vertical distance traveled along a particle trajectory (X, Y) over a given time ΔT, averaged over an ensemble starting from different positions (X_0, Y_0),

$$V_L = \frac{1}{\Delta T} \int_0^{\Delta T} dt\; V_p(X, Y, X_0, Y_0) \tag{11}$$

The bulk settling velocity V_B is defined on the basis of averaged time for particles to cross two different vertical levels separated by ΔY. If the characteristic length scale $(\widetilde{A}/V_T)$ of particle trajectories is much less than the characteristic horizontal vertical separation of vortices, then it is possible to show that $V_B \approx V_L$.

Computations of Davila and Hunt (2001) illustrate how the settling velocity (say V_L) is affected by $\overline{V}_T$ and F_p. When $\overline{V}_T > 1$, there are no equilibrium points, and thus no void and accumulation regions, leading to rather monotonic descent of particles. In this case the particles can show slightly larger settling speeds up to about $F_p = 3 - 5$, beyond which the settling speed slightly decreases.

The most striking effect of turbulence arises when there are equilibrium points (or empty regions), wherein $\overline{V}_T < 1$. In this case, for $F_p < F_p^*$, where $F_p^* \approx 1$, when the particles have lesser inertia, the particles follow fluid parcels along the fast tracks in the fringes of empty regions, leading to enhanced settling velocity. For finite F_p, as $\overline{V}_T$ increases, this enhancement of settling decreases because of the increasing dominance of buoyancy forces that tend to reduce the influence of the background flow. As $F_p > F_p^*$, there is a sharp reduction in the increase (above V_T) of settling velocity with F_p, and for $F_p > F_p^{**}$, $F_p^{**} \approx 10$, even a reduction of the settling velocity below V_T can be expected. This owes to the fact that with increasing F_p the inertia also increases, and thus particles lose their tendency to follow the fluid. This may cause particles to cross trajectories.

The results of an experiment conducted to verify the above arguments are shown in Figure 4. The experiments were conducted by generating 'nearly' isotropic turbulence in a mixing box, with two oscillating grids placed on either sides of the tank (Srdic et al., 1996). Note the variation of bulk settling velocity with F_p for the $a = 1$ case. The normalized V_B increases with F_p for small F_p, achieves a maximum and then reduces below unity. This observation is in qualitative agreement with the concepts described above. Although it is not possible to directly compare the predictions of the model with the experiments due to differences in flow configurations, a qualitative comparison is reasonable given that the experimental flow possesses some features, such as coherent vortex structures, assumed in the theoretical model. There have been a few previous experimental studies on settling of particles in turbulent fluids (Murray 1970; Nielsen 1992). In the former study the turbulence was generated using a horizontal bar grid oscillating vertically whereas in the latter two perforated plates oscillated in the vertical direction was used. Turbulence was not measured in either study, but inferred using oscillating grid/plate parameters. Both of the studies indicated that weak turbulence ($\overline{V}_T = V_T/\sigma > 1$) may have a delaying effect on settling whereas strong turbulence $\overline{V}_T < 1$ increases the settling velocity, where σ is the r.m.s. velocity. V_T/σ was the only parameter considered in both of these studies, but we expect the results to depend on F_p. In all, the experimental results are qualitatively consistent with the theoretical arguments presented above.

4 Applications of Particle Laden Flows

In this section, a few practical applications of particle driven flows are discussed in the context of urban air quality and visibility. Obviously, the simple fluid dynamic models described above cannot be directly used for practical predictions of particle laden flows in the atmosphere, given the complexity of natural flows subjected to ever changing chemical, physical and biological conditions of the atmosphere. Modeling systems that include flow and turbulence, two phase dynamics and chemical changes need to be used in this regard.

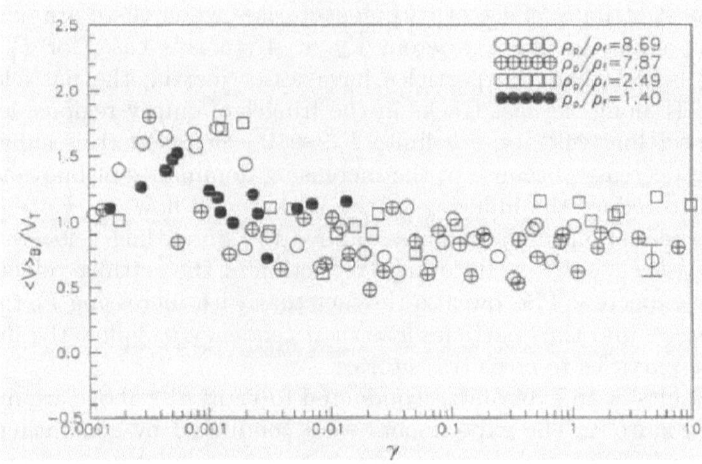

Fig. 4. A plot of normalized bulk settling velocity as a function of the rescaled Stokes number. The parameters are defined in the text. The experiments were conducted by releasing particles of different properties into a fluid with sustained turbulence introduced by oscillating grids.

Inherent to such complex models are a host of assumptions on particulate matter and their precursor emissions, nature of the sources and their interaction with the flow, secondary particle formation, chemical transformations, deposition and biogenic effects. Prediction of the composition and morphology of particles also enables us to estimate the visibility reduction through an aerosol-optical model. Below, we will first deal with the modeling of episodic particulate matter events using a 3-D air quality model for the case of a pair of cities in the U.S./Mexico border, which are amongst the most polluted of the U.S. A visibility calculation for the Phoenix urban area for typical days of winter and summer are presented next.

4.1 Particulate Matter Modeling for the U.S./Mexico Border

The U.S./Mexican border region consists of arid or semi-arid largely rural and/or agricultural land with irregularly spaced small and large twin cities across the border. These borderlands are characterized by a diverse mix of topography, ecology and human activity. As a result of fast demographic and economic growth, the U.S./Mexico border regions, particularly the twin cities across the border, have received increased attention with regard to environmental degradation and human health concerns. One of the major pollutants in the borderlands is the PM. It has been shown that episodic events contribute to many of the violations of the U.S. EPA's PM_{10} National Ambient Air Quality Standards (NAAQS) in the area. A number of factors, for example, topography, local anthropogenic sources, mesoscale transport, resuspension, frontal turbulence, prevailing wind speed and direction, daily cycle of the atmospheric boundary layer and traffic density have been highlighted as the drivers of PM episodes. To understand and quantify the causal factors leading to high PM episodes that frequent the border twin-city area of Douglas/Agua Prieta, air quality modeling with high grid resolution was performed.

The Models-3/CMAQ modeling platform was employed to calculate the PM concentration. This predictive tool consists of three integrated elements: the Pennsylvania State University (PSU)/NCAR Mesoscale (Generation 5) model for meteorology (MM5), Sparse Matrix Operator Kernel Emissions (SMOKE) model for emission processing, and the Community Multiscale Air Quality (CMAQ) model. Douglas and Agua Prieta, are located in southeastern Arizona and northeastern Sonora, within the Sulfur Springs valley, and are characterized by complex topography and arid desert climatology. Because the sources are close together, substantial gradients exist within these relatively small cities, thus requiring fine grid simulations. Therefore, typical pollution inventories available in the modeling context ($\sim$ 4 km resolution) cannot be utilized for simulations, and a special pollution inventory with 0.5 km resolution was constructed (Reynolds et al. 2002).

MM5 modeling was performed using nested domains, down-sizing from the parent domain of 54 km grid resolution to the domain of interest 44 x 32 km^2 (1 km resolution). Meteorological inputs for 44 x 32 km^2 CMAQ simulations with 0.5 km grid spacing were obtained via linear interpolation of wind fields from the 1-km grid simulation, considering the putative notion that MM5 modeling paradigms and parameterizations may not be suitable for grid sizes below 1 km. Owing to the lack of available observations, CMAQ default values were used as initial conditions. To minimize the impact of this selection, a spin-up time of three days was employed. Time-dependent boundary conditions for the study domain were derived from one-way nested CMAQ simulations over the larger domain, which also encapsulate the effects of long-range transport of $PM_{2.5}$ and gaseous precursors (U.S. EPA 2004). Nine PM events were selected for modeling from a special measurement program conducted over the area by the Arizona Department of Environmental Quality during January 1999 through mid-February 2000.

The model predictions were evaluated against observation the data by computing conventional statistical measures, and performance measures were found to be in acceptable ranges compared to other air quality models and studies. Arid desert conditions make the area more vulnerable to wind-blown dust, and thus some of the observed PM_{10} NAAQS exeedances were related to high wind speeds. The other high episodes occurred under stagnant conditions with low wind speeds. A severe limitation of the sophisticated CMAQ/MM5/SMOKE modeling system was clearly identified during the simulations, which is the inability of the system to deal with wind-blown dust at high and time-dependent wind speeds. The predicted 24-hr averaged PM_{10} concentrations are shown in Figure 5. More hot spots with PM_{10} over 500 $\mu g/m^3$ are concentrated in the vicinity of emissions sources on a low wind day (02/23) than on a high wind day (12/02). The CMAQ, lacking a parameterization for enhanced dust entrainment during high wind events, predicted lesser particle concentrations than the observations for high wind days. The model particles are solely provided by the emission inventory, but not by dust entrainment.

By conducting three groups of sensitivity simulations with carefully selected boundary conditions, it was possible to conclude that the regional contribution to high PM events over the area is insignificant, and that high PM episodes are dominated by local conditions. The simulations showed only a small contribution of secondary particles to total PM, which are most likely to be regional rather than local. Considering all, the observed high PM events appear to be a result of entrainment of soil dust from unpaved/paved roads and wind-blown dust at high wind speeds. These results are consistent with the chemical analyses of PM measurements in the area. Based on our study, most of the PM_{10} exeedances took place on the Mexican side, mainly contributed by unpaved roads. The exeedances in the U.S. side occurred only at a site close to the border when winds were strong and southerly (Choi et al. 2006).

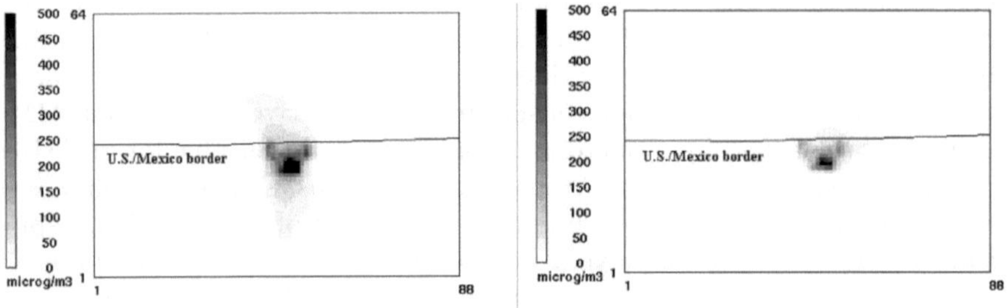

Fig. 5. CMAQ-simulated 24-hr averaged PM_{10} concentrations on the design (simulation) day of 02/23/1999 (left panel; this is a low wind day). The right panel shows a high wind day 12/2/1999.

4.2 Visibility Modeling

Visibility is the observer visual range, which is the greatest distance at which a large black object can just be seen against the horizon sky. Visibility impairment is probably the most easily recognizable effect of air pollution, which is caused by the scattering and absorption of light by particles and gases in the air (U.S. EPA 1999). In order to simulate visibility in the Urban Phoenix area, the MM5/CMAQ/SMOKE modeling system was employed. The SMOKE emission model was processed to produce speciated, gridded, hourly emissions, using EPA 1999 National Emissions Inventory. The MM5 was simulated to produce gridded, hourly meteorological fields. The CMAQ aerosol module calculates visibility indices (deciview and light extinction coefficient β_{ext}) using two different methods: an approximation to Mie theory and a mass reconstruction technique. The Mie algorithm estimates the light extinction efficiency due to aerosols based on the particle effective radius and refractive index (Wiscombe 1980). On the other hand, the mass reconstruction method is an empirical approach, and calculates β_{ext} from aerosol species mass concentrations, using (12) below (Malm et al. 1994),

$$\begin{aligned}\beta_{ext} = {} & 0.003 f_4(RH)\left[ammonium + sulfate + nitrate\right] \\ & + 0.004\left[all\ organic\ species\right] \\ & + 0.01\left[elemental\ carbon\right] \\ & + 0.001\left[unspeciated\ PM_{2.5}\right] \\ & + 0.0006\left[coarse\ particles\right] + 0.01 \end{aligned} \tag{12}$$

The brackets denote species mass concentration in $\mu g/m^3$. The coefficients represent scattering efficiencies, except for light absorbing carbon, which instead has absorption efficiency. The function $f_4(RH)$ is a relative humidity correction. Since ammonium sulfate and ammonium nitrate swell and increase in size at relative humidities above about 70

$$V_r = 3.91/\beta_{ext} \quad ; \quad deciview = 10\ln(\beta/0.01) \tag{13}$$

where V_r in km and β_{ext} in km^{-1} (Pitchford and Malm 1994).

According to the observations, visibility in winter is worse than any other season. It is thought that lower mixing height makes pollutants concentrate near the surface, and lower temperature provides more favorable condition to aerosol formation in winter than in other seasons. As an illustration, therefore, the worst and best visibility days in 2001 were simulated for the Phoenix area. December 18 was the worst visibility day recorded in that year, and August 11 registered the best visibility. Both days had a similar relative humidity (RH) around 40 50

Simulations were performed for a 4 day period including a 3-day spin-up time for each case. Hence, the simulation period for the worst visibility day, December 18, was 15-19 December, and that for the best visibility day, August 11, was 9-12 August 2001. The model seems to underestimate particle concentrations, and accordingly under-predict visibility degradation. Using Mie theory, the visibility degradation was under-predicted by a factor of two, and using the mass reconstruction method, by a factor of three. It should also be noted that CMAQ visibility calculations are somewhat hampered by uncertainty in coarse particle emissions due to the absence of wind blown dust. At this point, a complete evaluation of the model cannot be performed because of the lack of detailed observations. However, the relative magnitude of the modeled value for the worst day to that of the best day appears to be consistent with the observations, even though the absolute deciview values are far off from the observed values. Figure 6 displays simulated visual ranges at noon time for both days.

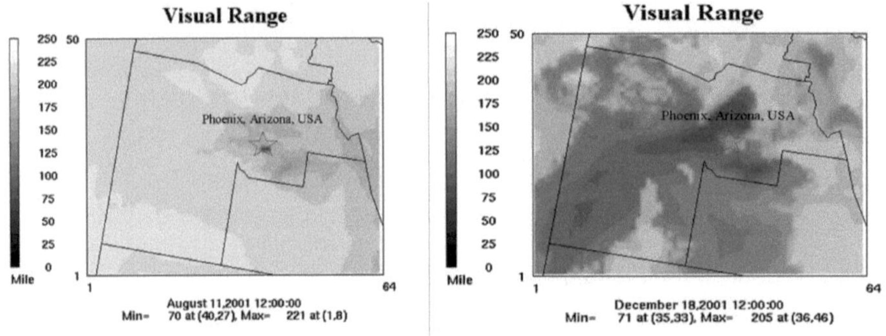

Fig. 6. Simulated visual range Vr at noon on August 11 and December 18, 2001. The visibility in winter days is smaller due to the high particle concentrations induced by more stable air-flows and smaller mixing heights.

References

[1] Bridgeman H (1990) Global Air Pollution. Belhaven Press

[2] Casamitjana X, Colomer J, and Fernando HJS (2000) Development of a turbid layer in conical basins and its applications to Lake Banyoles. Phys Limnol 27: 233-237

[3] Choi Y-J, Hyde P, and Fernando HJS (2006) Modeling of episodic particulate matter events using a 3-D air quality modeling system with fine grid: Applications to apair of cities in the U.S./Mexico border. Atmos Env 40: 5181-5201

[4] Colomer J and Fernando HJS (1999) Resuspension of sediments by multiple vertical jets. ASCE J Hydraulic Eng 125 (7): 765-770

[5] Colomer J and Fernando HJS (1996) Resuspension of a particle bed by a round vertical jet. J Env Eng 122(9): 864-869

[6] Davila J and Hunt JCR (2001) Settling of small particles near vortices and in turbulence. J Fluid Mech 440: 117-145

[7] Ghosh S, Davila JA, Hunt JCR, Srdic A, Fernando HJS, and Jonas PR (2005) How turbulence enhances coalescence of settling particles with applications to rain in clouds. Philosophical Transactions, Proceedings of the Royal Society, In Press (paper: 04PA0103), doi:10.1098, 29 pages

[8] Hinze JO (1975) Turbulence, 2nd Edition. McGraw Hill, NY p. 460

[9] Hunt JCR, Perkins RJ and Fung JCH (1994) Problems in modeling disperse two-phase flows. Appl Mech Rev 47(6pt.2): 49-60
[10] Malm WC, Sisler JF, Huffman D, Eldred RA, and Cahill TA (1994) Spatial and seasonal trends in particle concentrations and optical extinction in the United States. J Geophys Res 99:1347-1370
[11] Maxey M and Riley JJ (1983) Equation of motion for a small rigid sphere in a non-uniform flow. Phys Fluids 26: 883-889
[12] McCave I (1984) Size spectra and aggregation of suspended particles in the deep ocean. Deep Sea Res 31: 329-352
[13] Murray SP (1970) Settling velocities and vertical diffusion of particles in turbulent water. J Geophys Res 75: 1647-1654
[14] Neves MJ and Fernando HJS (1995) Sedimentation of particles from jets discharged by ocean outfalls - a theoretical and laboratory study. Water Sci and Technol 133-140
[15] Nielsen P (1992) Effects of turbulence on the settling velocity of isolated suspended particles. 11th Australasian Fluid Mechanics Conference, University of Tasmania, Hobart, Australia, pp. 179-182
[16] Pitchford ML, and Malm WC (1994) Development and application of a stnadrad visual index. Atmos Environ 28: 1049-1054
[17] Reynolds S, Meszler D, Causley M, Arons D, Acosta G, Diem J, and Jones R (2002) Emissions inventories for Douglas, Arizona and Agua Prieta, Sonora, Mexico (Final report for Arizona Department of Environment Quality)
[18] Srdic A, Fernando HJS and Montenegro LM (1996) Generation of nearly isotropic turbulence using two oscillating grids. Exp in Fluids 20: 395-397
[19] Tchen CM (1947) Mean value and correlation functions connected with the motion of small particles suspended in a turbulent fluid. Ph.D. Thesis, Delft
[20] US EPA (1999) Visibility monitoring guidance. EPA-454-R-99-003. US Environmental Protection Agency
[21] US EPA (2004) The Particle pollution report: Current understanding of air quality and emissions through 2003. EPA-454-R-04-002. US Environmental Protection Agency
[22] Wiscombe WJ (1980) Improved Mie scattering algorithms. Appl Opt 19: 1505-1509

ERCOFTAC SERIES

1. A. Gyr and F.-S. Rys (eds.): *Diffusion and Transport of Pollutants in Atmospheric Mesoscale Flow Fields*. 1995 ISBN 0-7923-3260-1
2. M. Hallbäck, D.S. Henningson, A.V. Johansson and P.H. Alfredsson (eds.): *Turbulence and Transition Modelling*. Lecture Notes from the ERCOFTAC/IUTAM Summerschool held in Stockholm. 1996 ISBN 0-7923-4060-4
3. P. Wesseling (ed.): *High Performance Computing in Fluid Dynamics*. Proceedings of the Summerschool held in Delft, The Netherlands. 1996 ISBN 0-7923-4063-9
4. Th. Dracos (ed.): *Three-Dimensional Velocity and Vorticity Measuring and Image Analysis Techniques*. Lecture Notes from the Short Course held in Zürich, Switzerland. 1996 ISBN 0-7923-4256-9
5. J.-P. Chollet, P.R. Voke and L. Kleiser (eds.): *Direct and Large-Eddy Simulation II*. Proceedings of the ERCOFTAC Workshop held in Grenoble, France. 1997 ISBN 0-7923-4687-4
6. A. Hanifi, P.H. Alfredson, A.V. Johansson and D.S. Henningson (eds.): *Transition, Turbulence and Combustion Modelling*. 1999 ISBN 0-7923-5989-5
7. P.R. Voke, N.D. Sandham and L. Kleiser (eds.): *Direct and Large-Eddy Simulation III*. 1999 ISBN 0-7923-5990-9
8. B.J. Geurts, R. Friedrich and O. Métais (eds.): *Direct and Large-Eddy Simulation IV*. 2001 ISBN 1-4020-0177-0
9. R. Friedrich, B.J. Geurts and O. Métais (eds.): *Direct and Large-Eddy Simulation V*. Proceedings of the Fifth International ERCOFTAC Workshop on Direct and Large-Eddy Simulation, held at the Munich University of Technology, August 27–29, 2003. 2004 ISBN 1-4020-2032-5
10. E. Lamballais, R. Friedrich, B.J. Geurts and O. Métais (eds.): *Direct and Large-Eddy Simulation VI*. Proceedings of the Sixth International ERCOFTAC Workshop on Direct and Large-Eddy Simulation, held at the University of Poitiers, September 12–14, 2005. 2006 ISBN 1-4020-4909-9
11. B.J. Geurts, H. Clercx and W. Uijttewaal (eds.): *Particle-Laden Flow: From Geophysical to Kolmogorov Scales*. 2007 ISBN 978-1-4020-6217-9

MIX
Papier aus verantwortungsvollen Quellen
Paper from responsible sources
FSC® C105338

If you have any concerns about our products,
you can contact us on
ProductSafety@springernature.com

In case Publisher is established outside the EU,
the EU authorized representative is:
Springer Nature Customer Service Center GmbH
Europaplatz 3, 69115 Heidelberg, Germany

Printed by Libri Plureos GmbH
in Hamburg, Germany